Mech

D0635527

Mechanics

Second edition

P. Smith

Department of Mathematics
University of Keele
Staffordshire

R. C. Smith

Open University, Milton Keynes
Buckinghamshire

JOHN WILEY & SONS

Chichester · New York · Brisbane · Toronto · Singapore

Other Wiley Editorial Offices

John Wiley & Sons, Inc, 605 Third Avenue,
New York, NY 10158–0012, USA

Jacaranda Wiley Ltd, G.P.O. Box 859, Brisbane,
Queensland 40001, Australia

John Wiley & Sons (Canada) Ltd, 22 Worcester Road,
Rexdale, Ontario M9W 1L1, Canada

John Wiley & Sons (SEA) Pte Ltd, 37 Jalan Pemimpin #05–04,
Block B, Union Industrial Building, Singapore 2057

Library of Congress Cataloging-in-Publication Data:

Smith, Peter, 1935–
  Mechanics / P. Smith, R. C. Smith.—2nd ed.
    p.   cm.
  R. C. Smith's name appears first on earlier ed.
  Includes bibliographical references.
  ISBN 0 471 92737 6
  1. Mechanics.   I. Smith, R. C. (Ralph Charles)   II. Title
QA805.S673   1990
531—dc20   90-31722
                                                      CIP

British Library Cataloguing in Publication Data:

Smith P. (Peter), *1935–*
  Mechanics, – 2nd ed
  I. Title   II. Smith, R. C. (Ralph Charles), *1930–*   III.
  Series
  531

ISBN 0 471 92737 6

Typeset by Techset Composition Ltd, Salisbury
Printed and bound in Great Britain by
Biddles Ltd, Guildford and King's Lynn

Contents

Foreword to the Second Edition

The first edition of *Mechanics* was one of a series of books on *Introductory Mathematics for Scientists and Engineers*. It proved to be one of the most popular texts in the series and has been widely used for teaching mechanics. Its success is due to the authors' skill in imparting the subject in a highly readable fashion and spicing the text with many models of practical situations. Nevertheless, after numerous years of good service, the first edition was beginning to look a little long in the tooth. There has been a shift of emphasis from the analytical to the computational in the treatment of equations. That has led to the present fascination with dynamical systems through the discovery of the phenomenon of chaos and the realisation that important effects were obscured or eliminated by earlier approximations.

The authors have, in this revised edition, brought the content of their book up-to-date without losing the clarity of exposition for which the first edition was so noteworthy. The text now recognises that readers expect to have easy access to powerful computational tools and to be regularly making calculations regarded as commonplace today but which were unthinkable when the first edition was published. Some more advanced material has been added for the reader who wants to go a little further. In particular, the reader can have a taste of what happens when non-linearities take control, producing a plethora of unexpected effects. So teachers and students should find this revision as valuable an asset as they found the old one.

I wish this new edition well and hope that it enjoys the success it so richly deserves.

D. S. JONES
Department of Mathematics
University of Dundee

Preface to the
Second Edition

This book is designed to give students a thorough grounding in particle dynamics and elementary rigid body dynamics. The contents were originally developed from classes given to first and second year undergraduates at the University of Keele, in joint degree programmes in which Mathematics was a major component. Over the years since the first edition was published, we have discovered that the book has also been adopted by many lecturers in Physics and Engineering for courses in dynamics.

With the considerable current interest in non-linear systems, it seemed an appropriate time to revise the previous edition, in order to incorporate some of the background required in the study of what has now become known as dynamical systems, at least as they arise from Newtonian mechanics. With this aim in view we have written a completely new chapter on non-linear dynamics which includes material on the methods of phase plane analysis, non-linear conservative and damped systems and Poincaré maps, leading to examples of mechanical systems which display period doubling and chaotic outputs. This material is preceded by a completely revised and enlarged chapter on the linear theory of oscillations which also now includes normal modes of multiple systems.

The introduction contains a brief survey of the development of mechanics, together with some general comments on the art of successful modelling of dynamical problems. The text is designed to accommodate students with varied backgrounds and, with this mind, the book starts with a chapter on vectors. Elementary analytical and numerical techniques for solving ordinary differential equations have now been placed in an Appendix. Most of the basic principles of Mechanics are covered in Chapters 2, 3 and 5. Certainly in a first course, most of this material, together with the applications in Chapter 4, would form the nucleus of a possible syllabus. The remaining chapters on rocket motion, linear oscillations, orbits, non-linear dynamics and rotating axes are largely self-contained and may be included or not in a course as time permits, although usually we would expect that non-linear dynamics would be preceded by the linear theory of oscillations. The book contains over 360 examples and problems, with many worked through in detail in the text. Answers and comments to the problems are provided at the end of the book.

A new edition is always a conflict between, on the one hand, a wish to change everything, which would produce a completely new book, and, on the other hand, the preservation of all existing chapters with additional material simply tacked on to the end of the book, greatly increasing its length. We have attempted to achieve a balance between these pressures. We have been particularly assisted in this endeavour by many comments from readers, reviewers, students and correspondents since the publication of the original edition. For many years the text was a Set Book for an Open University course on Mechanics and Applied Calculus, and Course Material and Units were designed around the book. This resulted in our approach to dynamics being subjected to a detailed textual analysis by the Course Team. We are grateful to all these contributors, many unknown to us, who have helped us, often unwittingly, in this revision.

The book has a different appearance since almost all the figures have been re-drawn. Many solutions to actual problems are now either directly computer drawn or based on plotter outputs. The comparative ease with which solutions to initial value problems in ordinary differential equations can be solved, and displayed and manipulated graphically, means that many dynamical problems which were difficult or inaccessible analytically, can now be usefully investigated with the aid of microcomputers. Software is still so machine- or language-dependent and time-sensitive that we have decided not to include programs in the book. However, to this end, the Appendix does contain some elementary numerical methods for solving initial value problems in ordinary differential equations.

PETER SMITH
RALPH C. SMITH
February, 1990

An Introduction to the Study
of Mechanics

MOTIVATION

The modern theoretical study of mechanics has a long history dating from the early part of the Seventeenth century. The first major contribution to particle dynamics was made by Galileo (1564–1642) with his elementary theory of the motion of a projectile. The fundamental importance of Galileo's approach lay in his attempt to relate theory and experimental evidence; at the time this was a new concept of scientific method. Prior to this breakthrough scientific theories of mechanics were put forward solely on this basis of reason and the authoritative statement. Galileo's writings had a profound effect on subsequent developments. The work of Galileo was continued and developed further by Huygens (1629–95).

From this early incomplete theory of particle dynamics Newton (1642–1727), in his great work *Principia* (1687), established particle dynamics as a scientific and mathematical discipline. Newton introduced the notions of force and mass and set forth his laws of motion (now known as Newton's laws of motion) for particle motion substantially in the form that appears in modern textbooks of dynamics. Significantly, he was able to use his laws to show that they were consistent with the numerical data on planetary motion obtained earlier by Kepler (1571–1630) and Brahë (1546–1601), usually expressed in Kepler's laws of planetary motion (see Chapter 8). For this purpose he needed to propound a law of gravitation and secondly, perhaps more significant for the study of dynamics in particular and in the wider context of mathematics, he made the first tentative steps towards the discovery of mathematical calculus, a powerful analytic tool in applied mathematics. For these obvious reasons the subject has come to be known as Newtonian mechanics.

Newton's principles were extended to rigid-body dynamics by Euler (1736), d'Alembert (1743) and by Lagrange (1788) in his famous *Mécanique Analytique*.

This short summary of the early history of the subject serves as a prototype for the modern conception of mechanics. The aim is to be able to predict the behaviour of the phenomena on the basis of certain postulates. But the postulates can only be tested by experimental evidence and clearly only a limited number of experiments are available to us. In the light of experience many of our fundamental assumptions must be modified. Thus whilst Newtonian mechanics is appropriate to the laboratory experiment or the motion of the planets in the solar system, significant departures from Newton's theories appear in very large and very small scale phenomena. For this reason relativistic and quantum mechanics have taken over in these two spheres. Applied mathematics must be a continuous interplay between experiment and experience on the one hand and fundamental hypothesis on the other. Hence dynamics has not developed as a completely deductive discipline based on certain consistent axioms. The study of dynamics requires a firm grasp of the intuitive meaning of such physical quantities as force, mass, energy etc., rather than seeing them as components of a strictly mathematical argument, and therein lies one of the difficulties of applied mathematics: the need to continuously assess and interpret the results of the mathematics involved in solving a problem.

SPACE AND TIME

The fundamentals of mechanics involve assumptions which are often left unsaid. Newtonian mechanics assumes that space is Euclidean, and this seems a natural view of space. It assumes that there exists an absolute reference system relative to which all other points in space can be described. Time is supposed continuous and measurable according to an absolute time scale so that observers at different positions in space can synchronise their clocks with this time scale. An event has an absolute position and takes place at an absolute time.

The simple intuitive notions of space and time have been highly successful, for example, in predicting the orbital paths of the planets, with the exception of a few minor deviations which cannot be explained by the Newtonian theory of dynamics. Only by rejecting the principles of absolute space and time through Einstein's relativistic theory can these discrepancies be accounted for. However, the principles of Newtonian mechanics have that element of simplicity and logical clarity which make them the best introduction to mechanics. Whilst relativistic theories can, in principle, give finer detail it is given at the expense of greater analytical difficulty.

UNITS

The international standard of length is the *metre* defined as a certain number of wavelengths of a radiation from krypton. It used to be defined by a standard metre rod made of platinum and iridium and kept at a specified temperature in a suburb of Paris. The unit of time is the second which has two definitions. The astronomical definition is $1/31\ 556\ 925 \cdot 9747$ of the tropical year 1900 (the tropical year is the time interval between successive south–north transits by the Sun across the Earth's equatorial plane). The atomic definition is a given number of periods of oscillations of the Caesium 133 atom.

NEWTON'S LAWS OF MOTION

The laws of motion arise partially as the answers to certain questions. Both historically and philosophically the right questions to ask taxed many scholars for many years. With hindsight we can express our ideas with greater awareness of the consequences of the postulates. Our considerations here are kept deliberately general: greater detail will be found in later chapters. First we ask: what happens to a body which is completely isolated in space from all other material objects? We answer this question by postulating that the body continues in a state of rest or of uniform motion in a straight line. The object will only deviate from this state if it is subject to some external influence or force. This is essentially Newton's first law and contains a qualitative definition of force.

We next ask: what quantitative effect does a force have on an object? Since velocity is a relative concept (that is, an isolated observer cannot detect his own velocity), we must suppose that force is independent of velocity. However, in Newtonian mechanics *acceleration* can be measured. Newton's second law of motion states that the force acting on a body is proportional to the acceleration induced in the body, the constant of proportionality being the (inertial) mass of the body, a measure of the quantity of material of which it is composed. This law is a simple *linear* relation between force and acceleration, and it is not inconsistent with the first law.

These two laws do not form a complete basis for dynamics. Consider a simple coil spring which has been compressed. The internal compressive force will endeavour to bring the spring back to its natural length and its external effect will be experienced at the two ends of the spring. Newton's third law postulates how the compressive reaction will be distributed at the two ends. We assume that action and reaction are equal and opposite: in our illustration the spring exerts equal and opposite forces at both ends. For the same reason we suppose that the Earth exerts the same gravitational pull on the Moon as the Moon does on the Earth.

These are the basic postulates of Newtonian mechanics which are assumed to hold universally. However, they must be backed up by hypotheses concerning the constitutive nature of the force which appears in Newton's second law. We cannot solve the equation

$$\text{force} = \text{mass} \times \text{acceleration}$$

until we specify the nature of the force. Returning to the illustration above
we must specify how the force varies with, for example, the extension or
contraction of the spring. In a compressed linear spring the constitutive law is
taken to be

$$\text{force} \propto \text{contraction},$$

but it is obvious that not all springs behave in this way for large contractions.
A modified law may have to be devised for large contractions perhaps involving
further parameters.

The nature of gravitational force must be specified and Newton's law of
universal gravitation states that the gravitational attraction between any two
bodies is proportional to their masses and inversely proportional to the square
of the distance between them. The test of this law can be found in the large
amount of data collected on the positions of the planets in the solar system.
The only significant discrepancies detected so far can be accounted for by
relativistic theories.

Much of the elegance of Newtonian mechanics lies in its dependence on
simple premises. More complicated dynamical theories may be capable of giving
finer detail and wider application but this is often at the expense of more
elaborate mathematical analysis. Many problems are capable of solution within
the terms of reference of the Newtonian system and they can serve as models
for further physical situations.

MODELLING

This book is in large part devoted to particle dynamics but this is not to
say that we restrict ourselves to 'particles'. Mathematically a particle can be
considered as a point in space which has finite mass, or physically as an object
of negligible dimension but with finite mass. Whilst the former definition is an
idealisation, the latter definition may be a useful approximation to a real
situation depending on whether we can justify the adjective 'negligible'. The
Earth's diameter is approximately 6400 km and the radius of the Earth's orbit
about the Sun is about 1.5×10^8 km and in certain problems it is useful to
think of the Earth as a particle moving about the Sun since its dimensions are
so small compared with the dimension of its orbit. However, if we are interested
in computing eclipses between the Earth, Moon and Sun, the diameters of the
Earth and Moon are of particular importance.

As we shall see later, the centre of mass of a rigid body behaves like a particle
whose mass equals the mass of the body. Thus the application of particle
dynamics is not restricted simply to particles but does have application to
bodies. Whilst we can determine the position of the centre of mass of a rigid
body using the principles of particle dynamics, we cannot discuss the orientation

of the body about its centre of mass. We can track a rocket leaving the Earth but we cannot discuss how it is tumbling about its centre of mass.

The art of modelling is often a difficult compromise. In constructing a model for a real situation, one is attempting to achieve simplicity whilst still retaining what one hopes are the significant features of the problem. In practice this often entails *ad hoc* estimates of the relative weight one must attach to various effects which may be present. To some extent this is overcome by examining 'control' problems where various effects are isolated in a single problem. For example, a car may be considered as a particle if we are interested in its acceleration and velocity, a box on four springs if we are concerned with its damping characteristics, or a box on four springs which are attached to four wheels if we are interested in its more general dynamical behaviour.

The real test of a good model lies in tolerable agreement between the theoretical and experimental results. However, it is not always possible to test agreement for reasons of expense. We can find the maximum load which a bridge can tolerate by simply adding weights until the bridge collapses. Clearly non-destructive testing is cheaper, and a theoretical assessment of the bridge's characteristics together with a sufficient margin of safety between the theoretical maximum load and the permitted load is a more reasonable approach to the problem.

MATHEMATICAL ANALYSIS AND APPROXIMATION

All discussion of fundamental postulates and modelling loses much of its point if it leads to heavy or intractable mathematical analysis. An important feature of particle dynamics from a teaching point of view is that many problems can be constructed which do have simple analytic answers. However, much criticism has been levelled at the artificiality of the traditional sort of examination question which has achieved a conventionality of its own. The sphere rolling down the inclined face of a wedge which can slide on a rough horizontal plane has taxed the ingenuity of teacher and student alike for many decades. Indeed some textbooks of dynamics approach the subject as a self-contained discipline with no apologies to the reader. The ideal problem would state: model the following situation and solve the resulting equations. This would be unfair on the beginner since what may seem to be reasonable assumptions may lead to a system of equations which cannot be readily solved. We have attempted to include at the end of each chapter a rather mixed collection of problems which cover the theory and application without resorting too much to the artificial type of exercise cited above.

The art of approximation depends greatly on experience. The applied mathematician tries to keep one eye on the practical implications of an approximation and the other on the mathematics which it entails. The most elegant theory is the one which supplies useful information with the minimum of mathematical analysis.

Differential equations play a key role in dynamics and because of this we have included an appendix on analytic and numerical methods of solving ordinary differential equations. Most problems in dynamics lead to one or more ordinary differential equations, and it seems appropriate to include the necessary theory in an introductory book on dynamics. Differential equations can be usefully classified into two categories, those which are linear and the rest, which are non-linear. There are a number of general methods of solving certain types of linear ordinary differential equations. On the other hand, non-linear ordinary differential equations can only usually be solved analytically, if at all, by the use of miscellaneous *ad hoc* techniques. For this sound practical reason applied mathematics has developed a heavy emphasis towards the linear or approximately linear phenomena.

DYNAMICAL SYSTEMS AND COMPUTING

As we have just remarked, many simplified models in dynamics lead to linear differential equations, but we need to remain aware that in almost all applications the actual governing equations are non-linear. Until recent times it was felt that Newtonian mechanics might be complicated but, given sufficiently powerful mathematical techniques and computing hardware, it should be possible to predict the future behaviour of any system. Although there remain many outstanding difficult theoretical questions such as the analytic solution of the *n*-body problem, interest in the subject from a research point of view had become less active.

The introduction of powerful and accessible computing facilities in the past 20 years has changed the situation. The numerical solution of relatively simple dynamical systems began to cast doubt on the universal predictive capacity of Newtonian theories. It was discovered that quite 'simple' model systems such as the forced pendulum, the bouncing ball and the gyroscope are *deterministic* (that is, their future states are the solutions of differential equations and not probability models), but that computer simulations in some parametric states could only predict outcomes over the short term. To some extent this notion cut across received wisdom in Newtonian mechanics, although Henri Poincaré (1854–1912) at the beginning of this century was aware of this possibility.

Such systems are said now to exhibit *chaos*, and one way in which it displays itself is in extreme sensitivity to initial data. Of course, not all parametric states of a dynamical system display this behaviour, and sometimes orderly states can co-exist with chaotic ones. The Earth continues to spin on its axis; and the Earth predictably orbits the Sun. However, all is not so orderly in the solar system. It has recently been observed from the *Voyager* probe that *Hyperion*, a satellite of Saturn, tumbles in what appears to be a random manner. Its future attitude a few months hence is effectively not predictable. Newton's laws of motion still apply but since any initial data can only be observed to within certain errors, it seems unlikely that a sufficiently accurate fix on the satellite can ever be found to predict its long term future orientation. Chaotic outcomes are usually bounded and irregular, and effectively random within this state. This leads to interesting questions about the stability of the spin and orbits of the planets, and of the whole solar system. Do we know whether the orientation of the Earth is close to a chaotic state which could be triggered by a small parametric change?

Chaotic behaviour limits weather prediction, may cause turbulence in fluid flow and arises in the logistic difference equation in population dynamics and in many other applications in subjects such as biology, economics and medicine. The simulation of these chaotic states has depended on the ability to compute, at high speed, numerical solutions of ordinary differential and difference equations, and much can now be achieved with relatively simple programs run on microcomputers. The mixture of mathematical analysis and computer experimentation is now becoming an important feature in the study of dynamical systems.

1
Vectors

1.1 DEFINITIONS

In the study of mechanics it is useful to distinguish between quantities which have magnitude only and those with which we can associate both magnitude and direction in space. The former are called *scalars* and include, for example, mass, length and temperature, and the latter are called *vectors*, of which velocity and acceleration are examples. A vector can be thought of as a directed quantity with specific rules of addition which we will define in the next section.

Geometrically, a vector can be represented in three-dimensional space by a straight line, say AB in Figure 1.1. In this form we express the vector as a *directed line segment* written as $\overline{AB}$ in the direction of the vector from point A to point B as signified by the order of the letters. The magnitude of the directed line segment is its length AB. Our definition of a vector specifies only the direction and magnitude of a directed line segment but not its position in space. Thus $\overline{CD}$ in Figure 1.1 is also a representation of the same vector as $\overline{AB}$. In fact, any directed line segment parallel to $\overline{AB}$ and with the same length as $\overline{AB}$ is a representative of the vector. We can write $\overline{AB} = \overline{CD}$ without ambiguity. Hence $\overline{AB}$ stands for the whole collection of directed line segments rather than just one individual one. These vectors which are not anchored to any particular point in space are known as *free vectors*.

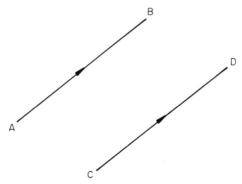

Figure 1.1   Directed line segments.

Notwithstanding this description of vectors, certain vectors have the additional constraint of being associated with particular points (not necessarily fixed) in space. These are known as *fixed vectors*. The actual manipulation of vectors, whether free or fixed, is not materially affected by their status but it is as well to be aware of the distinction between the two.

1.2 ADDITION OF VECTORS

Consider the triangle ABC in Figure 1.2. In the triangle $\overline{AB}$, $\overline{AC}$, $\overline{BC}$ represent vectors. We define the *sum* of $\overline{AB}$ and $\overline{BC}$ to be $\overline{AC}$ and we write

$$\overline{AB} + \overline{BC} = \overline{AC}.$$

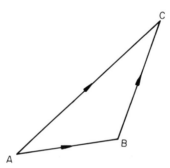

Figure 1.2   Triangle law of addition for vectors.

(It is important to observe the order in which the 'letters' are written down.) One consequence of this law is the following:

$$\overline{AB} + \overline{BA} = \overline{AA} \tag{1}$$

where $\overline{AA}$ is defined as the zero or null vector. Thus (1) may be written as

$$\overline{BA} = -\overline{AB}.$$

This tells us that if we write the letters of a vector in the reverse order, we must introduce a negative sign in any additive processes we are considering. Care must therefore be exercised when writing down any vector equation to ensure that all the vectors are written down in the correct sense. From our earlier remarks about free vectors, this law of addition will apply to any three vectors which can be translated to form a triangle as in Figure 1.2.

This property, that vectors add according to the triangle law, is what distinguishes a vector from a directed quantity as such. There are some directed quantities, a rotation for example, which do not combine like vectors and so we cannot represent them by vectors.

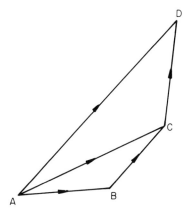

Figure 1.3 Extended triangle law.

The triangle law can be readily extended to the addition of three or more vectors. For example in Figure 1.3

$$\overline{AD} = \overline{AC} + \overline{CD}$$
$$= \overline{AB} + \overline{BC} + \overline{CD}.$$

1.3 AN ALTERNATIVE NOTATION

The geometrical notation is not particularly adaptable to the requirements of dynamics. The alternative notation in which bold-face type, **b**, represents a vector and the ordinary italic, *b*, a scalar is more convenient. On the blackboard or in written work the form $\underline{b}$ can be used to distinguish vectors from scalars. Always remember that a representative of a vector **b** can be interpreted mentally as a straight line joining two points Q and R. The magnitude of **b** is denoted by |**b**| (called the modulus of **b**) or simply *b*. The vector −**b** is a vector in the opposite direction to, but with the same magnitude as **b**. Geometrically it will be $\overline{RQ}$.

In the application of the triangle law to the sum **a** + **b** we draw **a** and **b** as the two sides of a triangle in sense indicated by the arrows in Figure 1.4. The sum is represented by the third side **c**.

Vectors obey the following rules which can be deduced from the triangle law:

(i) $\mathbf{a} + \mathbf{b} = \mathbf{b} + \mathbf{a}$ commutative rule

(ii) $\mathbf{a} + (\mathbf{b} + \mathbf{c}) = (\mathbf{a} + \mathbf{b}) + \mathbf{c}$ associative rule

(iii) $m(\mathbf{a} + \mathbf{b}) = m\mathbf{a} + m\mathbf{b}$; $(mn)\mathbf{a} = m(n\mathbf{a})$; $(m + n)\mathbf{a} = m\mathbf{a} + n\mathbf{a}$

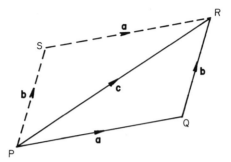

Figure 1.4   Commutative rule for addition.

for any real numbers m and n. The first may be verified by inspection of Figure 1.4. The left-hand side of (i) is represented in the figure by the path PQR whilst the right-hand side is represented by the dashed path PSR. Since PQRS is a parallelogram, both of these by the triangle law are equal to **c**. The proofs of (ii) and (iii) are left as exercises for the reader. In (iii) m**a** is vector of magnitude $m|$**a**$|$ in the direction of **a** if m is positive, and in the opposite direction to **a** if m is negative. By $-$**a** we mean (-1)**a**.

The zero or null vector which we defined geometrically in Section 1.2 can now be written **0**. Thus

$$\mathbf{a} - \mathbf{a} = \mathbf{0}.$$

1.4 SCALAR AND VECTOR PRODUCTS

There are two products which can be usefully associated with vectors in three dimensions—the so-called scalar and vector products. We start by introducing 'physical' definitions of them.

The *scalar product* of two vectors **a** and **b** is a scalar defined as the product of a, b and the cosine of the smaller angle θ between the vectors (see Figure 1.5). We denote it by **a** $\cdot$ **b** (called **a** dot **b**). Thus

$$\mathbf{a} \cdot \mathbf{b} = ab \cos \theta, \qquad 0 \le \theta \le \pi.$$

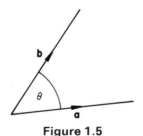

Figure 1.5

As shown in Figure 1.5 the angle θ between two vectors **a** and **b** is defined as the smaller angle between the vectors when drawn from the same point.

Example 1   *Show that* $\mathbf{a} \cdot \mathbf{b} = \mathbf{b} \cdot \mathbf{a}$.

By definition

$$\mathbf{a} \cdot \mathbf{b} = ab \cos \theta$$
$$= ba \cos \theta = \mathbf{b} \cdot \mathbf{a}.$$

Example 2   *Show that* $\mathbf{a} \cdot (\mathbf{b} + \mathbf{c}) = \mathbf{a} \cdot \mathbf{b} + \mathbf{a} \cdot \mathbf{c}$.

We note that, by definition

$$\mathbf{a} \cdot \mathbf{b} = a(b \cos \theta)$$
$$= a \times (\text{magnitude of the projection of } \mathbf{b} \text{ on } \mathbf{a}).$$

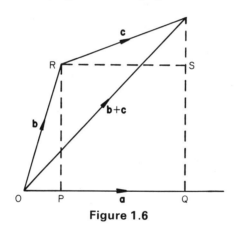

Figure 1.6

Using Figure 1.6

$$\mathbf{a} \cdot \mathbf{b} + \mathbf{a} \cdot \mathbf{c} = a \times \text{OP} + a \times \text{RS}$$
$$= a \times \text{OP} + a \times \text{PQ}$$
$$= a \times \text{OQ}$$
$$= a \times (\text{projection of } \mathbf{b} + \mathbf{c} \text{ on } \mathbf{a})$$
$$= \mathbf{a} \cdot (\mathbf{b} + \mathbf{c}).$$

We note that if m is a number, then

$$m(\mathbf{a} \cdot \mathbf{b}) = (m\mathbf{a}) \cdot \mathbf{b} = \mathbf{a} \cdot (m\mathbf{b})$$

by the definition of scalar product.

If $\mathbf{a} \cdot \mathbf{b} = 0$ and neither $\mathbf{a}$ nor $\mathbf{b}$ is a null vector, then $\cos \theta$ must vanish, implying that $\theta = \frac{1}{2}\pi$ which means the vectors are perpendicular.

The *vector* (or *cross*) *product* of $\mathbf{a}$ and $\mathbf{b}$ is a *vector* $\mathbf{c}$ whose magnitude is $ab \sin \theta$ where θ is the angle between $\mathbf{a}$ and $\mathbf{b}$ such that $0 \le \theta \le \pi$. The direction of $\mathbf{c}$ is perpendicular to both $\mathbf{a}$ and $\mathbf{b}$ and such that $\mathbf{a}$, $\mathbf{b}$ and $\mathbf{c}$ form a right-hand system of vectors. The right-hand screw rule applied to $\mathbf{a}$, $\mathbf{b}$, $\mathbf{c}$ states that a right-handed screw rotation along $\mathbf{c}$ turns from $\mathbf{a}$ to $\mathbf{b}$ through the angle θ (Figure 1.7). This is a conventional definition: in vectors and mechanics we adopt the right-hand rather than the left-hand convention. We write the product as

$$\mathbf{c} = \mathbf{a} \times \mathbf{b} \quad \text{(said as } \mathbf{a} \text{ vec } \mathbf{b}, \text{ or } \mathbf{a} \text{ cross } \mathbf{b})$$

where

$$|\mathbf{c}| = |\mathbf{a} \times \mathbf{b}| = ab \sin \theta.$$

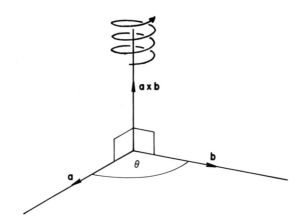

Figure 1.7   Vector product and right-hand screw convention.

Example 3   Show that $\mathbf{a} \times \mathbf{b} = -\mathbf{b} \times \mathbf{a}$. *Unlike the scalar product, the vector product is not commutative.*

The vector $\mathbf{a} \times \mathbf{b}$ has magnitude $ab \sin \theta$ and a direction determined by the right-hand screw rule. The vector $\mathbf{b} \times \mathbf{a}$ also has magnitude $ab \sin \theta$ but its direction is opposite to $\mathbf{a} \times \mathbf{b}$ by a second use of the rule (Figure 1.8). If two vectors are equal in magnitude but opposite in direction, their sum vanishes and it therefore follows that

$$\mathbf{a} \times \mathbf{b} = -\mathbf{b} \times \mathbf{a}.$$

This example indicates that care must be taken when writing down vector products.

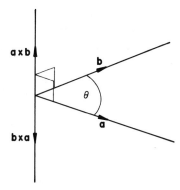

Figure 1.8 Non-commutative property of vector product.

We quote the following two results without proof:

(i) $\mathbf{a} \times (\mathbf{b} + \mathbf{c}) = \mathbf{a} \times \mathbf{b} + \mathbf{a} \times \mathbf{c}$,

(ii) $m(\mathbf{a} \times \mathbf{b}) = (m\mathbf{a}) \times \mathbf{b} = \mathbf{a} \times (m\mathbf{b})$.

If $\mathbf{a} \times \mathbf{b} = 0$ and neither $\mathbf{a}$ nor $\mathbf{b}$ is a null vector then $\sin \theta$ must vanish, implying that $\theta = 0$ or π. The two vectors are therefore parallel, pointing in the same direction if $\theta = 0$ and in opposite directions if $\theta = \pi$.

The vector product has a geometrical interpretation in that its magnitude is the area of the parallelogram constructed on the representatives of $\mathbf{a}$ and $\mathbf{b}$ drawn from the same point.

1.5 UNIT VECTORS AND RECTANGULAR AXES

A *unit vector* is simply a vector of unit magnitude. For any non-zero vector $\mathbf{b}$, $\mathbf{b}/b$ is a unit vector in the same direction as $\mathbf{b}$. Unit vectors provide a useful means of specifying directions in space.

An important set or triple of unit vectors is associated with the Cartesian or rectangular frame reference. A frame of reference is necessary in order to describe the location of a point in space in relation to other points. The simplest frame is the rectangular Cartesian one and it consists of three mutually perpendicular straight lines or axes Ox, Oy, Oz intersecting in the point O, called the origin (Figure 1.9). We again choose to use a right-handed system in the sense that a right-handed screw rotation along the z-axis turns from Ox to Oy through a right angle: if Oy and Oz in Figure 1.9 are in the plane of the paper Ox will point at the reader.

Any point P can be related to axes by a triple of numbers a, b, c whose magnitudes are the perpendicular distances of P from each of the planes Oyz, Oxz, Oxy. From Figure 1.9 these are equal respectively to OQ, OR, OS. We

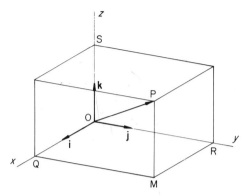

Figure 1.9   Rectangular Cartesian coordinate system: unit vectors.

now adopt the convention that a, b, c measured in the directions Ox, Oy, Oz are positive whilst a, b, c measured from O in the opposite directions are negative. The numbers (a, b, c) are the *coordinates* of P.

We may associate unit vectors with the directions of the axes. These are customarily written **i**, **j**, **k** and are in the directions Ox, Oy, Oz respectively. Using the triangle law of addition in Figure 1.9

$$\overline{OP} = \overline{OQ} + \overline{QM} + \overline{MP}$$
$$= a\mathbf{i} + b\mathbf{j} + c\mathbf{k}.$$

Since P is any point it is implied that any vector can be expressed in terms of these Cartesian unit vectors and, as we shall see, this representation is particularly important in computations involving vectors. The directed line segment $\overline{OP}$ drawn from O is called the *position vector* of P with respect to O and is usually denoted by **r**. Thus **r** is a *fixed* vector given by

$$\mathbf{r} = a\mathbf{i} + b\mathbf{j} + c\mathbf{k}$$

which is also often written $\mathbf{r} = (a, b, c)$. The numbers (a, b, c) as well as being the coordinates of P are also referred to as the components of **r** in the appropriate context and are the projections of **r** onto the coordinate axes.

The magnitude of **r** is given by

$$|\mathbf{r}| = r = OP = (OQ^2 + QP^2)^{1/2}$$
$$= (OP^2 + QM^2 + MP^2)^{1/2}$$
$$= (a^2 + b^2 + c^2)^{1/2}.$$

Example 4   Write down the position vector of the point with coordinates $(1, -1, 2)$ and find its magnitude.

The position vector of this point is given by

$$\mathbf{r} = \mathbf{i} - \mathbf{j} + 2\mathbf{k}$$

and its magnitude is $|\mathbf{r}| = r = (1 + 1 + 4)^{1/2} = \sqrt{6}$.

Example 5   *Find the vector $\overline{AB}$ where A and B have coordinates $(-1, 3, 5)$ and $(2, -1, 3)$.*

By the triangle law

$$\overline{OB} = \overline{OA} + \overline{AB}$$
$$\overline{AB} = \overline{OB} - \overline{OA}.$$

The position vectors of A and B are given by

$$\overline{OA} = -\mathbf{i} + 3\mathbf{j} + 5\mathbf{k}$$
$$\overline{OB} = 2\mathbf{i} - \mathbf{j} + 3\mathbf{k}$$

and therefore

$$\overline{AB} = (2\mathbf{i} - \mathbf{j} + 3\mathbf{k}) - (-\mathbf{i} + 3\mathbf{j} + 5\mathbf{k})$$
$$= 3\mathbf{i} - 4\mathbf{j} - 2\mathbf{k}.$$

This vector can be thought of as the position vector of B *relative* to *A*.

Note that the vector equation

$$\mathbf{a} = \mathbf{b}$$

where $\mathbf{a} = a_1\mathbf{i} + a_2\mathbf{j} + a_3\mathbf{k}$ and $\mathbf{b} = b_1\mathbf{i} + b_2\mathbf{j} + b_3\mathbf{k}$ means that

$$a_1\mathbf{i} + a_2\mathbf{j} + a_3\mathbf{k} = b_1\mathbf{i} + b_2\mathbf{j} + b_3\mathbf{k}$$

which implies the three scalar equations

$$a_1 = b_1, \qquad a_2 = b_2, \qquad a_3 = b_3.$$

Since the Cartesian system is right-handed, the unit vectors **i**, **j**, **k** satisfy the following relations:

(i) $\mathbf{i} \cdot \mathbf{j} = \mathbf{j} \cdot \mathbf{k} = \mathbf{k} \cdot \mathbf{i} = 0$

(ii) $\mathbf{i} \cdot \mathbf{i} = \mathbf{j} \cdot \mathbf{j} = \mathbf{k} \cdot \mathbf{k} = 1$

(iii) $\mathbf{i} \times \mathbf{j} = \mathbf{j} \times \mathbf{j} = \mathbf{k} \times \mathbf{k} = 0$

(iv) $\mathbf{i} \times \mathbf{j} = \mathbf{k}, \mathbf{j} \times \mathbf{k} = \mathbf{i}, \mathbf{k} \times \mathbf{i} = \mathbf{j}$, (note the cycle order in **i**, **j**, **k**).

The scalar product of two vectors **a** and **b** in terms of their components is given by

$$\begin{aligned}\mathbf{a} \cdot \mathbf{b} &= (a_1\mathbf{i} + a_2\mathbf{j} + a_3\mathbf{k}) \cdot (b_1\mathbf{i} + b_2\mathbf{j} + b_3\mathbf{k}) \\ &= a_1b_1\mathbf{i} \cdot \mathbf{i} + a_1b_2\mathbf{i} \cdot \mathbf{j} + a_1b_3\mathbf{i} \cdot \mathbf{k} + \cdots \\ &= a_1b_1 + a_2b_2 + a_3b_3,\end{aligned} \qquad (2)$$

using Example 2 and (i) and (ii) above. In particular if $\mathbf{a} = \mathbf{b}$, we note that $|\mathbf{a}| = (\mathbf{a} \cdot \mathbf{a})^{1/2}$.

The vector product of two vectors is given by

$$
\begin{aligned}
\mathbf{a} \times \mathbf{b} &= (a_1\mathbf{i} + a_2\mathbf{j} + a_3\mathbf{k}) \times (b_1\mathbf{i} + b_2\mathbf{j} + b_3\mathbf{k}) \\
&= a_1b_1\mathbf{i} \times \mathbf{i} + a_1b_2\mathbf{i} \times \mathbf{j} + a_1b_3\mathbf{i} \times \mathbf{k} + a_2b_1\mathbf{j} \times \mathbf{i} + a_2b_2\mathbf{j} \times \mathbf{j} \\
&\quad + a_2b_3\mathbf{j} \times \mathbf{k} + a_3b_1\mathbf{k} \times \mathbf{i} + a_3b_2\mathbf{k} \times \mathbf{j} + a_3b_3\mathbf{k} \times \mathbf{k} \\
&= a_1b_2\mathbf{k} - a_1b_3\mathbf{j} - a_2b_1\mathbf{k} + a_2b_3\mathbf{i} + a_3b_1\mathbf{j} - a_3b_2\mathbf{i} \\
&= (a_2b_3 - a_3b_2)\mathbf{i} + (a_3b_1 - a_1b_3)\mathbf{j} + (a_1b_2 - a_2b_1)\mathbf{k} \qquad (3)
\end{aligned}
$$

by (iii) and (iv) above. For those readers familiar with determinants the vector product may be written symbolically as

$$
\mathbf{a} \times \mathbf{b} = \begin{vmatrix} \mathbf{i} & \mathbf{j} & \mathbf{k} \\ a_1 & a_2 & a_3 \\ b_1 & b_2 & b_3 \end{vmatrix}
$$

and evaluated as such.

In many textbooks on vectors (2) and (3) are used as the definitions of scalar and vector products and the 'physical' properties deduced from them.

Example 6   Find a formula for the angle between two vectors $\mathbf{a}$ and $\mathbf{b}$ in terms of their components.

By definition

$$
\mathbf{a} \cdot \mathbf{b} = ab \cos \theta
$$

where θ is the angle between the vectors. Thus, using (2)

$$
\cos \theta = \frac{\mathbf{a} \cdot \mathbf{b}}{ab} = \frac{a_1b_1 + a_2b_2 + a_3b_3}{(a_1^2 + a_2^2 + a_3^2)^{1/2}(b_1^2 + b_2^2 + b_3^2)^{1/2}}.
$$

Example 7   If $\mathbf{a} = \mathbf{i} + 2\mathbf{j} - \mathbf{k}$ and $\mathbf{b} = 2\mathbf{i} + \mathbf{j} + 3\mathbf{k}$ determine (i) $\mathbf{a} \cdot \mathbf{b}$, (ii) $\mathbf{a} \times \mathbf{b}$, (iii) $(\mathbf{a} \times \mathbf{b}) \times \mathbf{a}$.

(i) $\mathbf{a} \cdot \mathbf{b} = 2 + 2 - 3 = 1$.

(ii) $\mathbf{a} \times \mathbf{b} = [6 - (-1)]\mathbf{i} + (-2 - 3)\mathbf{j} + (1 - 4)\mathbf{k} = 7\mathbf{i} - 5\mathbf{j} - 3\mathbf{k}$.

(iii) Using (ii)

$$
\begin{aligned}
(\mathbf{a} \times \mathbf{b}) \times \mathbf{a} &= (7\mathbf{i} - 5\mathbf{j} - 3\mathbf{k}) \times (\mathbf{i} + 2\mathbf{j} - \mathbf{k}) \\
&= [5 - (-6)]\mathbf{i} + [-3 - (-7)]\mathbf{j} + [14 - (-5)]\mathbf{k} \\
&= 11\mathbf{i} + 4\mathbf{j} + 19\mathbf{k}.
\end{aligned}
$$

1.6 TRIPLE PRODUCTS

Example 7(iii) above is an illustration of a *triple product* between three vectors. A triple product between three vectors involving scalar and vector products can only be constructed in certain ways, which must be consistent with the original definitions. For example, $(\mathbf{b} \cdot \mathbf{c}) \times \mathbf{a}$ has no meaning since the term in parentheses is a scalar and its vector product with $\mathbf{a}$ is undefined. The product $\mathbf{a} \times \mathbf{b} \times \mathbf{c}$ is ambiguous since there is no indication of the order in which the products should be taken.

A product between three vectors involving a scalar and a vector product is called a *scalar triple product*. For example, $\mathbf{a} \cdot \mathbf{b} \times \mathbf{c}$ is a scalar triple product and the answer is a scalar. No brackets are needed since the product is unambiguous: the vector product must be evaluated first, followed by the scalar product. Geometrically, the absolute value of $\mathbf{a} \cdot \mathbf{b} \times \mathbf{c}$ is the volume of the parallelepiped constructed on the representatives of $\mathbf{a}$, $\mathbf{b}$, $\mathbf{c}$ drawn from the same point. This can be seen from

$$\mathbf{a} \cdot \mathbf{b} \times \mathbf{c} = abc \sin \theta \cos \phi$$

and Figure 1.10, using the definitions of scalar and vector product. If $\mathbf{a}$, $\mathbf{b}$ and $\mathbf{c}$ are all non-zero vectors, $\mathbf{a} \cdot \mathbf{b} \times \mathbf{c} = 0$ implies that the volume of the parallelepiped vanishes and that $\mathbf{a}$, $\mathbf{b}$ and $\mathbf{c}$ are coplanar, that is they lie in the same plane when drawn from the same point.

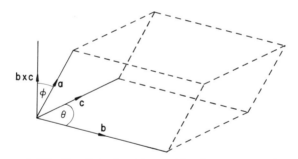

Figure 1.10 Parallelepiped formed by the vectors **a**, **b** and **c**.

With

$$\mathbf{a} = a_1\mathbf{i} + a_2\mathbf{j} + a_3\mathbf{k}, \quad \mathbf{b} = b_1\mathbf{i} + b_2\mathbf{j} + b_3\mathbf{k}, \quad \mathbf{c} = c_1\mathbf{i} + c_2\mathbf{j} + c_3\mathbf{k}$$

the scalar triple product in terms of Cartesian components becomes

$$\mathbf{a} \cdot \mathbf{b} \times \mathbf{c} = a_1(b_2c_3 - b_3c_2) + a_2(b_3c_1 - b_1c_3) + a_3(b_1c_2 - b_2c_1)$$

$$= \begin{vmatrix} a_1 & a_2 & a_3 \\ b_1 & b_2 & b_3 \\ c_1 & c_2 & c_3 \end{vmatrix}$$

in determinant form. The reader can verify directly, or by interchanging rows in the determinant, that

$$\mathbf{a} \cdot \mathbf{b} \times \mathbf{c} = \mathbf{c} \cdot \mathbf{a} \times \mathbf{b} = \mathbf{b} \cdot \mathbf{c} \times \mathbf{a}.$$

Note that the cyclic order is preserved.

The *vector triple product* involves two vector products between three vectors. Thus, $\mathbf{a} \times (\mathbf{b} \times \mathbf{c})$ is an example of a vector triple product. The brackets are now important since in general,

$$\mathbf{a} \times (\mathbf{b} \times \mathbf{c}) \neq (\mathbf{a} \times \mathbf{b}) \times \mathbf{c}.$$

This can be verified by a counter-example or by the following geometrical argument. Referring back to Figure 1.10, we recall that $\mathbf{b} \times \mathbf{c}$ is perpendicular to the plane of $\mathbf{b}$ and $\mathbf{c}$. Further $\mathbf{a} \times (\mathbf{b} \times \mathbf{c})$ will be perpendicular to $\mathbf{a}$ and $\mathbf{b} \times \mathbf{c}$, that is it must lie in the plane of $\mathbf{b}$ and $\mathbf{c}$. By a similar chain of reasoning $(\mathbf{a} \times \mathbf{b}) \times \mathbf{c}$ lies in the plane of $\mathbf{a}$ and $\mathbf{b}$. Except in the case when the resultant is in the direction $\mathbf{b}$, the inequality is demonstrated.

The vector triple product satisfies the following identity in terms of scalar products

$$\mathbf{a} \times (\mathbf{b} \times \mathbf{c}) = (\mathbf{a} \cdot \mathbf{c})\mathbf{b} - (\mathbf{a} \cdot \mathbf{b})\mathbf{c}.$$

This can be verified by expressing both sides in terms of their Cartesian components. We leave this as an exercise for the reader.

1.7 MOMENT OF A VECTOR

The moment $\mathbf{M}$ about the origin O of the representative of the vector $\mathbf{F}$ located at P (usually abbreviated by 'F at P') which has position vector $\mathbf{r}$ is defined as

$$\mathbf{M} = \mathbf{r} \times \mathbf{F}$$

If $\mathbf{F} = F_1\mathbf{i} + F_2\mathbf{j} + F_3\mathbf{k}$ and $\mathbf{r} = x\mathbf{i} + y\mathbf{j} + z\mathbf{k}$, then, in component form, the moment $\mathbf{M}$ is given by

$$\mathbf{M} = (yF_3 - zF_2)\mathbf{i} + (zF_1 - xF_3)\mathbf{j}$$
$$+ (xF_2 - yF_1)\mathbf{k}.$$

From the definition of the vector product

$$|\mathbf{M}| = |\mathbf{r}||\mathbf{F}| \sin(\pi - \theta) \qquad \text{(Figure 1.11)}$$
$$= |\mathbf{F}|OP \sin \theta$$
$$= |\mathbf{F}|d$$

where $(\pi - \theta)$ is the angle between the actual direction of the two vectors and d is the perpendicular distance from O onto the line through $\mathbf{F}$ at P.

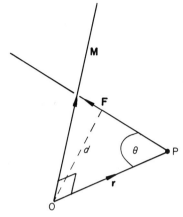

Figure 1.11 Moment **M** of **F** at P about O.

If the vectors $\mathbf{F}_1, \mathbf{F}_2, \ldots$, act at the points with position vectors $\mathbf{r}_1, \mathbf{r}_2, \ldots$, the total moment is the vector sum of the individual moments, that is

$$\mathbf{M} = \mathbf{r}_1 \times \mathbf{F}_1 + \mathbf{r}_2 \times \mathbf{F}_2 + \ldots.$$

1.8 SPHERICAL POLAR COORDINATES

Many other coordinate systems are available and useful for particular classes of problems. The coordinates r, θ, ϕ shown in Figure 1.12 are spherical polar coordinates. The point P can be considered as the intersection of a sphere

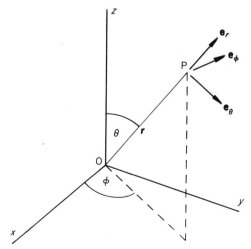

Figure 1.12 Spherical polar coordinates (r, θ, ϕ).

($r = $ constant), a cone ($\theta = $ constant) and a plane ($\phi = $ constant). For this reason spherical polar coordinates usually find their application in problems involving cones or spheres. For those familiar with geographical location on the surface of the earth, ϕ is the measure of longitude (from the Greenwich meridian on the surface of the Earth) and θ is the co-latitude (the latitude being measured from the equator on the Earth), where we have taken the origin at the centre of the Earth and z-axis through the North Pole. The relation between Cartesian and spherical polar coordinates can be worked out from Figure 1.12:

$$x = r \sin \theta \cos \phi \qquad y = r \sin \theta \sin \phi \qquad z = r \cos \theta.$$

We can associate unit vectors $\mathbf{e}_r$, $\mathbf{e}_\theta$, $\mathbf{e}_\phi$ with this coordinate system such that $\mathbf{e}_r$ is in the direction in which θ and ϕ are constant, with r increasing, etc. By analogy with the Earth's surface again, $\mathbf{e}_r$ is pointing vertically upward, $\mathbf{e}_\theta$ points to the south and $\mathbf{e}_s$ to the east. Unlike the Cartesian triple $\mathbf{i}$, $\mathbf{j}$, $\mathbf{k}$, the unit vectors $\mathbf{e}_r$, $\mathbf{e}_\theta$, $\mathbf{e}_\phi$ do not point in fixed directions in space. The unit vectors are right-handed and orthogonal so that

$$\mathbf{e}_r = \mathbf{e}_\theta \times \mathbf{e}_\phi, \qquad \mathbf{e}_\theta = \mathbf{e}_\phi \times \mathbf{e}_r, \qquad \mathbf{e}_\phi = \mathbf{e}_r \times \mathbf{e}_\theta.$$

1.9 PLANE VECTORS

Many problems occurring in later chapters will be essentially two-dimensional. In these cases we can locate any point in the plane of the two dimensions by reference to two rectangular axes Ox, Oy. If P is a point in the plane with coordinates (x, y) its position can be determined by the plane position vector

$$\mathbf{r} = x\mathbf{i} + y\mathbf{j}.$$

If vector products are included then the third dimension specified is required since the vector product of any two plane vectors will be in the direction $\mathbf{k}$ perpendicular to the Oxy plane.

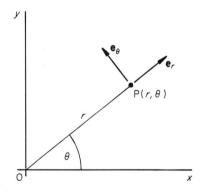

Figure 1.13 Plane polar coordinates.

A second useful representation of a point P is by *polar coordinates* (r, θ) where $r = |\mathbf{r}|$ and θ is the angle between the position vector and the positive direction of the axis (Figure 1.13). Clearly

$$x = r \cos \theta, \qquad\qquad y = r \sin \theta$$
$$r = (x^2 + y^2)^{1/2} \qquad \tan \theta = y/x.$$

With each of the coordinates r and θ we can associate unit vectors $\mathbf{e}_r$ and $\mathbf{e}_\theta$ as shown in Figure 1.13: $\mathbf{e}_r$ is called the *unit radial vector* and $\mathbf{e}_\theta$ the *unit transverse vector*. We observe that $\mathbf{r} = r\mathbf{e}_r$.

1.10 VELOCITY AND ACCELERATION

We wish now to consider derivatives of vectors and this can be most usefully illustrated through velocity and acceleration. Consider a particle P moving along a straight line such that its displacement x from a fixed point O of the line is given at time t by $x(t)$ (by this symbol we mean that x is a function of t or depends on t). At time t_1 the particle will have a displacement $x(t_1)$ (the point P$_1$ in Figure 1.14). Positive displacements occur to the right and negative to the left. In the time interval $t - t_1$, the particle will move a distance $x(t) - x(t_1)$. The *average velocity* $\bar{v}$ over this interval is defined by

$$\bar{v} = \frac{x(t) - x(t_1)}{t - t_1}. \tag{4}$$

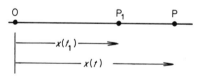

Figure 1.14

The *instantaneous velocity* v of the particle at time t is defined as the value of the right-hand side of (4) as t_1 approaches t. This is the mathematical notion of limit. Thus

$$v = \lim_{t_1 \to t} \frac{x(t) - x(t_1)}{t - t_1},$$

which is, by definition, the derivative of x with respect to t expressed as

$$v = \frac{dx}{dt} \quad \text{or } \dot{x}. \tag{5}$$

Speed is the magnitude of the velocity or $|\dot{x}|$. The use of the notation $\dot{x}$ (said as 'x dot') for dx/dt has a long history in mechanics.

Example 8   If the displacement, in metres, of a particle is given by

$$x = 12(t^2 + 2t)$$

where t is measured in seconds, find its velocity after three seconds.

At any time t

$$v = \frac{dx}{dt} = 12(2t + 2) \text{ metres per second (m/s)}. \tag{6}$$

At $t = 3$ s, $v = 12(6 + 2) = 96$ m/s.

Whilst velocity is rate of change of displacement, acceleration f is, in turn, the rate of change of velocity. Thus

$$f = \lim_{t_1 \to t} \frac{v(t) - v(t_1)}{t - t_1} = \frac{dv}{dt} = \dot{v} = \frac{d^2x}{dt^2} \quad \text{or } \ddot{x}$$

using (5): acceleration is the second time derivative of displacement.

Example 9   Show that the acceleration in Example 8 is constant at all times.

From (6)

$$\frac{dv}{dt} = 12(2) = 24 \text{ metres per second per second (m/s}^2\text{)}.$$

which is independent of time and constant.

An important alternative formula for acceleration exists if we consider speed depending on displacement, and use the rule for the derivative of a function of a function in calculus:

$$f = \frac{dv}{dt} = \frac{dv}{dx} \cdot \frac{dx}{dt} = v \frac{dv}{dx}$$

a form which will be used frequently in subsequent work.

When a particle moves along a space curve, as indicated in Figure 1.15, its position vector **r** will be dependent on time t. Let the particle be at Q at time

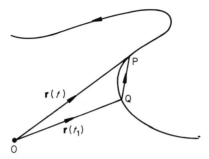

Figure 1.15 Position vector **r** of a particle describing a curve in space.

t_1 and at P at time t. We define the *average velocity* of the particle over the interval $t - t_1$ as

$$\frac{\overline{QP}}{t - t_1} = \frac{\overline{OP} - \overline{OQ}}{t - t_1} = \frac{\mathbf{r}(t) - \mathbf{r}(t_1)}{t - t_1}$$

by the triangle law. The instantaneous velocity **v** at time t is then the limit of this ratio as t_1 approaches t, or

$$\mathbf{v} = \lim_{t_1 \to t} \frac{\mathbf{r}(t) - \mathbf{r}(t_1)}{t - t_1}. \qquad (7)$$

Express $\mathbf{r}(t)$ and $\mathbf{r}(t_1)$ in terms of their components so that

$$\mathbf{r}(t) = x(t)\mathbf{i} + y(t)\mathbf{j} + z(t)\mathbf{k} \quad \text{and} \quad \mathbf{r}(t_1) = x(t_1)\mathbf{i} + y(t_1)\mathbf{j} + z(t_1)\mathbf{k}.$$

Equation (7) becomes

$$\mathbf{v} = \mathbf{i} \lim_{t_1 \to t} \frac{x(t) - x(t_1)}{t - t_1} + \mathbf{j} \lim_{t_1 \to t} \frac{y(t) - y(t_1)}{t - t_1} + \mathbf{k} \lim_{t_1 \to t} \frac{z(t) - z(t_1)}{t - t_1}$$

$$= \mathbf{i}\frac{dx}{dt} + \mathbf{j}\frac{dy}{dt} + \mathbf{k}\frac{dz}{dt}$$

$$= \frac{d}{dt}(x\mathbf{i} + y\mathbf{j} + z\mathbf{k}) = \frac{d\mathbf{r}}{dt} \quad \text{or} \quad \dot{\mathbf{r}} \qquad (8)$$

because **i**, **j**, **k** are fixed vectors independent of t. The velocity vector is the time derivative of the position vector, and as $\overline{QP}$ contracts **v** will become tangential to the curve at P. The magnitude of the velocity given by $|\mathbf{v}|$ is called the speed. This is consistent with our earlier definition of speed and agrees with it if we imagine the particle moving in a straight line parallel to the x-axis.

The *acceleration* **f** is the time derivative of the velocity

$$\mathbf{f} = \frac{d\mathbf{v}}{dt} = \frac{d^2\mathbf{r}}{dt^2} = \frac{d^2x}{dt^2}\mathbf{i} + \frac{d^2y}{dt^2}\mathbf{j} + \frac{d^2z}{dt^2}\mathbf{k} \quad \text{or} \quad \ddot{x}\mathbf{i} + \ddot{y}\mathbf{j} + \ddot{z}\mathbf{k}$$

using (8). Note that, in general, the acceleration is *not* tangential to the path described by the particle.

The rules governing the differentiation of the sums and products of vectors are similar to the familiar rules of calculus. If **a**, **b** and c are dependent on t, then

$$\frac{d}{dt}(\mathbf{a} + \mathbf{b}) = \frac{d\mathbf{a}}{dt} + \frac{d\mathbf{b}}{dt}$$

$$\frac{d}{dt}(\mathbf{a} \cdot \mathbf{b}) = \frac{d\mathbf{a}}{dt} \cdot \mathbf{b} + \mathbf{a} \cdot \frac{d\mathbf{b}}{dt}$$

$$\frac{d}{dt}(c\mathbf{a}) = \frac{dc}{dt}\mathbf{a} + c\frac{d\mathbf{a}}{dt}$$

$$\frac{d}{dt}(\mathbf{a} \times \mathbf{b}) = \frac{d\mathbf{a}}{dt} \times \mathbf{b} + \mathbf{a} \times \frac{d\mathbf{b}}{dt}.$$

These can be most easily verified by expressing **a** and **b** in terms of their components. In the last formula the order of the vector products is important.

Example 10 Show that the velocity of a particle is independent of the origin to which its position vector is referred.

Let O and O′ be two fixed origins with **a** the position vector of O′ relative to O. Let **r** and **r′** be the position vectors of P relative to O and O′ respectively (Figure 1.16).

Let **v** be the velocity of P so that

$$\mathbf{v} = \frac{d\mathbf{r}}{dt}.$$

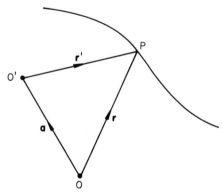

Figure 1.16 Position vector of P relative to a second origin O′.

By the triangle law $\mathbf{r} = \mathbf{a} + \mathbf{r}'$ consequently

$$\mathbf{v} = \frac{d}{dt}(\mathbf{a} + \mathbf{r}') = \frac{d\mathbf{a}}{dt} + \frac{d\mathbf{r}'}{dt}$$

but $\mathbf{a}$ is a constant vector (that is, constant in magnitude and direction) and therefore $d\mathbf{a}/dt = 0$ and

$$\mathbf{v} = \frac{d\mathbf{r}'}{dt}$$

which is the velocity of P relative to O'.

Example 11 Show that the path of a particle whose position vector at time t is given by

$$\mathbf{r} = a \cos \omega t \mathbf{i} + a \sin \omega t \mathbf{j} \quad (a, \omega \text{ are constants})$$

is a circle. Find the velocity and acceleration of the particle and show that $\mathbf{v} \cdot \mathbf{f} = 0$.

The components of the position vector are

$$x = a \cos \omega t, \qquad y = a \sin \omega t$$

and squaring and adding

$$x^2 + y^2 = a^2(\cos^2 \omega t + \sin^2 \omega t) = a^2.$$

This is the equation of a circle with centre at the origin and radius a.
   The velocity $\mathbf{v}$ is given by

$$\mathbf{v} = \frac{d\mathbf{r}}{dt} = \frac{d}{dt}(a \cos \omega t \mathbf{i} + a \sin \omega t \mathbf{j})$$

$$= -a\omega \sin \omega t \mathbf{i} + a\omega \cos \omega t \mathbf{j}.$$

The acceleration

$$\mathbf{f} = \frac{d\mathbf{v}}{dt} = \frac{d}{dt}(-a\omega \sin \omega t \mathbf{i} + a\omega \cos \omega t \mathbf{j})$$

$$= -a\omega^2 \cos \omega t \mathbf{i} - a\omega^2 \sin \omega t \mathbf{j}. \tag{9}$$

Now

$$\mathbf{v} \cdot \mathbf{f} = (-a\omega \sin \omega t \mathbf{i} + a\omega \cos \omega t \mathbf{j}) \cdot (-a\omega^2 \cos \omega t \mathbf{i} - a\omega^2 \sin \omega t \mathbf{j})$$
$$= a^2\omega^3 \sin \omega t \cos \omega t - a^2\omega^3 \cos \omega t \sin \omega t = 0.$$

Since the velocity vector is a tangent to the circle, this implies that the acceleration is radial. Its direction can be deduced from (9) which can be written in the form

$$\mathbf{f} = -\omega^2 \mathbf{r}.$$

Thus $\mathbf{f}$ is opposite in direction to the position vector and points towards the origin.

Example 12   A vector has constant magnitude but its direction varies with time. Show that its derivative is always perpendicular to itself.

Let $\mathbf{a}(t)$ denote the vector. We are given that $\mathbf{a} \cdot \mathbf{a} = |\mathbf{a}|^2 = $ constant. Differentiating this equation with respect to t, we get

$$\frac{d}{dt}(\mathbf{a} \cdot \mathbf{a}) = 0$$

or

$$\frac{d\mathbf{a}}{dt} \cdot \mathbf{a} + \mathbf{a} \cdot \frac{d\mathbf{a}}{dt} = 0$$

that is

$$\mathbf{a} \cdot \frac{d\mathbf{a}}{dt} = 0$$

the required result since the scalar product vanishes.

Physically the vector in this example can be interpreted as the position vector of a point which moves on the surface of a sphere. The velocity or rate of change of $\mathbf{a}$ must be perpendicular to $\mathbf{a}$.

In terms of plane polars we saw in Section 1.9 that

$$\mathbf{r} = r\mathbf{e}_r.$$

We wish now to find the components of velocity and acceleration in the directions of the unit vectors $\mathbf{e}_r$ and $\mathbf{e}_\theta$. The velocity

$$\mathbf{v} = \frac{d\mathbf{r}}{dt} = \frac{d}{dr}(r\mathbf{e}_r) = \frac{dr}{dt}\mathbf{e}_r + r\frac{d\mathbf{e}_r}{dt} \tag{10}$$

where it must be noted that $\mathbf{e}_r$ is not a constant vector. Referring back to Figure 1.13, it is clear that

$$\mathbf{e}_r = \cos\theta\mathbf{i} + \sin\theta\mathbf{j} \quad \mathbf{e}_\theta = -\sin\theta\mathbf{i} + \cos\theta\mathbf{j}.$$

Thus

$$\frac{d\mathbf{e}_r}{dt} = (-\sin\theta\mathbf{i} + \cos\theta\mathbf{j})\frac{d\theta}{dt} = \dot{\theta}\mathbf{e}_\theta$$

$$\frac{d\mathbf{e}_\theta}{dt} = (-\cos\theta\mathbf{i} - \sin\theta\mathbf{j})\frac{d\theta}{dt} = -\dot{\theta}\mathbf{e}_r.$$

(Note the significance of Example 12 in this context.) Equation (10) becomes

$$\mathbf{v} = \dot{r}\mathbf{e}_r + r\dot{\theta}\mathbf{e}_\theta$$

where $v_r = \dot{r}$ is the radial component and $v_\theta = r\dot{\theta}$ is the transverse component of the velocity.

The acceleration $\mathbf{f}$ is given by

$$\mathbf{f} = \frac{d\mathbf{v}}{dt} = \frac{d}{dt}(\dot{r}\mathbf{e}_r + r\dot{\theta}\mathbf{e}_\theta)$$

$$= \ddot{r}\mathbf{e}_r + \dot{r}\frac{d\mathbf{e}_r}{dt} + \dot{r}\dot{\theta}\mathbf{e}_\theta + r\ddot{\theta}\mathbf{e}_\theta + r\dot{\theta}\frac{d\mathbf{e}_\theta}{dt}$$

$$= \ddot{r}\mathbf{e}_r + \dot{r}\dot{\theta}\mathbf{e}_\theta + \dot{r}\dot{\theta}\mathbf{e}_\theta + r\ddot{\theta}\mathbf{e}_\theta - r\dot{\theta}^2\mathbf{e}_r$$

$$= (\ddot{r} - r\dot{\theta}^2)\mathbf{e}_r + (2\dot{r}\dot{\theta} + r\ddot{\theta})\mathbf{e}_\theta.$$

Here the radial component of the acceleration $f_r = \ddot{r} - r\dot{\theta}^2$ and the transverse component

$$f_\theta = 2\dot{r}\dot{\theta} + r\ddot{\theta} = \frac{1}{r}\frac{d}{dt}(r^2\dot{\theta}).$$

In particular, if a particle is describing a circle of radius a and centre at the origin, $r = a$ and the velocity and acceleration become

$$\mathbf{v} = a\dot{\theta}\mathbf{e}_\theta \qquad \mathbf{f} = -a\dot{\theta}^2\mathbf{e}_r + a\ddot{\theta}\mathbf{e}_\theta.$$

The speed $v = |\mathbf{v}| = a|\dot{\theta}|$ and the *radial* component of acceleration can be written $f_r = -v^2/a$, an important result. If we denote by ω the *angular rate* $\dot{\theta}$ at which the particle moves round the circle, then $f_r = -a\omega^2$. The angular rate $\dot{\theta}$ is measured in radians per second (there are π radians in $180°$ and 1 radian $= 57.296°$).

We can also introduce the concept of the angular velocity of a body. Consider a wheel spinning on its axle (Figure 1.17). The angle $\theta = \theta(t)$ is a measure of its angular displacement at any time t. The position vector of a point on the wheel is $r\mathbf{e}_r$ where r is a constant. The velocity of this point must be $r\dot{\theta}\mathbf{e}_\theta$. The *angular velocity* $\boldsymbol{\omega}$ is defined as $\dot{\theta}\mathbf{k}$ where $\mathbf{k}$ is a unit vector in the direction of the axle and its magnitude is the magnitude of the angular rate $|\dot{\theta}|$.

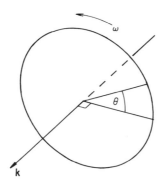

Figure 1.17 Wheel spinning on its axis at an angular rate ω.

This description refers only to rotation about a *fixed* axis. We observe that the velocity of a point of the wheel is

$$r\dot{\theta}\mathbf{e}_\theta = \boldsymbol{\omega} \times \mathbf{r}$$

a formula which can be generalised to variable angular velocity.

1.11 INTEGRATION OF VECTORS

If a vector **a** depends on t, then the interpretation of the integral of **a** with respect to t presents no special difficulty if we express **a** in terms of its components. Thus we may define

$$\mathbf{I} = \int_{t_1}^{t_2} \mathbf{a}(t)\, dt$$

$$= \int_{t_1}^{t_2} (a_1(t)\mathbf{i} + a_3(t)\mathbf{j} + a_3(t)\mathbf{k})\, dt$$

$$= \mathbf{i} \int_{t_1}^{t_2} a_1(t)\, dt + \mathbf{j} \int_{t_1}^{t_2} a_2(t)\, dt + \mathbf{k} \int_{t_1}^{t_2} a_3(t)\, dt$$

the scalar integrals having their usual meaning.

Example 13   Evaluate $\displaystyle\int_1^2 \mathbf{a}(t)\, dt$ *where* $\mathbf{a}(t) = t\mathbf{i} + \sin \pi t\mathbf{j} + \dfrac{1}{t}\mathbf{k}$.

$$\int_1^2 \mathbf{a}(t)\, dt = \int_1^2 \left(t\mathbf{i} + \sin \pi t\mathbf{j} + \frac{1}{t}\mathbf{k} \right) dt$$

$$= \left(\tfrac{1}{2}t^2\mathbf{i} - \frac{\cos \pi t}{\pi}\mathbf{j} + \ln t\mathbf{k} \right)_1^2$$

$$= \tfrac{1}{2}(4 - 1)\mathbf{i} - \frac{1}{\pi}(\cos 2\pi - \cos \pi)\mathbf{j} + (\ln 2 - \ln 1)\mathbf{k}$$

$$= \frac{3}{2}\mathbf{i} - \frac{2}{\pi}\mathbf{j} + \ln 2\mathbf{k}.$$

More elaborate integrals can be constructed in which the integrand contains scalar or vector products—for example

$$\int_{t_1}^{t_2} \mathbf{r} \cdot \mathbf{r}\, dt \qquad \int_{t_1}^{t_2} \mathbf{r} \times \frac{d\mathbf{r}}{dt}\, dt \qquad \int_{t_1}^{t_2} a(t)\mathbf{r} \times \frac{d^2\mathbf{r}}{dt^2}\, dt$$

where we write $a(t)$ to emphasise that it is not necessarily a constant, $\mathbf{r}$ we assume to be dependent on t. It is important to recognise whether an integral is a scalar or a vector. Above, the first integral is a scalar and the other two are vectors. In all cases we evaluate the product in the integrand and integrate the resulting scalar or vector accordingly.

Example 14   *Evaluate* $\displaystyle\int_0^\pi \mathbf{r} \cdot \ddot{\mathbf{r}} \, dt$ *where* $\mathbf{r} = \cos \omega t \mathbf{i} + \sin \omega t \mathbf{j}$.

$$\int_0^\pi \mathbf{r} \cdot \ddot{\mathbf{r}} \, dt = \int_0^\pi (\cos \omega t \mathbf{i} + \sin \omega t \mathbf{j}) \cdot (-\omega^2 \cos \omega t \mathbf{i} - \omega^2 \sin \omega t \mathbf{j}) \, dt$$

$$= -\omega^2 \int_0^\pi (\cos^2 \omega t + \sin^2 \omega t) \, dt$$

$$= -\omega^2 \int_0^\pi dt = -\pi \omega^2.$$

1.12   LINE INTEGRALS

We shall require to integrate variables along a given path in space. If $\mathbf{r}$ is the position vector of a point on the path C joining points P_1 and P_2 in Figure 1.18 and $\mathbf{r} + \delta\mathbf{r}$ is a neighbouring point, then, by the triangle law of addition, $\delta\mathbf{r}$ is a chord of the path. Now scalars and vectors may be functions of position as well as time. Thus $\phi(x, y, z)$ is a scalar dependent on position and is called a *scalar field* and $\mathbf{F}(x, y, z)$ is a vector dependent on position and is called a *vector field*, the vector being interpreted as

$$\mathbf{F}(x, y, z) = F_1(x, y, z)\mathbf{i} + F_2(x, y, z)\mathbf{j} + F_3(x, y, z)\mathbf{k}$$

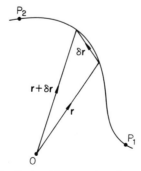

Figure 1.18   Path of integration for a line integral.

where F_1, F_2, F_3 are scalar functions. The temperature of the atmosphere is a scalar function of position at any instant of time whilst the local velocity in a flowing liquid is an example of a vector field. (Both scalar and vector fields may depend also on the fourth variable, time.) The integrals

$$\int_C \phi \, d\mathbf{r}, \qquad \int_C \mathbf{F} \cdot d\mathbf{r}, \qquad \int_C \mathbf{F} \times d\mathbf{r}$$

are examples of line integrals. In these integrals x, y, z take their values on the path C. The simplest way of specifying the curve C is to express the position vector in terms of a parameter t (usually time in mechanics), so that

$$\mathbf{r}(t) = x(t)\mathbf{i} + y(t)\mathbf{j} + z(t)\mathbf{k} \quad t_1 \le t \le t_2.$$

Thus the second integral can be written

$$\int_{t_1}^{t_2} \mathbf{f}(t) \cdot \frac{d\mathbf{r}}{dt} \, dt$$

where $\mathbf{f}(t) = \mathbf{F}[x(t), y(t), z(t)]$ and $\mathbf{r}(t_1)$ and $\mathbf{r}(t_2)$ are the position vectors of P_1 and P_2. The integral can now be evaluated in the usual way. The following example will supply the details.

Example 15   Evaluate $\int_0^P \mathbf{F} \cdot d\mathbf{r}$ *where* $\mathbf{F} = xy\mathbf{i} + yz^2\mathbf{j} + y^2z\mathbf{k}$, *O is the origin and P is* (1, 1, 1) *along*

(a) *the path* $x = t^2$, $y = t^3$, $z = t^4$,
(b) *the straight lines* (0, 0, 0) *to* (0, 0, 1) *and* (0, 0, 1) *to* (1, 1, 1).

The two paths are shown in Figure 1.19.

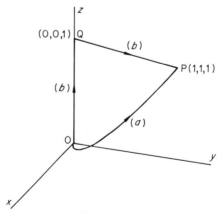

Figure 1.19

(*a*) Along this path

$$F = \mathbf{f}(t) = t^2 \cdot t^3 \mathbf{i} + t^3 \cdot t^8 \mathbf{j} + t^6 \cdot t^4 \mathbf{k}$$
$$= t^5 \mathbf{i} + t^{11} \mathbf{j} + t^{10} \mathbf{k}$$

and

$$\frac{d\mathbf{r}}{dt} = \frac{d}{dt}(t^2 \mathbf{i} + t^3 \mathbf{j} + t^4 \mathbf{k})$$

$$= 2t\mathbf{i} + 3t^2 \mathbf{j} + 4t^3 \mathbf{k}.$$

Thus

$$\mathbf{f}(t) \cdot \frac{d\mathbf{r}}{dt} = 2t^6 + 3t^{13} + 4t^{13} = 2t^6 + 7t^{13}.$$

At O, $t = 0$ and at P, $t = 1$ so that

$$\int_O^P \mathbf{F} \cdot d\mathbf{r} = \int_0^1 (2t^6 + 7t^{13}) \, dt = \frac{2}{7} + \frac{1}{2} = \frac{11}{14}.$$

(*b*) We can write

$$\int_O^P \mathbf{F} \cdot d\mathbf{r} = \int_O^Q \mathbf{F} \cdot d\mathbf{r} + \int_Q^P \mathbf{F} \cdot d\mathbf{r}.$$

Along OQ, $x = y = 0$ with the result that $\mathbf{F} = 0$ and the first integral on the right-hand side makes no contribution.

Along QP, $y = x$ and $z = 1$. Thus

$$\int_O^P \mathbf{F} \cdot d\mathbf{r} = \int_Q^P (xy \, dx + yz^2 \, dy + y^2 z \, dz)$$

$$= \int_0^1 (x^2 \, dx + x \, dx) \qquad (\text{putting } y = x \text{ and } z = 1)$$

$$= \int_0^1 (x^2 + x) \, dx \qquad (\text{using } x \text{ as the parameter})$$

$$= \frac{5}{6}.$$

We note that the value of the line integral between O and P depends on the path joining the points.

1.13 GRAD AND CURL

Suppose that $\phi(x, y, z)$ is a scalar dependent on position and defined in some region of space. The gradient (grad) of ϕ is a *vector* defined by

$$\text{grad } \phi = \frac{\partial \phi}{\partial x} \mathbf{i} + \frac{\partial \phi}{\partial y} \mathbf{j} + \frac{\partial \phi}{\partial z} \mathbf{k}$$

in terms of its Cartesian components. Grad ϕ will, in turn, be a *vector field*.

Given the scalar field $\phi(x, y, z)$, then the *equation* $\phi(x, y, z) = c$, a constant, will represent a surface in space. Consider a point P on the surface with position vector $\mathbf{r}$ and a neighbouring point Q on the surface with position vector $\mathbf{r} + \delta\mathbf{r}$. Since P and Q both lie on the surface (Figure 1.20),

$$0 = \phi(x + \delta x, y + \delta y, z + \delta z) - \phi(x, y, z)$$

$$= \frac{\partial\phi}{\partial x}\delta x + \frac{\partial\phi}{\partial y}\delta y + \frac{\partial\phi}{\partial z}\delta z$$

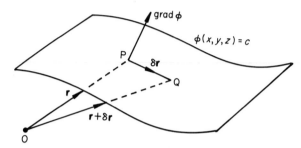

Figure 1.20   The gradient at a point on the surface $\phi = $ constant.

to the first order in $\delta x, \delta y, \delta z$. This equation can be written

$$\text{grad } \phi \cdot \delta\mathbf{r} = 0.$$

In the limit as $|\delta\mathbf{r}| \to 0$, $\delta\mathbf{r}$ will become a tangent to the surface at P. We conclude that geometrically grad ϕ is normal to the surface at P.

Example 16   *Find a unit vector normal to $xy^2z^3 = 1$ at $(1, 1, 1)$.*

We see that

$$\text{grad } (xy^2z^3) = y^2z^3\mathbf{i} + 2xyz^3\mathbf{j} + 3xy^2z^2\mathbf{k}.$$

At $(1, 1, 1)$, grad $(xy^2z^3) = \mathbf{i} + 2\mathbf{j} + 3\mathbf{k}$. A unit normal to the surface is given by

$$\mathbf{n} = \frac{\mathbf{i} + 2\mathbf{j} + 3\mathbf{k}}{(1 + 4 + 9)^{1/2}} = (\mathbf{i} + 2\mathbf{j} + 3\mathbf{k})/\sqrt{14}.$$

The *curl* of a vector field

$$\mathbf{V}(x, y, z) = V_1(x, y, z)\mathbf{i} + V_2(x, y, z)\mathbf{j} + V_3(x, y, z)\mathbf{k}$$

is defined by

$$\text{curl } \mathbf{V} = \left(\frac{\partial V_3}{\partial y} - \frac{\partial V_2}{\partial z}\right)\mathbf{i} + \left(\frac{\partial V_1}{\partial z} - \frac{\partial V_3}{\partial x}\right)\mathbf{j} + \left(\frac{\partial V_2}{\partial x} - \frac{\partial V_1}{\partial y}\right)\mathbf{k}$$

or

$$\begin{vmatrix} \mathbf{i} & \mathbf{j} & \mathbf{k} \\ \dfrac{\partial}{\partial x} & \dfrac{\partial}{\partial y} & \dfrac{\partial}{\partial z} \\ V_1 & V_2 & V_3 \end{vmatrix}$$

the latter being a symbolic determinant representation.

Example 17   If $\mathbf{V} = xyz\mathbf{i} + \cos y\mathbf{j} + x\,e^{-y}\mathbf{k}$, *find* curl $\mathbf{V}$.

$$\text{curl } \mathbf{V} = \left(\frac{\partial}{\partial y}(x\,e^{-y}) - \frac{\partial}{\partial z}(\cos y) \right)\mathbf{i} + \left(\frac{\partial}{\partial z}(xyz) - \frac{\partial}{\partial x}(x\,e^{-y}) \right)\mathbf{j}$$

$$+ \left(\frac{\partial}{\partial x}(\cos y) - \frac{\partial}{\partial y}(xyz) \right)\mathbf{k}$$

$$= -x\,e^{-y}\mathbf{i} + (xy - e^{-y})\mathbf{j} - xz\mathbf{k}.$$

The importance of the curl derives in part from the following result. If $\mathbf{V} = \text{grad } \phi$, that is, if $\mathbf{V}$ is the gradient of a scalar,

$$\text{curl } \mathbf{V} = \text{curl grad } \phi$$

$$= \left(\frac{\partial^2 \phi}{\partial y \partial z} - \frac{\partial^2 \phi}{\partial z \partial y} \right)\mathbf{i} + \left(\frac{\partial^2 \phi}{\partial z \partial x} - \frac{\partial^2 \phi}{\partial x \partial z} \right)\mathbf{j} + \left(\frac{\partial^2 \phi}{\partial x \partial y} - \frac{\partial^2 \phi}{\partial y \partial x} \right)\mathbf{k}$$

$$= \mathbf{0}$$

assuming ϕ is sufficiently smooth. In this case we say that $\mathbf{V}$ has associated with it a *scalar potential* ϕ. The converse, that curl $\mathbf{V} = 0$ implies the existence of ϕ such that $\mathbf{V} = \text{grad } \phi$, is also true but we shall omit the proof.

Example 18   If $\mathbf{V} = e^{-x}(-yz\mathbf{i} + z\mathbf{j} + y\mathbf{k})$, *show that* curl $\mathbf{V} = 0$ *and find the scalar potential.*

$$\text{curl } \mathbf{V} = \left(\frac{\partial}{\partial y}(y\,e^{-x}) - \frac{\partial}{\partial z}(z\,e^{-x}) \right)\mathbf{i} + \left(\frac{\partial}{\partial z}(-yz\,e^{-x}) - \frac{\partial}{\partial x}(y\,e^{-x}) \right)\mathbf{j}$$

$$+ \left(\frac{\partial}{\partial x}(z\,e^{-x}) - \frac{\partial}{\partial y}(-yz\,e^{-x}) \right)\mathbf{k}$$

$$= (e^{-x} - e^{-x})\mathbf{i} + (-y\,e^{-x} + y\,e^{-x})\mathbf{j} + (-z\,e^{-x} + z\,e^{-x})\mathbf{k}$$

$$= \mathbf{0}.$$

Thus **V** has a scalar potential ϕ such that

$$\frac{\partial \phi}{\partial x} = -e^{-x}yz, \qquad \frac{\partial \phi}{\partial y} = z\,e^{-x}, \qquad \frac{\partial \phi}{\partial z} = y\,e^{-x}.$$

Integrating each of these with respect to x, y, z respectively

$$\phi = -\int e^{-x}\,yz\,dx + f_1(y, z) = e^{-x}\,yz + f_1(y, z)$$

$$\phi = \int z\,e^{-x}\,dy + f_2(z, x) = e^{-x}\,yz + f_2(z, x)$$

$$\phi = \int y\,e^{-x}\,dz + f_3(x, y) = e^{-x}\,yz + f_3(x, y)$$

where f_1, f_2, f_3 are arbitrary functions. We now match f_1, f_2, f_3 in the three equations. Obviously we choose $f_1 = f_2 = f_3 = C$, a constant, which gives the answer as $\phi = yz\,e^{-x} + C$. Note that the scalar potential always contains an arbitrary additive constant.

Returning to the line integral

$$I = \int_{P_1}^{P_2} \mathbf{F} \cdot d\mathbf{r}$$

considered in Section 1.12, we observe that if the integral is independent of the path joining P_1 and P_2, then the line integral around a closed path is zero since if P_1AP_2 and P_1BP_2 are two different paths between P_1 and P_2,

$$I = \int_{P_1AP_2BP_1} \mathbf{F} \cdot d\mathbf{r} = \int_{P_1AP_2} \mathbf{F} \cdot d\mathbf{r} + \int_{P_2BP_1} \mathbf{F} \cdot d\mathbf{r}$$

$$= \int_{P_1AP_2} \mathbf{F} \cdot d\mathbf{r} - \int_{P_1BP_2} \mathbf{F} \cdot d\mathbf{r} = 0.$$

As a corollary to this there is a theorem in vector calculus, which we shall not prove here, which states the following:

*If **F** is well-behaved in a simply connected region R, then* $\displaystyle\int_C \mathbf{F} \cdot d\mathbf{r} = 0$ *implies* curl $\mathbf{F} = \mathbf{0}$ *and conversely, where* **C** *is any closed path in the region.* A *simply connected* region means that all paths within the region can be deformed into each other. This would be true for a path, or an elastic band for example, inside a pipe but not true of the region exterior to the pipe since an elastic band looped around the pipe could not be deformed into one not looped around the pipe. The two conditions, that **F** is well-behaved and that the region is simply connected, can be important for it is possible, if these are not satisfied, to have curl $\mathbf{F} = \mathbf{0}$ but $\int_C \mathbf{F} \cdot d\mathbf{r} \neq 0$ for some closed contours C, as is seen in Example 3 of Chapter 5.

Vector fields which have the property that the line integral is independent of the path are said to be *conservative* and since this property is nearly always implied by curl $\mathbf{F} = \mathbf{0}$, we have a much simpler test to apply than the almost impossible task of showing that the value of a line integral does not change whatever path we choose.

A vector field $\mathbf{V}$ is tangential at each point to a curve or *vector line* in space. For a conservative vector field these vector lines are everywhere normal to a family of surfaces where members are of the form $\phi = $ constant.

Exercises

1. Show that if G divides the line $P_1 P_2$ in the ratio λ_2 to λ_1 then the position vector of G is given by

$$\mathbf{r}_G = \frac{\lambda_1 \mathbf{r}_1 + \lambda_2 \mathbf{r}_2}{\lambda_1 + \lambda_2}$$

where $\mathbf{r}_1$ and $\mathbf{r}_2$ are the position vectors of P_1 and P_2, respectively.

2. A ship sails 10 km due north, 4 km N 60° E and then 7 km due east. Find the position vector of the final position of the ship relative to its initial position. What is the distance of the ship from its initial position?

3. If $\mathbf{a} = \mathbf{i} + 2\mathbf{j} + \mathbf{k}$, $\mathbf{b} = -\mathbf{i} + \mathbf{k}$ and $\mathbf{c} = 3\mathbf{i} + \mathbf{j} - \mathbf{k}$, evaluate

(i) $\mathbf{a} \cdot \mathbf{b}$ (iii) $\mathbf{a} \cdot (\mathbf{b} \times \mathbf{c})$
(ii) $\mathbf{a} \times \mathbf{b}$ (iv) $\mathbf{a} \times (\mathbf{b} \times \mathbf{c})$.

4. Find a unit vector perpendicular to $\mathbf{i} + \mathbf{j} + \mathbf{k}$ and $\mathbf{i} + 3\mathbf{j} - \mathbf{k}$.

5. An equation $ax + by + cz = d$ represents a plane in Cartesian coordinates. Show that $(\mathbf{r} - \mathbf{A}) \cdot \mathbf{B} = 0$, where $\mathbf{r} = x\mathbf{i} + y\mathbf{j} + z\mathbf{k}$ and $\mathbf{A}$ and $\mathbf{B}$ are constant vectors, is the vector equation of a plane. Interpret the vectors $\mathbf{A}$ and $\mathbf{B}$.

6. Show that both the equations

$$\mathbf{r} = \mathbf{c}t + \mathbf{d} \quad \text{and} \quad (\mathbf{r} - \mathbf{d}) \times \mathbf{c} = \mathbf{0}$$

represent straight lines through the point with position vector $\mathbf{d}$ in the direction of the vector $\mathbf{c}$.

7. Show that the position vector $\mathbf{r}$ of an aircraft circling above an airfield at height h in a circle of radius a with constant speed V can be expressed as

$$\mathbf{r} = a[\mathbf{i} \cos (Vt/a) + \mathbf{j} \sin (Vt/a)] + h\mathbf{k}$$

in terms of the time t. Find the velocity and acceleration vectors of the aircraft.

8. The position vector of a particle is given by

$$\mathbf{r} = a \cos \omega t \mathbf{i} + a \sin \omega t \mathbf{j} + bt\mathbf{k}$$

where a, ω and b are constants. Sketch the path taken by the particle. Find the velocity and acceleration vectors.

9. The position vector of a particle is given by

$$\mathbf{r} = a \cos \omega t \sin \Omega t \mathbf{i} + a \sin \omega t \sin \Omega t \mathbf{j} + a \cos \Omega t \ \mathbf{k}$$

where a, ω and Ω are constants. Show that the particle moves on a sphere of radius a. Find the velocity of the particle and show that its magnitude is given by $a(\Omega^2 + \omega^2 \sin^2 \Omega t)^{1/2}$. Deduce that the minimum speed must occur at the highest and lowest points of the sphere and that the maximum speed occurs where the path of the particle cuts the horizontal plane through the centre of the sphere.

10. A particle describes a path with position vector

$$\mathbf{r} = a \cos \omega t \mathbf{i} + b \sin \omega t \mathbf{j}.$$

Show that

    (i) the path is the ellipse

$$\frac{x^2}{a^2} + \frac{y^2}{b^2} = 1$$

    (ii) the acceleration is directed towards the origin

    (iii) $\displaystyle\int_t^{t+t_1} \mathbf{r} \times d\mathbf{r} = \omega abt_1 \mathbf{k}$

and interpret this result.

11. The vectors $\mathbf{a}$ and $\mathbf{b}$, and the scalar c are functions of the single variable t. By expressing $\mathbf{a}$ and $\mathbf{b}$ in terms of their components prove that

    (i) $\displaystyle\frac{d}{dt}(\mathbf{a} \cdot \mathbf{b}) = \frac{d\mathbf{a}}{dt} \cdot \mathbf{b} + \mathbf{a} \cdot \frac{d\mathbf{b}}{dt}$

    (ii) $\displaystyle\frac{d}{dt}(c\mathbf{a}) = \frac{dc}{dt} \mathbf{a} + c \frac{d\mathbf{a}}{dt}$

    (iii) $\displaystyle\frac{d}{dt}(\mathbf{a} \times \mathbf{b}) = \frac{d\mathbf{a}}{dt} \times \mathbf{b} + \mathbf{a} \times \frac{d\mathbf{b}}{dt}.$

12. If $\mathbf{a} = t^2\mathbf{i} + (2t + 1)\mathbf{j} + t\mathbf{k}$ and $\mathbf{b} = (t - 1)\mathbf{i} - t\mathbf{j} + \mathbf{k}$ find

    (i) $\displaystyle\frac{d}{dt}(\mathbf{a} + \mathbf{b})$ (ii) $\displaystyle\frac{d}{dt}(\mathbf{a} \cdot \mathbf{b})$ (iii) $\displaystyle\frac{d}{dt}(\mathbf{a} \times \mathbf{b}).$

13. The polar coordinates of a particle are given by $r = e^t$, $\theta = t$. Find the radial and transverse components of its velocity and acceleration.

14. If $\mathbf{r} = t^2\mathbf{i} - t\mathbf{j} + t^3\mathbf{k}$, O is the origin and P is the point $(1, -1, 1)$, evaluate the following line integrals

(i) $\displaystyle\int_O^P \mathbf{r} \cdot d\mathbf{r}$ (iii) $\displaystyle\int_O^P \dot{\mathbf{r}} \cdot d\mathbf{r}$

(ii) $\displaystyle\int_O^P \mathbf{r} \times d\mathbf{r}$ (iv) $\displaystyle\int_O^P \dot{\mathbf{r}} \times d\mathbf{r}$.

15. If $\mathbf{F} = x\mathbf{i} + yz\mathbf{j} + xyz\mathbf{k}$, evaluate

(i) $\displaystyle\int_C \mathbf{F} \cdot d\mathbf{r}$ (ii) $\displaystyle\int_C \mathbf{F} \times d\mathbf{r}$

where C is the curve $x = t$, $y = t^2$, $z = t$ joining the origin to $(1, 1, 1)$.

16. If $\phi = e^{xyz}$, find grad ϕ.

17. Which of the following vector fields has a scalar potential?

   (i) $yz\mathbf{i} + zx\mathbf{j} + xy\mathbf{k}$ (iii) $f(r)\mathbf{r}$
   (ii) $y\mathbf{i} + z\mathbf{j} + x\mathbf{k}$ (iv) $x\mathbf{r}$.

18. A point moves with constant speed V along the cardioid $r = a(1 + \cos\theta)$. Show that the magnitude of the angular velocity is $(V/2a)\sec\frac{1}{2}\theta$ and that the radial component of the acceleration is constant.

2

Kinematics: Geometry of Motion

2.1 INTRODUCTION

Kinematics is concerned with the description of the motion of a body or system of bodies without consideration of the laws of motion or the precise nature of the forces which are acting. It is connected only with the geometry of mechanics and for this reason it is independent of the laws of motion of classical mechanics. It may, for example, be concerned with the geometry of a gear mechanism, a cam on a rotating shaft or the effects of a given acceleration on a mechanical system. The following sections will indicate some of the problems which are within the scope of kinematics.

2.2 RELATIVE VELOCITY

We shall introduce the subject by considering two particles A and B moving along the same straight line with displacements $x_A(t)$ and $x_B(t)$ from some fixed origin O of the line (see Figure 2.1). The velocities of A and B are $v_A = dx_A/dt$ and $v_B = dx_B/dt$. The displacement of B relative to A is $x_B - x_A$ and the velocity of B relative to A is the time derivative of this displacement or $v_B - v_A$: it is the velocity that B appears to have when observed from A.

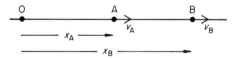

Figure 2.1   Relative velocity for motion in a straight line.

Example 1   Suppose that the straight line in Figure 2.1 represents a road and that B is a car which leaves O at 10.00 am travelling at a constant speed of 50 km/h (kilometres per hour). A second car leaves O at 10.15 am and travels at a constant speed of 60 km/h in the same direction as B. Find where and when A overtakes B.

With x_A and x_B as the two displacements, we know that $dx_B/dt = 50$, $dx_A/dt = 60$. These two *differential equations* have elementary *solutions*

$$x_B = 50t + C, \qquad x_A = 60t + D$$

which can be verified by substitution. Here C and D are two constants to be determined by the *initial conditions*. Measuring the time from 10.00 am and noting that time is measured in hours and distance in kilometres, the initial conditions are

$$\text{at } t = 0 \qquad x_B = 0$$
$$\text{at } t = \tfrac{1}{4} \qquad x_A = 0.$$

Thus $C = 0$ and $0 = 60 \times \tfrac{1}{4} + D$ or $D = -15$. The required solutions are

$$x_B = 50t, \qquad x_A = 60t - 15.$$

Car A overtakes B when their relative displacement is zero, that is when $x_B - x_A = 0$, or

$$50t - (60t - 15) = 0$$

or $t = 1\tfrac{1}{2}$ hours. At 11.30 am and at 75 km from O, A overtakes B.

Figure 2.2 shows a graphical solution of the problem in the (t, x) plane. With suitably scaled axes, lines with slopes 50 and 60 are drawn from points P and Q on the t axis. The point of intersection of the lines indicates the time and location where A overtakes B.

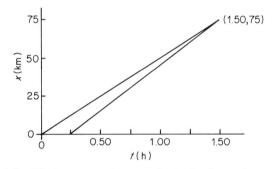

Figure 2.2   Displacement–time graphs for both cars in Example 1.

The ideas of one-dimensional relative motion can be readily generalized to motion in three dimensions. Let $r_A(t)$ and $r_B(t)$ be the position vectors at time t of two particles A and B with respect to a fixed frame of reference with origin O. By the triangle law (see Figure 2.3),

$$\overline{AB} = r_B - r_A = R, \text{ say.}$$

The vector $\mathbf{R}$ is the position vector of B relative to A. The velocity $\mathbf{V}$ of B *relative* to A is defined as the time derivative of $\mathbf{R}$:

$$\mathbf{V} = \frac{d\mathbf{R}}{dt} = \frac{d\mathbf{r}_B}{dt} - \frac{d\mathbf{r}_A}{dt}.$$

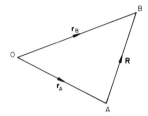

Figure 2.3   Relative position vector.

The relative velocity **V** is the velocity which B appears to have when observed from A. In a similar manner the acceleration **f** of B relative to A is defined by

$$\mathbf{f} = \frac{d^2\mathbf{R}}{dt^2} = \frac{d^2\mathbf{r_B}}{dt^2} - \frac{d^2\mathbf{r_A}}{dt^2}.$$

Looked at another way, relative motion (in this context) is concerned with measuring displacement, velocity and acceleration relative to a frame with origin A (Figure 2.4) whose axes Ax', Ay', Az' remain always parallel to Ox, Oy, Oz of the fixed frame as A accelerates. In other words frame A can be *translated* into the fixed frame *without rotation*. Rotating axes will be discussed in Chapter 10.

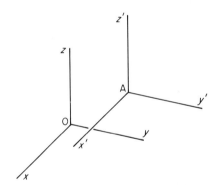

Figure 2.4   Translated frame of reference.

Example 2   To a cyclist travelling due north along a straight road at 16 km/h the wind appears to come from the east. If he increases his speed to 30 km/h it appears to blow from the north-east. Find the speed and direction of the wind.

Select a two-dimensional Cartesian frame on the Earth's surface with **i** pointing east and **j** north. Let the velocity of the wind be $\mathbf{v_w} = v_1\mathbf{i} + v_2\mathbf{j}$. If $\mathbf{v_c}$ is the velocity of the cyclist, the velocity of the wind relative to the cyclist is $\mathbf{v_w} - \mathbf{v_c}$. In the first case $\mathbf{v_c} = 16\mathbf{j}$ and

$$\mathbf{v_w} - \mathbf{v_c} = v_1\mathbf{i} + v_2\mathbf{j} - 16\mathbf{j} = v_1\mathbf{i} + (v_2 - 16)\mathbf{j}.$$

The wind appears to come from the east so that the j-component of $\mathbf{v}_w - \mathbf{v}_c$ must vanish. Hence $v_2 = 16$. In the second case $\mathbf{v}_c = 30\mathbf{j}$ and

$$\mathbf{v}_w - \mathbf{v}_c = v_1\mathbf{i} + v_2\mathbf{j} - 30\mathbf{j} = v_1\mathbf{i} + (v_2 - 30)\mathbf{j}$$
$$= v_1\mathbf{i} - 14\mathbf{j}.$$

Since the wind appears to blow from the north-east the i- and j-components must be equal. Hence $v_1 = -14$.

The wind speed is therefore $|\mathbf{v}_w| = (16^2 + 14^2)^{1/2} = \sqrt{452} = 21.2$ km/h approximately, and comes from a direction θ where $\tan\theta = -8/7$, that is from $49°$ south of east approximately.

Example 3 A ship is steaming due north at 20 km/h and a second ship, 10 km north-west of it initially, is steaming due east at 10 km/h. Find the shortest distance between the ships subsequently.

Take the origin O at the initial position of the first ship, A, with the x- and y-axes east and north. The initial positions and velocities of A and the second ship B are shown in Figure 2.5. The velocity of A is $20\mathbf{j}$ and that of B is $10\mathbf{i}$. The velocity of A relative to B is $(20\mathbf{j} - 10\mathbf{i})$ and the direction of this vector is shown by the dotted line drawn from the origin in Figure 2.5. This is the path which A appears to take when viewed from B. Clearly A and B are at their closest distance when A reaches position C, the foot of the perpendicular from B to the relative path. Thus the required distance

$$BC = 10\sin(\pi/4 - \alpha) = 10(\sin\tfrac{1}{4}\pi \cos\alpha - \cos\tfrac{1}{4}\pi \sin\alpha)$$

$$= \frac{10}{\sqrt{2}}(\cos\alpha - \sin\alpha)$$

where $\tan\alpha = \tfrac{1}{2}$. Therefore $BC = \dfrac{10}{\sqrt{2}}\left(\dfrac{2}{\sqrt{5}} - \dfrac{1}{\sqrt{5}}\right) = \sqrt{10} = 3.16$ km.

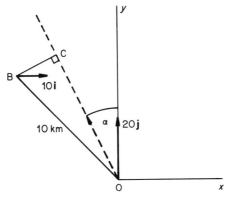

Figure 2.5

We can also find the time which elapses before A reaches C. The speed of A relative to B is $(10^2 + 20^2)^{1/2} = 10\sqrt{5}$ km/h and the distance

$$AC = (10^2 - 10)^{1/2} = 3\sqrt{10} \text{ km.}$$

Thus the ships reach their closest distance after $\dfrac{3\sqrt{10}}{10\sqrt{5}} = \dfrac{3\sqrt{2}}{10} = 0.42$ hours.

Note that the problem can be solved graphically by measuring BC in Figure 2.5, using a suitable scale.

Example 4   The runway of an airfield faces west. A helicopter, flying north at a height of 2 km and at a speed of 360 km/h on a path which passes over a point 2 km west of the runway end, is spotted $2\sqrt{2}$ km horizontally south-west of the runway end. At this time an aircraft takes off from the airfield and climbs at an angle of 45° with a constant speed of 480 km/h. Show that the helicopter and aircraft are in danger of collision.

Take a Cartesian frame with its origin at the runway end, **i** pointing west, **j** south and **k** vertically as shown in Figure 2.6.

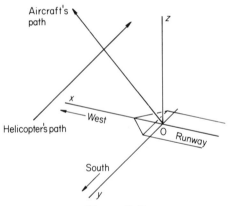

Figure 2.6

As we shall see the two machines considered as particles do not actually meet but we shall show that they pass at a distance which is within the dimensions of both. If two aircraft pass within several tens of metres of each other there is a risk of collision.

We shall employ an alternative method to that used in example 3. If $\mathbf{v}_H$ and $\mathbf{v}_A$ are the velocities of the helicopter and aircraft, we have

$$\mathbf{v}_H = \frac{d\mathbf{r}_H}{dt} = -360\mathbf{j}, \qquad \mathbf{v}_A = \frac{d\mathbf{r}_A}{dt} = 240\sqrt{2}(\mathbf{i} + \mathbf{k}).$$

Integrating these simple differential equations, we find that

$$\mathbf{r}_H = -360t\mathbf{j} + \boldsymbol{\alpha}, \qquad \mathbf{r}_A = 240\sqrt{2}t(\mathbf{i} + \mathbf{k}) + \boldsymbol{\beta}$$

where α and β are constant vectors. Initially ($t = 0$), the helicopter is at $2\mathbf{i} + 2\mathbf{j} + 2\mathbf{k}$ and the aircraft is at the origin. Thus

$$\alpha = 2\mathbf{i} + 2\mathbf{j} + 2\mathbf{k}, \qquad \beta = 0$$

and

$$\mathbf{r_H} = 2\mathbf{i} + 2(1 - 180t)\mathbf{j} + 2\mathbf{k}, \qquad \mathbf{r_A} = 240\sqrt{2}t(\mathbf{i} + \mathbf{k}).$$

At time t the distance D between them is given by

$$D^2 = |\mathbf{r_A} - \mathbf{r_H}|^2 = 8(120\sqrt{2}t - 1)^2 + 4(1 - 180t)^2$$
$$= 4(90\,000t^2 - 120(4\sqrt{2} + 3)t + 3). \qquad (1)$$

We now determine when D is a minimum by putting $dD/dt = 0$. This occurs when

$$0 = 180\,000t - 120(4\sqrt{2} + 3)$$

or when

$$t = \frac{4\sqrt{2} + 3}{1500} = 0.005\,77 \text{ hours.}$$

Substituting for t back into (1), the minimum distance

$$D_M = 0.08 \text{ km or } 80 \text{ m}$$

approximately. This distance is an insufficient margin of safety.

2.3 CONSTANT ACCELERATION

Many problems in mechanics involve or can be approximated to by constant acceleration; for example projectiles move vertically under gravity with constant acceleration (Section 4.2) and a car accelerating along a road can be approximately considered as a particle accelerating uniformly. For these reasons it is useful to obtain the formulae relating velocity, displacement and time for constant acceleration along a straight line.

Consider a particle P moving in a straight line with constant acceleration f. If x is its displacement from a fixed origin O on the line

$$\frac{d^2x}{dt^2} = f. \qquad (2)$$

If v is the velocity, the first integral of (2) gives

$$v = \frac{dx}{dt} = \int f \, dt + A = ft + A$$

where A is the constant of integration. With the initial velocity at $t = 0$ given as u, $A = u$ and

$$v = u + ft. \qquad (3)$$

We can integrate this equation a second time since (3) is equivalent to

$$\frac{dx}{dt} = u + ft$$

and obtain

$$x = \int (u + ft)\, dt + B = ut + \tfrac{1}{2} ft^2 + B$$

where B is a further constant of integration. Suppose we measure distance from the position the particle occupied initially, that is take $x = 0$ when $t = 0$. Then $B = 0$ and

$$x = ut + \tfrac{1}{2} ft^2. \tag{4}$$

A third formula connecting velocity and displacement can be found by using the following identity

$$\frac{d^2x}{dt^2} = \frac{dv}{dt} = \frac{dx}{dt} \cdot \frac{dv}{dx} = v\frac{dv}{dx}.$$

We can therefore rewrite equation (2) as

$$v\frac{dv}{dx} = f$$

and integrate it, giving

$$\int v\, dv = \int f\, dx + C$$

or

$$\tfrac{1}{2}v^2 = fx + C$$

where C is a constant. With the initial conditions specified above, $v = u$ at $x = 0$, so that

$$v^2 = u^2 + 2fx. \tag{5}$$

Equations (3), (4) and (5) are very important formulae and the reader will find it helpful to remember them. We emphasise, however, that they are valid only for constant acceleration in a straight line.

Example 5   A car can maintain a constant acceleration of 2.5 m/s². Find the time taken for the car to reach 90 km/h from rest and the distance it travels in this time.

In the notation above we are given that $u = 0$ and

$$v = \frac{90 \times 1000}{3600} = 25 \text{ m/s.}$$

Also $f = 2.5$ m/s^2 and we require x and t. From (3)

$$t = \frac{v - u}{f} = \frac{25}{2.5} = 10 \text{ s.}$$

From (5)

$$x = \frac{v^2 - u^2}{2f} = \frac{25 \times 25}{2 \times 2.5} = 125 \text{ m.}$$

Thus the car reached 90 km/h in 10 s and 125 m.

Example 6   A train has a maximum speed of 72 km/h which it can achieve at an acceleration of 0.25 m/s^2. With its brakes fully applied the train has a deceleration (negative acceleration) of 0.5 m/s^2. What is the shortest time that the train can travel between stations 8 km apart if it stops at both stations?

The journey can be broken into three stages: the acceleration, the uniform speed and the deceleration. We first determine the distance covered by the train during its periods of acceleration and deceleration and, by subtraction, how far it travels at uniform speed. The total time can be found by summing the times over the three stages.

(i) acceleration. In the usual notation $f = 0.25$ m/s^2, $u = 0$ and

$$v = 72 \text{ km/h} = \frac{72 \times 1000}{60 \times 60} = 20 \text{ m/s.}$$

We require x and t. From (5)

$$x = \frac{v^2 - u^2}{2f} = \frac{20 \times 20}{2 \times 0.25} = 800 \text{ m}$$

and from (3)

$$t = \frac{v - u}{f} = \frac{20}{0.25} = 80 \text{ s.}$$

(ii) deceleration. We now require the distance before the second station where the brakes must be applied to bring the train to rest at the station. The acceleration is now $f = -0.5$ m/s^2. The initial speed $u = 20$ m/s and $v = 0$. By (5)

$$x = \frac{v^2 - u^2}{2f} = \frac{-20 \times 20}{-2 \times 0.5} = 400 \text{ m}$$

and by (3)

$$t = \frac{v - u}{f} = \frac{-20}{-0.5} = 40 \text{ s}$$

The brakes must be applied 400 m before the second station.

(iii) uniform speed. The train must travel at a constant speed of 72 km/h for $8000 - 800 - 400 = 6800$ m. This takes a time of $6800/20 = 340$ s.
Thus the minimum time for the journey is $340 + 80 + 40 = 460$ s.

2.4   VARIABLE ACCELERATION

Acceleration can be experienced by a ride in a lift or elevator, or in a car which suddenly brakes. In the former example the rate of change of velocity can be felt but the destination can only really be decided by watching the lights against the floors of the building. If we can measure the acceleration of a body (a device which does this is known as an *accelerometer*) can we also calculate the velocity and displacement of the body from this information? Such an instrument can be used in an aircraft to compute the height changes induced by up and down currents in the atmosphere, or the position of a submarine when conventional location finding is not available.

The acceleration of a particle or point with position vector **r** and velocity **v** is given by the vector $d^2\mathbf{r}/dt^2$ or $d\mathbf{v}/dt$. If the acceleration is a known function then it could depend on the position of the particle, its velocity and time. Thus, in this case, **r** would satisfy the *vector differential equation*

$$\frac{d^2\mathbf{r}}{dt^2} = \ddot{\mathbf{r}} = \mathbf{f}(\mathbf{r}, \dot{\mathbf{r}}, t) \tag{6}$$

where **f** is a known vector function.

Example 7   A point with position vector **r** *experiences an acceleration* **r** × **ṙ**. *Write down the Cartesian components of its acceleration.*

The acceleration is given by

$$\ddot{\mathbf{r}} = \mathbf{r} \times \dot{\mathbf{r}}.$$

The components of the vector product (Section 1.4) imply that

$$\ddot{x} = y\dot{z} - z\dot{y}, \quad \ddot{y} = z\dot{x} - x\dot{z}, \quad \ddot{z} = x\dot{y} - y\dot{x}.$$

Thus we are left with three ordinary scalar differential equations for x, y and z. Notice that x, y and z appear in each equation so that we must solve the equations *simultaneously* for the coordinates in terms of t.

Let us look at the one-dimensional case. Here the point can be assumed to be moving along the x axis with its acceleration $\ddot{x}$ dependent on its displacement x, velocity $\dot{x}$, and the time t. Thus

$$\ddot{x} = f(x, \dot{x}, t). \tag{7}$$

Equation (7) is a second-order differential equation for x in terms of t.

Example 8   A point moves along a straight line with acceleration which is proportional to its velocity. Find its velocity and displacement if their initial values are respectively $v_0 > 0$ *and* x_0 *at* $t = 0$.

Let k be the constant of proportionality. Then

$$\ddot{x} = k\dot{x} \quad \text{or} \quad \dot{v} = \ddot{x} = kv.$$

This is a separable equation with solution

$$\int \frac{dv}{v} = k \int dt + C.$$

Thus

$$\ln|v| = kt + C \quad \text{or} \quad v = e^C\, e^{kt}$$

since $v_0 > 0$ is given. Hence

$$v = \dot{x} = v_0\, e^{kt}$$

which implies

$$x = v_0 \int e^{kt}\, dt + D = \frac{v_0}{k}\, e^{kt} + D.$$

From the initial conditions, $D = x_0 - (v_0/k)$ with the conclusion that the displacement is given by

$$x = x_0 + \frac{v_0}{k}\,(e^{kt} - 1).$$

If $k > 0$ $(k < 0)$ then the velocity and displacement exhibit exponential growth (decay).

Example 9 Repeat the previous example but assume an acceleration law in which the acceleration is proportional to the square of the speed.

In this case

$$\ddot{x} = k\dot{x}^2 \quad \text{or} \quad \dot{v} = kv^2.$$

The equation is still separable but with solution

$$\int \frac{dv}{v^2} = k \int dt + C,$$

which can be integrated to give

$$-\frac{1}{v} = kt + C \quad \text{or} \quad v = -\frac{1}{kt + C}.$$

From the same initial data as in Example 8, we deduce $C = -1/v_0$. Hence

$$\dot{x} = -v_0/(v_0 kt - 1).$$

Thus, separating the variables again, we obtain

$$x = -v_0 \int \frac{dt}{v_0 kt - 1} + D = -\frac{1}{k} \ln|v_0 kt - 1| + D.$$

From the initial condition on x it follows that $D = x_0$, and that

$$x = x_0 - \frac{1}{k} \ln |v_0 kt - 1|.$$

However, we must carefully define the time domains for which these answers are valid. For example, if $k > 0$ and $v_0 > 0$, then $\dot{x} \to \infty$ and $x \to \infty$ as $t \to 1/v_0 k$. Thus the solutions 'blow up' at a finite time (often referred to as *finite time blow-up*) even though solutions exist beyond the blow-up time. Figure 2.7 shows a typical solution for the case in which $x_0 = -10$, $v_0 = 1$ and $k = 0.25$.

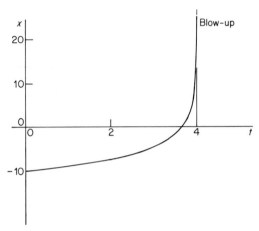

Figure 2.7 Solution showing blow-up at time $t = 4$ in Example 9.

This example makes the point that the forces which generate the acceleration (as we shall see in the next chapter) cannot grow too rapidly without causing catastrophic consequences.

Usually the output from an accelerometer would display the acceleration as a function of time in the form of data, or as an acceleration–time graph assuming that we are still considering the one-dimensional case. Thus the acceleration would be given by $\ddot{x} = f(t)$ with $f(t)$ a function fitting the data. It may take the form that $f(t_i) = f_i$, $i = 0, 1, 2, \ldots, N$. In other words the acceleration is obtained by measurements over the time interval (t_0, t_N) at the time steps $t_0, t_1, \ldots, t_N$. The velocity–time relation will require the integration of the acceleration–time relation, whilst the displacement–time graph will require yet a further integration. Both these integrations can only be performed numerically.

Rather than expressing the acceleration in terms of data, consider the case in which $f(t)$ is a given function. In particular, let us examine the problem in which

$$\ddot{x} = f(t) = a \sin (t + b \sin t) \tag{8}$$

where a and b are constants. Since $f(t + 2\pi) = f(t)$ for *all* t, the acceleration is a *periodic* function of time with period 2π. In order to find the displacement $x(t)$ we would need to integrate this equation twice. However, analytically the prospect of obtaining a closed form of solution seems unlikely in this case. Hence our only hope of further progress lies with numerical integration. Since this is a stepwise process in terms of the time t, it is rather like using data in a 'real' problem. We usually write (8) as the first-order system of differential equations

$$\dot{x} = v \quad \dot{v} = f(t) = a \sin (t + b \sin t).$$

One cycle of the acceleration on the interval $(0, 2\pi)$ is shown in Figure 2.8(a): this may be thought of as the *input* of the system. A numerical solution or *output* $x(t)$ for the initial conditions $x(0) = 0$, $\dot{x}(0) = -1.107$ is shown in Figure 2.8(b). These particular initial conditions lead exceptionally to a periodic solution, and were found by experimenting with the initial data. Note, however, that not all solutions of (8) are periodic.

It is generally a routine matter these days to display equations such as (8) for a wide variety of acceleration functions. Simple numerical methods of solving initial-value problems for first-order systems of differential equations enable solutions to be displayed at computer terminals. Some simple techniques for numerical integration are discussed in Appendix A.

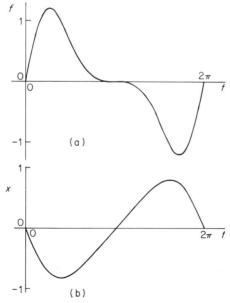

Figure 2.8 (a) Input and (b) periodic output for the system given by (8) with $a = 1.2$, $b = 1$ and the special initial conditions $x(0) = 0$, $\dot{x}(0) = -1.107$.

2.5 PURSUIT PROBLEMS

Pursuits occur when a *pursuer* is attempting to catch or overtake a *target*. This could be a predator chasing a prey, as in the case of a dog chasing a rabbit, or a missile locked on to an aircraft. We have to make a number of assumptions about the speeds of both participants and the strategy followed by the pursuer. Let $\mathbf{r_B}(t)$ represent the position vector of the target B, and let $\mathbf{r_A}(t)$ be the position vector of the pursuer A (see Figure 2.9). It is assumed that A only knows the current and past positions of B. The most obvious strategy for A is to move always towards the target's current position. The position of B relative to A is $\mathbf{R} = \mathbf{r_B}(t) - \mathbf{r_A}(t)$ by the triangle law. The velocity of A is $\mathbf{v_A}(t) = \dot{\mathbf{r}}_A(t)$, and the direction in which A moves will be given by $\mathbf{v_A}(t)$ and this must be parallel to $\mathbf{r_B}(t) - \mathbf{r_A}(t)$ according to our strategy. We can express this by the vector product condition

$$\mathbf{v_A}(t) \times (\mathbf{r_B}(t) - \mathbf{r_A}(t)) = \mathbf{0}. \qquad (9)$$

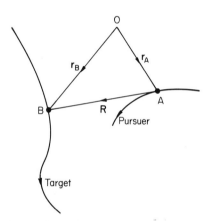

Figure 2.9 Pursuit problem: the target's path need not be a plane curve.

Whilst equation (9) is necessary to determine the pursuit curve, it contains insufficient information to derive the path. In addition we need to specify the speed of A. Equation (9) implies that

$$\dot{\mathbf{r}}_A(t) = \mathbf{v_A}(t) = v_A(t)(\mathbf{r_B}(t) - \mathbf{r_A}(t))/|\mathbf{r_B}(t) - \mathbf{r_A}(t)| \qquad (10)$$

where $v_A(t) = |\mathbf{v}_A(t)|$ is the speed of A. If we specify this speed as a function of t, then (10) provides a vector differential equation for the position vector $\mathbf{r_A}$. In some cases it is more convenient to use the relative position vector $\mathbf{R}(t) = \mathbf{r_B}(t) - \mathbf{r_A}(t)$ in which case (10) can be replaced by

$$\dot{\mathbf{R}}(t) = \dot{\mathbf{r}}_B(t) - v_A(t)\mathbf{R}(t)/R(t). \qquad (11)$$

Capture occurs whenever $\mathbf{R} = \mathbf{0}$.

Consider a pursuit problem which takes place in the (x, y) plane. Let $\mathbf{r_B} = x_B\mathbf{i} + y_B\mathbf{j}$ and $\mathbf{R} = X\mathbf{i} + Y\mathbf{j}$. The components of (11) supply the simultaneous differential equations

$$\dot{X} = \dot{x}_B - v_A(t)X/(X^2 + Y^2)^{1/2} \tag{12}$$

$$\dot{Y} = \dot{y}_B - v_A(t)Y/(X^2 + Y^2)^{1/2} \tag{13}$$

for the relative pursuit coordinates (X, Y).

Example 10   In a pursuit problem the target's path is given by $x_B = ut + x_0$ $y_B = y_0$. The pursuer starts from the origin at time $t = 0$ and moves towards the target with constant speed v. Find and solve the equations of the pursuit curve.

In Equations (12) and (13), $\dot{x}_B = u$, $\dot{y}_B = 0$ and $v_A(t) = v$. Hence

$$\dot{X} = u - vX/(X^2 + Y^2)^{1/2} \quad \dot{Y} = -vY/(X^2 + Y^2)^{1/2}.$$

In this example it is more convenient to introduce polar coordinates (R, θ) defined by $X = R \cos \theta$, $Y = R \sin \theta$. Thus

$$\dot{X} = \dot{R} \cos \theta - R \sin \theta, \quad \dot{\theta} = u - v \cos \theta \tag{14}$$

$$\dot{Y} = \dot{R} \sin \theta + R \cos \theta, \quad \dot{\theta} = -v \sin \theta. \tag{15}$$

Solving (14) and (15) for $\dot{R}$ and $\dot{\theta}$, we obtain

$$\dot{R} = u \cos \theta - v, \qquad R\dot{\theta} = -u \sin \theta.$$

The equation of the path can be found by integrating the ratio

$$\frac{dR}{d\theta} = \frac{\dot{R}}{\dot{\theta}} = \frac{(u \cos \theta - v)R}{-u \sin \theta}.$$

This is a first-order separable differential equation with solution

$$\int \frac{dR}{R} = \int \left(-\cot \theta + \frac{v}{u} \operatorname{cosec} \theta\right) d\theta + C$$

where C is a constant. Thus

$$\ln R = -\ln |\sin \theta| - \frac{v}{u} \ln |\operatorname{cosec} \theta + \cot \theta| + C$$

or

$$R = A|\operatorname{cosec} \theta|/|\operatorname{cosec} \theta + \cot \theta|^{v/u} \qquad A = e^C. \tag{16}$$

From the initial conditions we infer that, at time $t = 0$,

$$R = (x_0^2 + y_0^2)^{1/2}, \quad \cos \theta = x_0/(x_0^2 + y_0^2)^{1/2}, \quad \sin \theta = y_0/(x_0^2 + y_0^2)^{1/2}.$$

Hence the constant in (16) is given by

$$A = |x_0 + (x_0^2 + y_0^2)^{1/2}|^{v/u}/|y_0|^{(v-u)/v} \qquad (y_0 \neq 0).$$

If $y_0 = 0$, then $\theta = 0$ or π and the pursuit path is the x axis.

Capture takes place if $R = 0$ can occur in (16). In fact, if $v \leq u$ this will not take place as we might reasonably expect since the pursuer's speed never exceeds that of the target.

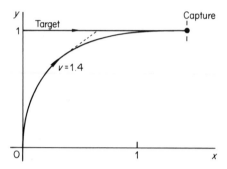

Figure 2.10 Pursuit curve and capture in Example 10 with $x_0 = 0$, $y_0 = 1$ and $v = 1.4$.

Figure 2.10 shows a typical computed path for this example with $u = 1$, $x_0 = 0$, $y_0 = 1$ and $v = 1.4$. As is often the case it is easier to plot the path by numerical solution of the simultaneous differential equations (12) and (13) rather than from the exact solution. Analytic solutions are rare anyway for the pursuit equations. Plane pursuit problems are particularly suited to interactive graphics facilities. Figure 2.11 shows the pursuit curve for a circular target path $x_B =$

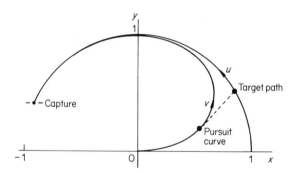

Figure 2.11 Computed pursuit curve and capture for target path $x_B = u \cos \omega t$, $y_B = u \sin \omega t$ with $u = 1$, $\omega = 1.2$, $x_0 = 1$ and $y_0 = 0$. The pursuit curve starts from the origin with $v = 1.32$ at $t = 0$.

$u \cos \omega t$, $y_B = u \sin \omega t$ with the pursuer starting from the origin at $t = 0$ with speed $v \geq u$. The data used are $u = 1$, $\omega = 1.2$, $x_0 = 1$, $y_0 = 0$ and $v = 1.32$.

Exercises

1. A ship steaming north at 12 km/h passes a fixed buoy at 12.00 hours. A second ship steaming east at 16 km/h passes the same buoy at 12.50 hours. At what time are the two ships closest together and what is then their distance apart?

2. A ship leaves a port A and sails due west at a speed of 32 km/h. One hour later an aeroplane takes off from an airfield 160 km due south of A to intercept the ship. If the aeroplane's speed is 360 km/h, find the direction in which it should be flown. Where will the ship be when the aeroplane flies over it?

3. An aeroplane is to fly from a point A to an airfield B 300 km due north of A. If a wind of 30 km/h blows from the north-west, find the direction the plane should be pointing and the time taken to reach B if the speed of the plane relative to still air is 150 km/h.

4. A ship A equipped with radar having a range of 75 km leaves port and steams due north at 10 km/h. At the same time a second ship 150 km north-east of A is steaming due west. Show that this ship will come within the range of A's radar if its speed is between 2.7 and 37.3 km/h approximately.

5. An aircraft is flying due north at 340 km/h relative to the ground and finds that the wind appears to be blowing from north $\theta°$ east where $\tan \theta = \frac{1}{4}$. It then turns and flies east at the same speed and finds that the wind is coming from east $\theta°$ south. Obtain the speed and direction of the wind.

6. In steady rain, raindrops fall at about 16 km/h. On the basis of this figure draw up a table of values of angles of raindrop streaks on train windows which would enable a passenger in the train to estimate his speed in units of 10 km/h.

7. An aeroplane has a speed v and a flying range (out and back) of R in calm weather. Prove that in a north wind of speed n $(n < v)$ its range is

$$\frac{R(v^2 - n^2)}{v(v^2 - n^2 \sin^2 \phi)^{1/2}}$$

in a direction whose true bearing from north is ϕ. What is the maximum value of this range and in what directions may it be attained?

8. An aircraft has a cruising speed of 400 km/h and a safe maximum range of 3200 km in still air when fully loaded with 100 passengers. For each empty seat the range can be increased by 10 km. How many passengers can the aircraft safely carry for a flight of 3200 km, flying into a headwind of 50 km/h?

9. In Example 4 find how long the aircraft should wait on the runway before taking off in order that the minimum distance between it and the helicopter should be at least 2 km.

10. A fly crawls with constant speed v along the radial spoke of a wheel which is rotating with constant angular speed ω. If the fly starts from the centre of the wheel find the actual velocity of the fly and the path on which it moves relative to a fixed frame of reference.

11. A rocket is fired vertically from a point on the Earth's equator with a speed of 8000 km/h. Taking account of the Earth's rotation, find the actual firing speed relative to a *fixed* equatorial plane.
    The rocket continues its course in a straight line with constant speed. Define suitable unit vectors in the fixed equatorial plane and find the position vectors of the rocket and the launching site relative to the centre of the Earth and the rocket relative to the launching site. What is the distance between the launching site and the rocket after four hours have elapsed? (Radius of the Earth's equator = 6400 km.)

12. A car accelerates uniformly from rest to 100 km/h in 15 s. Find (a) the acceleration, (b) the distance travelled in this time, (c) its speed after 10 s.

13. A point is moving in a straight line with constant acceleration. It is seen to move 15 m in the first second and 65 m in the sixth second. How far does it travel in the fourth second?

14. Two trains start 6 minutes apart and attain their maximum speeds of 120 km/h in 2 km. Assuming constant acceleration show that the first has travelled 10 km before the second starts and that they run 12 km apart at maximum speed.

15. Telegraph poles at the side of a railway track are spaced at intervals of 50 m. A passenger in a train observes that successive poles pass at time intervals of 3 s and 2.8 s. Calculate the acceleration of the train (assuming it to be uniform).

16. Two particles A and B are moving with uniform positive accelerations f and $3f$ along parallel straight paths in the same direction. At a certain instant they are level and their speeds are then $2v$ and v respectively. Show that the particles will again draw level after a time v/f has elapsed, and find their velocities.
    If B now moves with constant velocity, show that the particles will draw level for a third time when they have both covered a total distance $21v^2/2f$.

17. A car is fitted with three gear ratios such that, within the speed ranges indicated below, the car can maintain the maximum uniform accelerations given on the right-hand side:

first gear:	0–40 km/h	3 m/s²
second gear:	30–80 km/h	4 m/s²
third gear:	60–120 km/h	3 m/s²

At what speeds should the driver change gear in order to reach 120 km/h in the shortest possible time? What is this time?

18. The maximum acceleration of a train is α, the maximum retardation is β and its maximum speed is V. Show that it cannot run a distance a from rest to rest in a time less than

$$\left(\frac{2a(\alpha + \beta)}{\alpha\beta}\right)^{1/2} \quad \text{if} \quad a \leq \frac{V^2(\alpha + \beta)}{2\alpha\beta}$$

or less than

$$\frac{V}{2}\left(\frac{1}{\alpha} + \frac{1}{\beta}\right) + \frac{a}{V} \quad \text{if} \quad a > \frac{V^2(\alpha + \beta)}{2\alpha\beta}.$$

19. Approximately, the stopping distance of a car travelling at v km/h is given by the expression $\frac{5}{24}\left(v + \frac{v^2}{32}\right)$ m, the first term representing the 'reaction' distance of the driver and the second the distance through which the brakes are actually applied. Two cars are 35 m apart when the first car, travelling at 45 km/h, makes an emergency stop. The driver of the second car has no prior warning of this hazard. Show that a collision must occur if the speed of the second car exceeds 81 km/h.

20. A closed racing track consists of two semi-circular roads of radii 0.5 km joined by two straight roads each of length 4 km. What is the shortest time in which a racing car can travel once round the track if its maximum speed on the straight is 250 km/h and on the curved sections is 180 km/h, and its maximum acceleration and retardation are both 4 m/s²?

21. A wheel rolls without slipping along a straight line so that its centre moves with constant velocity V. Find the actual velocity of any point of the wheel and show that its direction is perpendicular to the straight line joining the point to the point of contact of the wheel and line.

22. The crankshaft of an engine of a car is connected through a four-speed gearbox to the rear axle. If the overall gear-ratios between engine and axle are

first gear: 20:1
second gear: 15:1
third gear: 10:1
top gear: 5:1

and the axle drives a wheel 60 cm in diameter, find the speed of the car in the four gears when the engine is turning at 3000 r.p.m.

23. Show that the hour and minute hands of a clock cross every 12/11 hours.

24. Figure 2.12 shows a sliding valve V being driven by a circular cam C of radius a which rotates with constant angular speed ω about an eccentric axis through a point O distance b from its centre. Find the displacement of the valve from O at any time t.

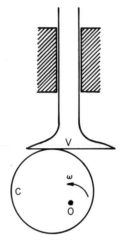

Figure 2.12   Cam and valve.

25. Figure 2.13 shows the cross-section of a simple model for a piston P driving a crankshaft AB. If AB = 2 cm and the connecting rod BP = 4 cm and the crankshaft rotates at a constant rate of 3000 revolutions per minute (r.p.m.), show that

$$AP = 2 \sin \omega t + 2(3 + \sin^2 \omega t)^{1/2} \text{ cm}$$

where $\omega = 100 \, \pi$ rad/s, and time is measured from $\theta = 0$.

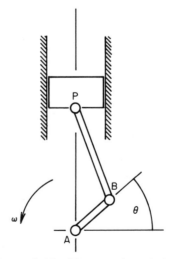

Figure 2.13   Piston and crankshaft.

26. Find the displacement $x(t)$ when $\ddot{x} = \sin t$ subject to the initial conditions $x(0) = x_0$, $\dot{x}_0(0) = v_0$. Show that the displacement will always be periodic if $v_0 = -1$.

27. The system $\ddot{x} = f(t)$ has the stepped acceleration input given by

$$f(t) = \begin{matrix} 0 & t < t_0 \\ 1 & t_0 \le t \le t_1 \\ 0 & t > t_1 \end{matrix}$$

where $t_1 > t_0 > 0$. Let $x(0) = \dot{x}(0) = 0$. Find the velocity and displacement of the system for $t > 0$. Sketch the input acceleration and output displacement.

28. The acceleration of a particle is given by the power-law relation $\dot{v} = kv^\alpha$ for $v > 0$, where v is the velocity of the particle and $k > 0$ and $\alpha > 0$ are constants. If $v(0) = v_0 > 0$ find the values of α for which the system exhibits blow-up in a finite time, and the times at which it occurs.

29. Find all solutions of the differential equation

$$\ddot{x} = x\dot{x}$$

Show that with the exception of the singular solution $x = 0$, no solutions are sustainable for all time.

30. A person rows a boat across a straight river of width d, always rowing in the direction of a point directly opposite to his or her starting point. He or she can row the boat at a constant speed V and the river flows with constant speed w. Show that the differential equation of the path is

$$\frac{dy}{dx} = \frac{y - k(x^2 + y^2)^{1/2}}{x}, \qquad k = \frac{w}{v},$$

where the origin of coordinates is on the opposite bank and y points downstream. Solve this equation and find the path of the boat. Show that the boat will reach the opposite bank if $k < 1$.

Rather than using analytic techniques compute some paths directly from the expressions for $\dot{x}$ and $\dot{y}$ for a range of values of k, say $k = 0.6, 0.8, 1.0, 1.2$. Display the paths on screen or plot the results.

31. Unlike the previous model let us now assume that the water speed in the river is not constant. In a straight river of width $2a$ the water flows parallel to the banks and its surface speed is $V_0(1 - x^2/a^2)$ where x is measured from the mid-line of the river (this is a *parabolic* velocity distribution: the water is at rest at the banks and has a maximum value V_0 on the mid-line). A person, who can maintain a speed V in still water, swims across the river directly towards the opposite bank. Find where he or she reaches the opposite bank.

32. A rabbit is running with speed v along a straight hedge. A dog, which is at a distance d from the hedge and which runs at speed $2v$ sees the rabbit when it is at its nearest

point to the dog. The dog then chases the rabbit by running in such a direction that it is always pointing at the moving rabbit. By using suitable polar coordinates relative to the rabbit to find the path of the dog, show that the rabbit runs a distance $\frac{2}{3}d$ before it is caught. The definite integral

$$\int_0^{\pi/2} \frac{d\theta}{(1 + \sin \theta)^2} = \frac{2}{3}$$

should be useful.

3

Principles of Mechanics

3.1   INERTIAL FRAMES OF REFERENCE

Newtonian mechanics presupposes that we can specify a 'fixed' or inertial frame of reference which is unaccelerated and non-rotating. For most phenomena within the solar system it is usual to take a frame which is fixed relative to the background of stars. This inertial frame can be imagined to have its origin at the Sun with its axes determined by the directions of certain stars (this ignores small perturbations of the Sun due to the motion of the planets about it). Such a hypothesis seems reasonable for local solar phenomena and most time intervals in which we are interested because of the remoteness of the stars (and consequently their small gravitational pull on the Sun) and their relatively slow change of orientation.

Whilst such a *sidereal* frame of reference is appropriate for studying the orbits of the planets or the path of an interplanetary rocket, it is not convenient for describing events which occur in the laboratory or in the immediate vicinity of the Earth's surface. A frame of reference fixed in the Earth would be more useful. However, since the Earth rotates on its polar axis and revolves about the Sun, a *terrestrial* frame will translate and rotate relative to the sidereal frame.

Let us calculate some typical relative accelerations of the two frames. Imagine a terrestrial frame of reference with origin fixed at the centre of the Earth, one axis towards the North Pole and two axes moving with the equatorial plane of the Earth. Approximately, the Earth describes a circular orbit about the Sun of radius $a = 1.5 \times 10^{11}$ m in a time of 3.16×10^7 s (one year). The orbital speed v of the Earth is therefore given by

$$v = \frac{2\pi \times 1.5 \times 10^{11}}{3.16 \times 10^7} = 3.0 \times 10^4 \text{ m/s.}$$

The acceleration of the Earth towards the Sun (see Section 1.10) is given by

$$\frac{v^2}{a} = 0.0060 \text{ m/s}^2.$$

In addition the Earth spins about its polar axis once every 8.616×10^4 s (why is this less than 24 hours?) which means that each point of the Earth describes

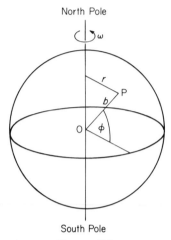

North Pole

ω

South Pole

Figure 3.1

a circle about the axis. Consider a point P on the Earth's surface with latitude $\phi°$ (latitude is the angle between the radius to P and the equatorial plane; see Figure 3.1). The angular speed of the Earth is

$$\omega = \frac{2\pi}{8.616 \times 10^4} = 7.29 \times 10^{-5} \text{ rad/s}.$$

The acceleration of P due to the spin of the Earth will vary with the latitude of P. If r is the distance of P from the polar axis, the acceleration of P towards the axis is

$$r\omega^2 = b \cos \phi \cdot \omega^2$$
$$= 0.0339 \cos \phi \text{ m/s}^2.$$

approximately where b, the mean radius of the Earth, is taken to be 6.37 $\times 10^6$ m. The acceleration induced by the spin varies between zero at the poles and 0.339 m/s² at the equator. (This discussion ignores the oblateness of the Earth and slight variations of the Earth's spin.)

We may compare the two figures 0.0060 m/s² and 0.0339 m/s² with the *acceleration due to gravity* near the Earth's surface. This is the acceleration with which a body falls freely towards the Earth due to the Earth's gravitational pull. It is approximately 9.81 m/s² (its value is of course not constant since the Earth is not perfectly spherical). Both accelerations are small compared with this typical acceleration due to gravity. For this reason we can take the Earth as an inertial frame for local phenomena on the Earth.

It must be emphasised, however, that it is not always safe to assume that small accelerations effective over long intervals of time can be ignored. A small sustained acceleration can produce significant effects.

3.2   LAWS OF MOTION

In the late seventeenth century Newton, following work by Galileo and others, proposed three laws of motion for particles. Mechanics is largely based on these. His first law states that every body continues in its state of rest or uniform motion in a straight line unless acted upon by an external force. Whilst this law is implied by Newton's second law which follows, it adds a qualitative meaning to the notion of force. We introduce the principles of mechanics through the fiction of the particle.

The fundamental empirical law of mechanics for an isolated particle subject to a force $\mathbf{F}$ is that there exists an inertial frame of reference in which

$$\mathbf{F} = m\mathbf{f} \tag{1}$$

where $\mathbf{f}$ is the acceleration of the particle and m is its inertial mass, a positive scalar magnitude independent of position and time which can be associated with the particle. We shall say more about comparative masses later. Equation (1) is *Newton's second law of motion*: force = mass × acceleration.

Note that (1) is a *vector* equation which implies that the acceleration is always in the direction of the force (this does not mean that the particle is necessarily moving in the direction of the force: for example, the Earth revolves about the Sun but its acceleration is directed towards the Sun in the direction of the gravitational force). If no force acts, $\mathbf{F} = 0$ and $\mathbf{f} = \dot{\mathbf{v}} = 0$. Then $\mathbf{v} = \mathbf{c}$, a constant vector, and the particle moves with constant velocity. Integrating $\mathbf{v} = \dot{\mathbf{r}} = \mathbf{c}$ again

$$\mathbf{r} = \mathbf{c}t + \mathbf{d}$$

where $\mathbf{d}$ is a further constant vector. The particle moves along a straight line (Exercise 6, Chapter 1) with constant velocity confirming Newton's first law.

A particle may be subject simultaneously to several external forces $\mathbf{F}_1$, $\mathbf{F}_2$, ..., $\mathbf{F}_n$ (Figure 3.2) each producing accelerations $\mathbf{f}_1$, $\mathbf{f}_2$, ..., $\mathbf{f}_n$ so that

$$\mathbf{F}_1 = m\mathbf{f}_1, \quad \mathbf{F}_2 = m\mathbf{f}_2, \ldots, \quad \mathbf{F}_n = m\mathbf{f}_n.$$

Adding these equations

$$\mathbf{F} = \mathbf{F}_2 + \mathbf{F}_2 + \cdots + \mathbf{F}_n = m(\mathbf{f}_1 + \mathbf{f}_2 + \cdots + \mathbf{f}_n) = m\mathbf{f}$$

where $\mathbf{F}$ is the total external force and $\mathbf{f}$ is the resultant acceleration. The relation above emphasises the *linear* character of Newton's law since $\mathbf{F}$ and $\mathbf{f}$ are simple vector sums of the separate forces and accelerations.

The vector $\mathbf{p} = m\mathbf{v}$ is called the *linear momentum* of the particle and if we assume the mass remains constant

$$\mathbf{F} = \frac{d\mathbf{p}}{dt}$$

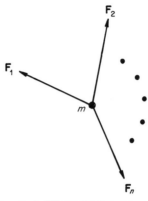

Figure 3.2

that is, the force is rate of change of linear momentum. If the force vanishes the linear momentum is *conserved*.

Newton's second law of motion is not sufficient in itself to describe the motion of two or more particles since particles may exercise mutual forces on one another. We must add a third postulate which defines the magnitude and direction of these forces. *Newton's third law of motion* asserts that if two particles exert forces on one another, these forces are equal in magnitude and opposite in direction and act along the line joining the two particles.

For example, if the Earth and the Moon are considered as two particles in isolation, the Moon experiences a gravitational attractive force in the direction of the Earth, and the Earth an equal force in the direction of the Moon.

With this simple postulate we can now examine the motion of each particle separately since we know the directions of the forces exerted on the particle by any other particles. Suppose we have two particles A and B with masses m_A and m_B subject only to a mutual force of attraction (see Figure 3.3). By Newton's third law if the force on A due to B is $\mathbf{F}$ then the force on B due to A must be $-\mathbf{F}$. Furthermore both forces must act along the line joining A and B, this

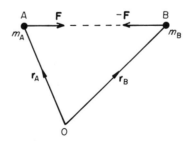

Figure 3.3 The direction of mutual forces acting on two particles.

property being expressed by the vector equation

$$\mathbf{F} \times (\mathbf{r_A} - \mathbf{r_B}) = \mathbf{0}.$$

Considering each particle in isolation from the other, the equations of motion of A and B are respectively

$$\mathbf{F} = m_A \ddot{\mathbf{r}}_A, \qquad -\mathbf{F} = m_B \ddot{\mathbf{r}}_B$$

where $\mathbf{r_A}$ and $\mathbf{r_B}$ are their position vectors. Once the nature of $\mathbf{F}$ is known these equations can be solved, in principle, to find the separate motions of A and B.

The mutual forces between particles can be used to compare masses of particles. Consider a spring with a particle attached to each end and compressed (the system is assumed to be far removed from the influence of other forces). The spring will exert the same force on each particle by Newton's third law. On release we measure the acceleration a_1 and a_2 of the two particles (we assume that motion takes place only in the line of the spring). If F is the magnitude of the force exerted by the spring on each particle, then

$$m_2 a_2 = F = m_1 a_1$$

where m_1 and m_2 are the masses of the two particles. Thus

$$m_2 = \frac{a_1}{a_2} m_1$$

and m_2 can be measured in terms of a standard mass m_1. The international standard is the prototype kilogram (kg), a platinum–iridium cylinder, and masses are compared against this standard. The gram is 10^{-3} kg.

We have now introduced the third fundamental dimension—mass—to add to length and time. The rationalized system of metric units known as the Système International d'Unités (S.I.) is now in general international use. The S.I. units are derived from six base units, three of which concern us in this book: they are the metre (m), the kilogram (kg) and the second (s) as the basic length, mass and time. They form the basis of a *coherent* system of units in that the product or quotient of two or more unit quantities in the system is the unit of the resultant quantity. Thus the unit of force is the newton (N) which represents force of one metre-kilogram/(second)^2.

It is appropriate to note here the difference between the colloquial and scientific uses of the words 'weight' and 'weigh'. It is convenient in practice to compare masses by weighing. A body on the Earth's surface experiences a force of attraction due to the gravitational influence of the earth. This force of attraction is called the *weight* of the body. This force will vary slightly from place to place on the Earth's surface, due to the shape of the Earth. A mass of 1 kg will, if allowed to fall freely, accelerate at about 9.81 m/s^2. Standard gravitational acceleration is actually taken as $g = 9.80665$ m/s^2. Thus the Earth exerts a force of 9.81 N on the mass. The weight of the mass is therefore 9.81 N which can also be described as 1 kg force. There is much to be said for using the coherent unit and avoiding the use of units which contain g.

Colloquially, when it is said that a person weighs 80 kg it means that their weight, as explained above, is the same as the weight of a mass of 80 kg.

The S.I. units are *preferred*, but it is often more convenient to use decimal multiples and sub-multiples of these. In calculations, however, S.I. units themselves should be used to avoid errors. Decimal multiples of 10 raised to a power which is a multiple of 3 are recommended. Some decimal multiples which do not obey the rules of 10^3 and 10^{-3} are, however, accepted. Thus the centimetre (cm) and the litre $= 1$ dm^3 are accepted S.I. units. Some multiples and prefixes are given in the table below:

Factor	Prefix	Symbol
10^6	mega	M
10^3	kilo	k
10	deca	da
10^{-1}	deci	d
10^{-2}	centi	c
10^{-3}	milli	m
10^{-6}	micro	μ

It should be noted that S.I. units contain one slight inconsistency in that the unit of mass is the *kilo*gram. From the table 1000 N is 1 kN.

The S.I. unit for plane angle, the *radian* (rad), is called a supplementary unit.

Many imperial units are still in extensive use, their values in terms of S.I. units are listed below:

1 ft	$= 0.3048$ m
1 mile	$= 1.609\,344$ km
1 ft^2	$= 0.092\,9030$ m^2
1 ft^3	$= 0.028\,3168$ m^3
1 mile/h	$= 0.447\,04$ m/s
1 pound (lb)	$= 0.453\,592\,37$ kg
1 poundal	$= 0.138\,255$ N
1 UK gallon	$= 4.546\,09$ dm^3

Example 1   Find the constant thrust required to accelerate uniformly an aircraft, which has mass 10^5 kg, from rest to a speed of 180 km/h in 20 s assuming no air resistance.

If the initial thrust is the value found in the first part and this is increased at the constant rate of 6×10^3 N/s, find the time taken to achieve the same speed and the distance covered in that time.

Thrust is an alternative term for force.

In the notation of Section 2.3, the aircraft must have an acceleration

$$f = \frac{v - u}{t}$$

where

$$v = 50 \text{ m/s}, \quad u = 0, \quad t = 20 \text{ s}.$$

Thus

$$f = 2.5 \text{ m/s}^2.$$

By Newton's second law, the required thrust

$$= mf = 2.5 \times 10^5 \text{ N}.$$

For the second part the initial thrust is 2.5×10^5 N and the thrust at any time t

$$= (2.5 \times 10^5 + 6 \times 10^3 t) \text{N}$$

where t is in seconds.

Therefore by Newton's second law

$$10^5 \frac{dv}{dt} = 2.5 \times 10^5 + 6 \times 10^3 t$$

that is

$$\frac{dv}{dt} = 2.5 + 0.06t.$$

Integrating

$$v = 2.5t + 0.03t^2 + C$$

where $C = 0$ because $v = 0$ when $t = 0$.

When $v = 50$ m/s, t is given by

$$50 = 2.5t + 0.03t^2$$

that is

$$0.03t^2 + 2.5t - 50 = 0.$$

We require the positive root of this equation, which is $t = \dfrac{50}{3}$ s.

To find the distance covered we must integrate

$$v = \frac{dx}{dt} = 2.5t + 0.03t^2.$$

This gives

$$x = \frac{2.5t^2}{2} + 0.01t^3$$

since again the constant of integration is zero from the initial condition. Substituting $t = \dfrac{50}{3}$ s we find that the distance covered is approximately 392 m.

Example 2   The meteorologist measures pressure in millibars (1 millibar = 10^2 N/m^2). In laboratory experiments pressures are usually expressed in millimetres (mm) of mercury whose density (mass per unit volume) is 1.359×10^4 kg/m^3. Normal atmospheric pressure is taken to be 760 mm of mercury. Find the corresponding pressure in terms of millibars.

Pressure is force per unit area.
Suppose that the cross-sectional area of an inverted tube containing mercury is A m^2. Then

The mass of 1 m^3 of mercury	$= 1.359 \times 10^4$ kg.
The weight of 1 m^3 of mercury	$= 1.359 \times 10^4 \times 9.81$ N.
The weight of 0.760 A m^3 of mercury	$= 1.359 \times 10^4 \times 9.81 \times 0.760$ A N.
The force exerted over 1 m^2	$= 1.359 \times 10^4 \times 9.81 \times 0.760$ N/m^2.
Pressure	$= 1.359 \times 10^2 \times 9.81 \times 0.760$ millibars.
	$= 1013$ millibars.

3.3   CENTRE OF MASS

Given two particles P_1 and P_2 with masses m_1 and m_2, their *mass-centre* G is defined to be the point between P_1P_2 such that

$$P_1G = \frac{m_2}{m_1 + m_2} P_1P_2$$

(see Figure 3.4). Obviously

$$GP_2 = \frac{m_1}{m_1 + m_2} P_1P_2.$$

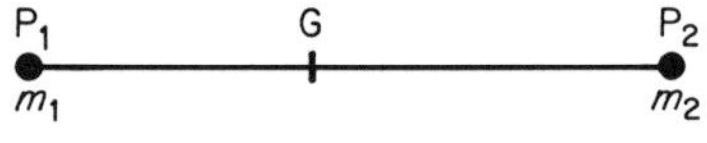

Figure 3.4   The mass-centre G divides P_1P_2 in the ratio $m_2 : m_1$.

The definition can easily be translated into vector notation. Let the particles have position vectors $\mathbf{r}_1$ and $\mathbf{r}_2$ (see Figure 3.5). The point G with position vector

$$\bar{\mathbf{r}} = \frac{m_1\mathbf{r}_1 + m_2\mathbf{r}_2}{m_1 + m_2} \qquad (2)$$

is then the mass-centre. It is easy to verify that P_1P_2 is divided by G in the ratio given above.

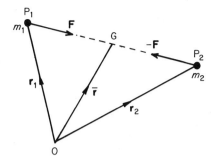

Figure 3.5 The position vector $\bar{r}$ of the mass-centre of two particles.

The importance of the mass-centre is revealed by considering again a two-particle system with mutual forces as in Figure 3.5. The equations of motion for the separate particles are

$$\mathbf{F} = m_1\ddot{\mathbf{r}}_1, \qquad -\mathbf{F} = m_2\ddot{\mathbf{r}}_2$$

so that, after the elimination of $\mathbf{F}$

$$m_1\ddot{\mathbf{r}}_1 + m_2\ddot{\mathbf{r}}_2 = \mathbf{0}. \qquad (3)$$

Differentiating equation (2) twice with respect to time, we get

$$(m_1 + m_2)\ddot{\bar{\mathbf{r}}} = m_1\ddot{\mathbf{r}}_1 + m_2\ddot{\mathbf{r}}_2$$
$$= \mathbf{0}$$

by (3). Thus, since $m_1 + m_2 \neq 0$, $\ddot{\bar{\mathbf{r}}} = \mathbf{0}$ and the acceleration of the mass-centre vanishes. Its velocity $\dot{\bar{\mathbf{r}}}$ must be a constant vector which implies that *the mass-centre moves in a straight line with constant velocity*. If the mass-centre is initially at rest it remains so, irrespective of the separate motions of the particles.

Example 3   Using a two-particle model for the Earth–Moon system, show that the Earth and Moon revolve about a point approximately 4800 km from the centre of the Earth. Assume that the mass of the Earth is 81 times that of the Moon and that the distance between them is 3.9×10^5 km.

If m_1 is the mass of the Earth and m_2 the mass of the Moon and d is the distance between them, the distance of G from the centre of the Earth is

$$\frac{m_2 d}{m_1 + m_2} = d\left(\frac{m_1}{m_2} + 1\right)^{-1}$$
$$= \frac{3.9 \times 10^5}{81 + 1} \text{ km}$$
$$= 4800 \text{ km}.$$

Since the mean radius of the Earth is about 6400 km, the mass-centre of the Earth-Moon system lies within the Earth. The perturbation in the Earth's orbit can be detected as a small oscillation of about 27 days' duration, the lunar month.

The *linear momentum* $\mathbf{p}$ of two particles is defined as the sum of the momenta of each particle. Thus if $\mathbf{r}_1$ and $\mathbf{r}_2$ are the position vectors of the two particles, then

$$\mathbf{p} = m_1\dot{\mathbf{r}}_1 + m_2\dot{\mathbf{r}}_2$$
$$= (m_1 + m_2)\dot{\bar{\mathbf{r}}}$$

through Equation (2). The linear momentum of the two particles is therefore the same as that of a single (fictitious) particle whose mass is the total mass moving with the speed of the mass-centre.

Further, if $\mathbf{F}_1$ and $\mathbf{F}_2$ are the forces which act on the two particles m_1 and m_2, respectively, then the separate equations of motion are

$$\mathbf{F}_1 = m_1\ddot{\mathbf{r}}_1, \qquad \mathbf{F}_2 = m_2\ddot{\mathbf{r}}_2.$$

The addition of these two equations gives

$$\mathbf{F}_1 + \mathbf{F}_2 = m_1\ddot{\mathbf{r}}_1 + m_2\ddot{\mathbf{r}}_2 = (m_1 + m_2)\ddot{\bar{\mathbf{r}}}$$

with the result that the mass-centre moves as though all the forces act there on a particle whose mass is the total mass. Since any mutual forces between the particles are equal and opposite, $\mathbf{F}_1 + \mathbf{F}_2$ represents the total *external* force on the system.

3.4   MULTI-PARTICLE SYSTEMS

It is a straightforward matter to extend the results so far derived for two-particle systems to systems consisting of n particles. Suppose the n particles have masses m_i, position vectors $\mathbf{r}_i$ and are subject to forces (mutual and external) $\mathbf{F}_i$ with $i = 1, 2, \ldots, n$. The *mass-centre* G of the system is defined to be the point with position vector

$$\bar{\mathbf{r}} = \sum_{i=1}^{n} m_i\mathbf{r}_i \bigg/ \sum_{i=1}^{n} m_i$$

where the usual summation notation is used.

The *linear momentum* $\mathbf{p}$ of the system is defined by

$$\mathbf{p} = \sum_{i=1}^{n} m_i\mathbf{v}_i$$

where $\mathbf{v}_i = \dot{\mathbf{r}}_i$.

For each particle of the system the second law of motion gives

$$\mathbf{F}_i = m_i\ddot{\mathbf{r}}_i \qquad i = 1, 2, \ldots, n \tag{4}$$

where $\ddot{\mathbf{r}}_i$ is the acceleration of the ith particle. Summing all these equations, we obtain

$$\sum_{i=1}^{n} \mathbf{F}_i = \sum_{i=1}^{n} m_i \ddot{\mathbf{r}}_i = \frac{\mathrm{d}}{\mathrm{d}t} \left(\sum_{i=1}^{n} m_i \mathbf{v}_i \right)$$

$$= \dot{\mathbf{p}}$$

$$= \frac{\mathrm{d}^2}{\mathrm{d}t^2} \left(\sum_{i=1}^{n} m_i \mathbf{r}_i \right)$$

$$= \frac{\mathrm{d}^2}{\mathrm{d}t^2} \left[\left(\sum_{i=1}^{n} m_i \right) \bar{\mathbf{r}} \right] \quad \text{(by the definition of } \bar{\mathbf{r}} \text{)}$$

$$= \left(\sum_{i=1}^{n} m_i \right) \ddot{\bar{\mathbf{r}}}.$$

We draw two conclusions. The total force $\sum_{i=1}^{n} \mathbf{F}_i$ is the *rate of change of the total linear momentum* and *the mass-centre of the system moves as a particle, whose mass is the total mass, subjected to the total force.* If the total force vanishes (for example, if the forces consist entirely of mutual actions and reactions between the particles) then $\sum_{i=1}^{n} \mathbf{F}_i = \mathbf{0}$ which implies that $\ddot{\bar{\mathbf{r}}} = \mathbf{0}$. In this case the mass-centre moves in a straight line with constant velocity or remains at rest.

Of course, knowledge of the motion of the mass-centre does not help in the description of the separate motions of the particles. Given the forces $\mathbf{F}_i$ in equation (4), the differential equations must be solved separately to achieve this.

3.5 MOMENT OF MOMENTUM

Consider a single particle of mass m, position vector $\mathbf{r}$ subject to a force $\mathbf{F}$. The linear momentum $\mathbf{p} = m\mathbf{v}$ where $\mathbf{v} = \dot{\mathbf{r}}$. The moment $\mathbf{h}_O$ of the linear momentum about the origin O is given by

$$\mathbf{h}_O = \mathbf{r} \times \mathbf{p} = m\mathbf{r} \times \mathbf{v}.$$

The vector $\mathbf{h}_O$ is called the *moment of momentum* or *angular momentum about* O. Taking the time derivative of $\mathbf{h}_O$, we have

$$\frac{\mathrm{d}\mathbf{h}_O}{\mathrm{d}t} = \frac{\mathrm{d}}{\mathrm{d}t} (m\mathbf{r} \times \mathbf{v}) = m\mathbf{v} \times \mathbf{v} + m\mathbf{r} \times \dot{\mathbf{v}}$$

$$= \mathbf{r} \times \mathbf{F}$$

since $\mathbf{v} \times \mathbf{v} = \mathbf{0}$ and $\mathbf{F} = m\dot{\mathbf{v}}$. The right-hand side is the moment of the force $\mathbf{F}$ about O. Thus *the moment of the force about* O *is the rate of change of the moment of momentum about* O.

Whilst the linear momentum is the vector whose magnitude and direction are independent of the fixed origin chosen, the moment of momentum vector varies with the choice of origin. In Figure 3.6, let O' be a second *fixed* origin with position vector $\mathbf{r}_{O'}$, and let the particle P have position vector $\mathbf{r}'$ relative to it. The moment of momentum about O' is

$$\mathbf{h}_{O'} = m\mathbf{r}' \times \mathbf{v}.$$

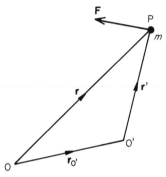

Figure 3.6

By the triangle law $\mathbf{r}' = \mathbf{r} - \mathbf{r}_{O'}$, so that

$$\mathbf{h}_{O'} = m(\mathbf{r} - \mathbf{r}_{O'}) \times \mathbf{v} = \mathbf{h}_O - m\mathbf{r}_{O'} \times \mathbf{v}$$

which clearly shows that generally $\mathbf{h}_O \neq \mathbf{h}_{O'}$. However, we still have

$$\frac{d\mathbf{h}_{O'}}{dt} = \frac{d}{dt}(m\mathbf{r}' \times \mathbf{v}) = \mathbf{r}' \times \mathbf{F},$$

the moment of the force about O'.

For a multi-particle system the moment of momentum is the sum of the individual moments. Let the particles have masses m_i, position vectors $\mathbf{r}_i$ referred to O and be subject to external forces $\mathbf{F}_i^{ext}$ and internal reactions $\mathbf{F}_i^{int}$, with $i = 1, 2, \ldots, n$. The total force on particle i is $\mathbf{F}_i^{ext} + \mathbf{F}_i^{int}$ and the moment of momentum

$$\mathbf{h}_O = \sum_{i=1}^{n} m_i \mathbf{r}_i \times \dot{\mathbf{r}}_i.$$

Taking the time derivative,

$$\frac{d\mathbf{h}_O}{dt} = \frac{d}{dt}\left(\sum_{i=1}^{n} m_i \mathbf{r}_i \times \dot{\mathbf{r}}_i\right) = \sum_{i=1}^{n} m_i \dot{\mathbf{r}}_i \times \dot{\mathbf{r}}_i + \sum_{i=1}^{n} m_i \mathbf{r}_i \times \ddot{\mathbf{r}}_i$$

$$= \sum_{i=1}^{n} m_i \mathbf{r}_i \times \ddot{\mathbf{r}}_i = \sum_{i=1}^{n} \mathbf{r}_i \times (\mathbf{F}_i^{ext} + \mathbf{F}_i^{int}).$$

Every internal force must be balanced by an equal and opposite force acting
in the same line with the result that the moments of these two forces must
cancel. Consequently

$$\frac{d\mathbf{h}_O}{dt} = \sum_{i=1}^{n} \mathbf{r}_i \times \mathbf{F}_i^{ext}$$

that is the moment of the external forces about the origin equals the rate
of change of moment of momentum about that point. If no external forces act
on the system, $\dot{\mathbf{h}}_O = \mathbf{0}$ and hence $\mathbf{h}_O =$ a constant vector, and we say that the
moment of momentum of the system is conserved.

The solar system (consisting of Sun, planets and moons) is an example of
such a situation since the only forces are the mutual gravitational attractions.
The moment of momentum vector is fixed in direction relative to the stars and
is consequently normal to an *invariant* plane called the *astronomical plane*. The
plane of the Earth's orbit (the ecliptic plane) is currently inclined at about 1.5°
to the astronomical plane.

Note that moment of momentum is required in the analysis of the *rotation*
of a body. We shall have more to say on this topic in Chapter 10. A particle
by its interpretation as a mass-point can have no sensible rotation. For this
reason the motion of a system of particles is completely described by equation
(4) of Section 3.4 without the introduction of moment of momentum.

3.6 GRAVITATIONAL FORCE

So far we have discussed force in general terms without considering the
nature of the force. One of the fundamental forces is gravity: any object in the
universe experiences a force due to the presence of any other object. This is the
Newtonian postulate of gravity which in detail states (in terms of particles) that
*any particle attracts any other particle with a force which is proportional to the
product of their masses and inversely proportional to the square of the distance
between them.* This force, of course, acts along the line joining the two particles.
Again the test of this postulate lies in its consistency with the laboratory and
visible evidence available to us.

This result can be expressed symbolically most conveniently in vector terms.
Let the particles A and B with masses m_A and m_B have position vectors $\mathbf{r}_A$ and
$\mathbf{r}_B$ in an inertial frame (see Figure 3.7). The vector

$$\overline{AB} = \mathbf{r}_B - \mathbf{r}_A$$

and the unit vector in the direction of $\overline{AB}$ is

$$\frac{\mathbf{r}_B - \mathbf{r}_A}{|\mathbf{r}_B - \mathbf{r}_A|}.$$

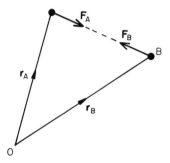

Figure 3.7   Mutual gravitational forces on two particles A and B.

According to Newton's law of gravitation, the force acting on A due to B is given by $\mathbf{F_A}$ where

$$\mathbf{F_A} = \frac{\gamma m_A m_B (\mathbf{r_B} - \mathbf{r_A})}{|\mathbf{r_B} - \mathbf{r_A}|^3}$$

in which γ, the constant of proportionality, is called the *gravitational constant* (the symbol G is also used in the literature). Although the gravitational force between two objects in a laboratory is very small, this force can be detected experimentally. Strictly, in the gravitational hypothesis the *gravitational mass* μ_B should be used instead of γm_B. The acceleration induced on A by B is then

$$\frac{\mu_B (\mathbf{r_B} - \mathbf{r_A})}{|\mathbf{r_B} - \mathbf{r_A}|^3}.$$

It is then a matter of experimental evidence (Poynting's or Cavendish–Boys' experiments) that μ_B/m_B is the same for all bodies. The value of γ is found to be 6.67×10^{-11} m^3/kg s^2 (at present the best estimates for γ are only accurate to three significant figures).

By Newton's third law the force on B due to A is given by

$$\mathbf{F_B} = -\mathbf{F_A} = \frac{\gamma m_A m_B (\mathbf{r_A} - \mathbf{r_B})}{|\mathbf{r_B} - \mathbf{r_A}|^3}.$$

Example 4   Estimate the mass of the Earth, given that the acceleration induced by the gravity of the Earth at its surface is 9.81 m/s^2, using a particle model for the Earth. (Use the following data: mean radius of the Earth = 6400 km. $\gamma = 6.7 \times 10^{-11}$ m^3/kg s^2.)

We shall consider the Earth as fixed in an inertial frame of reference. In our model the Earth is contracted to a particle and we consider the behaviour of a *unit* mass 6400 km from this point. The force required to produce an acceleration of 9.81 m/s^2 in a kg is

9.81 N. The gravitational force exerted on the unit mass is $\gamma M/a^2$ where M is the required mass and a is the radius of the Earth. Hence $9.81 = \gamma M/a^2$ or

$$M = \frac{9.81 \times (6400 \times 10^3)^2}{6.7 \times 10^{-11}} \text{ kg}$$

$$= 6.0 \times 10^{24} \text{ kg}.$$

This result does not take account of the oblateness of the Earth nor that the Earth has bulk but it compares very well with the currently adopted value of 5.98×10^{24} kg.

For spherical or nearly spherical bodies good estimates of the gravitational attraction *outside* the body can be obtained by replacing the body by a particle of equivalent mass at its centre. For arbitrary shapes the gravitational effects are much more complicated. The gravitational force *within* a body is also more elaborate and is beyond the scope of this book.

3.7   FIELDS OF FORCE

If at every point of a region of space a particle is acted upon by a force then there is said to exist a *field of force* in that region. The force will depend on position, in which case we would have $\mathbf{F}(x, y, z)$ in Cartesian coordinates, and possibly on time too in which case we would have $\mathbf{F}(x, y, z, t)$. The Earth is an example of the first form of the force field; it generates a gravitational field of force throughout space which with our particle model is directed towards the centre of the Earth. By the previous section this force obeys the inverse square law and takes the form, for a particle of mass m,

$$\mathbf{F} = -\frac{\gamma m M \mathbf{r}}{r^3} \tag{5}$$

if we treat the centre of the Earth as a fixed point and measure r from the centre of the Earth (M is the mass of the Earth).

The law of gravitation for a sphere can be replaced by a simpler one when we are dealing with phenomena which occur within a 'small' area of the Earth's surface (say about 300 km across and within a few tens of kilometres of the Earth's surface). Within such a region the gravitational force will be substantially constant and parallel. We can confirm this by two quick calculations.

Two points 300 km apart on the Earth's surface subtend an angle of 300/6400 radians or 2.7° at the centre of the Earth, if we take the radius of the Earth to be 6400 km. At the extremities of this distance the force directions will deviate from the parallel by about 2.7°.

Suppose at the surface of the Earth $g = 9.81$ m/s^2, then we have

$$g = \frac{\gamma M}{a^2} \tag{6}$$

where $a = 6400$ km. At a height of 150 km let the acceleration due to gravity be g' so that

$$g' = \frac{\gamma M}{(a + 150)^2}.$$

Division of this equation by (6) gives

$$\frac{g'}{g} = \frac{a^2}{(a + 150)^2} = \left(1 + \frac{150}{a}\right)^{-2}$$

$$\simeq 1 - \frac{2 \times 150}{a} = 0.95$$

using the first two terms of the binomial expansion. Therefore,

$$g' = 0.95 \times 9.81 = 9.3 \text{ m/s}^2.$$

Within the region indicated these two results show that the force is almost uniform in magnitude and direction. We can replace (5) by the uniform force field

$$\mathbf{F} = -\frac{\gamma m M}{a^2} \mathbf{k} = -mg\mathbf{k}$$

where $\mathbf{k}$ is the upward unit vector and $g = 9.81$ m/s^2. With a new frame of reference with origin on the Earth's surface and z-axis vertically upwards, Newton's second law gives

$$\ddot{\mathbf{r}} = -g\mathbf{k}$$

for the particle of unit mass. This equation separates into the three scalar equations

$$\ddot{x} = \ddot{y} = 0, \qquad \ddot{z} = -g \qquad (7)$$

which imply that the particle moves with constant velocity parallel to the surface of the Earth and with constant acceleration vertically. We shall examine motion in a uniform force field in the next chapter in the context of projectiles.

Other examples of fields of force are provided by a magnetic field and the force it exerts on a charged particle moving in it, or the wind in the atmosphere and its effect on an aircraft passing through it. Force fields may also depend on velocity or higher time-derivatives of displacement in addition to position and time.

3.8 RIGID BODIES

The results obtained so far tacitly assume that 'bodies' can be approximated to by particles. Indeed the basic laws of Newtonian mechanics refer only to particles. However it would not be realistic to exclude all consideration of bodies of finite dimension. Probably the simplest type of body is the *rigid body*. The rigidity can be expressed by the condition that the distance between any two points of the body remains constant, no matter what forces are applied to the body. By using this essentially geometric constraint, we can avoid examining the local forces or *stresses* within the body since the stresses in a rigid body must always be such as to maintain its rigidity. If the body is composed of deformable material and exhibits elastic or fluid properties, we must postulate the nature of the internal stresses and examine the consequent deformation suffered by the body. For many practical purposes the assumption of rigidity is a useful model. This is not to say, however, that all bodies which appear rigid can be treated as rigid bodies. Certain hard rubber balls show characteristics at variance with their predicted behaviour when a rigid body model is used.

The fundamental assumption concerning the nature of a rigid body is that every subdivision of it has mass. In other words we may cut the body into as many pieces as we please but each portion will have mass. This is contrary to the usual molecular concept of matter but is sufficient for most problems in the physical and engineering sciences where we are dealing with matter in bulk. In Figure 3.8, **r** is the position vector of a point P of a rigid body. The point P is enclosed by a small region of volume δV and mass δm. The ratio $\delta m/\delta V$ is the mean density of this region. The limit of this ratio as δV collapses to the point is the density at P. Symbolically, the density

$$\rho = \lim_{\delta V \to 0} \frac{\delta m}{\delta V}.$$

We note that ρ can be a function of position in the body.

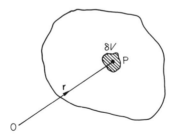

Figure 3.8 Position vector **r** of an element δV of the rigid body.

The mass-centre $\bar{\mathbf{r}}$ of the rigid body is defined by

$$M\bar{\mathbf{r}} = \int \mathbf{r} \, dm = \int \mathbf{r}\rho \, dV$$

where M is the total mass, and the integration is performed throughout the volume. Interpreting integration as a summation this definition is in line with that for multi-particle systems: it is the 'sum' of the product of elements of mass and **r**.

Example 5   Find the mass-centre of a solid hemisphere of uniform density ρ and radius a.

The radius perpendicular to the face is an axis of symmetry of the hemisphere. Since the density is uniform the mass-centre will lie on this line. The mass of the hemisphere is $\frac{2}{3}\pi a^3\rho$.

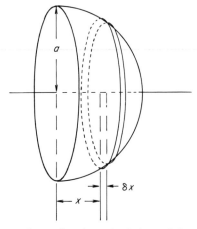

Figure 3.9   Integration scheme for the calculation of the mass-centre of a hemisphere.

Consider now a disc of thickness δx which is parallel to the face and distance x from it (see Figure 3.9). The mass of this disc is approximately $\rho\pi(a^2 - x^2)\delta x$. If $\bar{x}$ is the distance of the mass-centre from the face, we have

$$\bar{x}\cdot\tfrac{2}{3}\pi a^3\rho = \int_0^a \rho\pi(a^2 - x^2)x\ \mathrm{d}x$$

$$= \rho\pi\left[\frac{a^2x^2}{2} - \frac{x^4}{4}\right]_0^a$$

$$= \tfrac{1}{4}\rho\pi a^4.$$

Therefore $\bar{x} = 3a/8$.

The application of Newton's laws of motion for *particles* to rigid bodies presents serious technical difficulties since obviously a rigid body cannot be considered strictly speaking as an agglomeration of particles. It is possible to produce a plausible line of argument by imagining a rigid body subdivided into a large number of small elements of mass δm and then to replace these elements by particles of equal mass. This amounts to the assertion that a rigid body is a collection of particles held together by massless ties. This conflicts with our definition of a body having within it a continuous distribution of mass. In fact, in the same way that we had to redefine the mass-centre of a rigid body so must we also extend the Newtonian postulates and, in particular, the second law of motion.

The *linear momentum* **p** of a rigid body is defined by

$$\mathbf{p} = \int \dot{\mathbf{r}} \, dm = \int_V \dot{\mathbf{r}}\rho \, dV$$

the integration being performed throughout the body (refer back to Figure 3.8). Again this is 'sum' of 'mass × velocity' for each element δm and accords with the definition for a multi-particle system. Further

$$\mathbf{p} = \int_V \dot{\mathbf{r}}\rho \, dV = \frac{d}{dt}\int_V \mathbf{r}\rho \, dV = \frac{d}{dt}(M\bar{\mathbf{r}}) = M\dot{\bar{\mathbf{r}}}, \tag{8}$$

where the operations of integration and differentiation have been interchanged. This is a step which requires proof but we have not available the necessary mathematics to do this. For this reason we shall accept this operation as being a reasonable one. We conclude, as with the multi-particle system, that the linear momentum is the product of the total mass and the velocity of the mass-centre. However, we shall verify (8) in the following example.

Example 6   A wheel of radius a, mass M and density ρ rolls along a straight line with constant speed U. Verify that

$$\int_V \dot{\mathbf{r}}\rho \, dV = \frac{d}{dt}\int_V \mathbf{r}\rho \, dV = MU\mathbf{i}.$$

Assume that the wheel has negligible thickness so that the density is mass per unit area and the element of volume can be interpreted as an element of area. Let the x-axis be the line along which the wheel rolls with the z-axis vertical. If ω is the angular speed of the wheel, then $U = a\omega$. Consider at time t a point P of the wheel distance R from the centre and such that the radius through P makes an angle θ with the downward vertical (see Figure 3.10). Clearly the position vector of P is given by

$$\mathbf{r} = (s - R \sin \theta)\mathbf{i} + (a - R \cos \theta)\mathbf{k}$$

and, consequently,

$$\dot{\mathbf{r}} = (\dot{s} - R \cos \theta\dot{\theta})\mathbf{i} + R \sin \theta\dot{\theta}\mathbf{k} = \left(U - \frac{RU}{a}\cos \theta\right)\mathbf{i} + \frac{RU}{a}\sin \theta\mathbf{k}$$

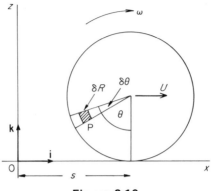

Figure 3.10

since R is a constant for a particular point P, $\dot{s} = U$ and $\dot{\theta} = \omega$. An element of area is $R\delta R\delta\theta$ (the shaded area in Figure 3.10). Thus

$$\int_V \dot{\mathbf{r}}\rho \, dV = \int_0^a \int_0^{2\pi} \left[\left(U - \frac{RU}{a} \cos\theta \right) \mathbf{i} + \frac{RU}{a} \sin\theta \, \mathbf{k} \right] \rho \, d\theta R \, dR$$

$$= \int_0^a \left[\left(U\theta - \frac{RU}{a} \sin\theta \right) \mathbf{i} - \frac{RU}{a} \cos\theta \, \mathbf{k} \right]_{\theta=0}^{\theta=2\pi} \rho R \, dR$$

$$= \int_0^a 2\pi\rho UR \, dR\mathbf{i} = 2\pi\rho U[\tfrac{1}{2}R^2]_{R=0}^{R=a}\mathbf{i}$$

$$= \pi a^2 \rho U\mathbf{i} = MU\mathbf{i}.$$

Similarly

$$\frac{d}{dt}\int_V \mathbf{r}\rho \, dV = \frac{d}{dt}\int_0^a \int_0^{2\pi} [(s - R\sin\theta)\mathbf{i} + (a - R\cos\theta)\mathbf{k}]\rho \, d\theta R \, dR$$

$$= \frac{d}{dt}\int_0^a (2\pi s\mathbf{i} + 2\pi a\mathbf{k})\rho R \, dR$$

$$= \frac{d}{dt}(\pi a^2 \rho s\mathbf{i} + \pi a^3 \rho\mathbf{k})$$

$$= \pi a^2 \rho \dot{s}\mathbf{i} = MU\mathbf{i}.$$

The forces which a rigid body may experience can be classified under two headings: the *body forces* which apply to every point of the body, as for example in the case of gravity, and the *applied surface forces* which may be either point loads, $\mathbf{F}_1, \mathbf{F}_2, \ldots$ in Figure 3.11, or pressure applied over a section of the surface (the hatched surface S in the figure). Suppose a rigid body is subject to a body force $\mathbf{Q}$ per unit mass, point loads $\mathbf{F}_1, \mathbf{F}_2, \mathbf{F}_3, \ldots, \mathbf{F}_n$ and pressure p over the

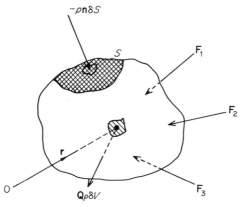

Figure 3.11 Rigid body subject to body force, surface pressure and point loads.

surface S. The total force $\mathbf{F}$ acting on the body (internal reactions in the body cancel one another out) is given by

$$\mathbf{F} = \int_V \mathbf{Q}\rho \; dV + \sum_{i=1}^n \mathbf{F}_i - \int_S p\mathbf{n} \; dS$$

where $\mathbf{n}$ is the unit outward vector normal to the surface of the body. The last term represents the total effect of the pressure over the surface S (remember that pressure is force per unit area and acts normal to the surface).

Newton's second law of motion modified for bodies asserts that the total force on the body equals the rate of change of linear momentum, that is

$$\mathbf{F} = \frac{d\mathbf{p}}{dt} = \frac{d}{dt} \int_V \dot{\mathbf{r}}\rho \; dV$$

$$= \frac{d}{dt}(M\dot{\mathbf{r}}) = M\ddot{\mathbf{r}} \tag{9}$$

by Equation (8). There results the important principle: the total force acting on a rigid body balances the product of total mass and the acceleration of the mass-centre. This statement reveals the added importance of particle dynamics since the mass-centre of a rigid body behaves as a particle.

Example 7   A rigid body moves in a uniform gravitational field. Show that the equation of motion of its mass-centre is

$$\ddot{\mathbf{r}} = -g\mathbf{k}$$

where $-\mathbf{k}$ is a unit vector in the direction of the force.

In this case the body force $\mathbf{Q}$ is $-g\mathbf{k}$ where g, the acceleration due to gravity, is constant. The total force

$$\mathbf{F} = -\int g\mathbf{k}\ \mathrm{d}m = -g\mathbf{k}\int \mathrm{d}m = -Mg\mathbf{k}.$$

By the equation of motion

$$M\ddot{\mathbf{r}} = -Mg\mathbf{k}$$

or $\ddot{\mathbf{r}} = -g\mathbf{k}$.

We note from (9) that if the total force $\mathbf{F}$ vanishes the linear momentum of the body is conserved and its mass-centre moves in a straight line with constant speed.

The position of the mass-centre at any time does not tell us anything about the *orientation* of the body about its mass-centre. Our second postulate for rigid bodies asserts that the moment of all the forces acting on the body about the origin of an inertial frame balances the rate of change of moment of momentum about the same point. The moment of momentum of a body subject to the forces indicated in Figure 3.11 is defined as

$$\mathbf{h} = \int_V \mathbf{r} \times \dot{\mathbf{r}}\rho\ \mathrm{d}V.$$

Again this is a natural extension of the definition for a multi-particle system. The moment of the forces acting on the body

$$\mathbf{M} = \int_V \mathbf{r} \times \mathbf{Q}\rho\ \mathrm{d}V + \sum_{i=1}^{n} \mathbf{r}_i \times \mathbf{F}_i - \int_S p\mathbf{r} \times \mathbf{n}\ \mathrm{d}S$$

where $\mathbf{r}_1, \mathbf{r}_2, \ldots$ are the position vectors of the points where the forces $\mathbf{F}_1, \mathbf{F}_2,$ ... act. Our second postulate says that

$$\mathbf{M} = \frac{\mathrm{d}\mathbf{h}}{\mathrm{d}t}. \tag{10}$$

Thus the *six* scalar equations governing the motion of a rigid body are contained in the two vector equations

$$\mathbf{F} = \frac{\mathrm{d}\mathbf{p}}{\mathrm{d}t}, \quad \mathbf{M} = \frac{\mathrm{d}\mathbf{h}}{\mathrm{d}t}. \tag{11}$$

The two postulates (due to Euler) for bodies replace Newton's three laws of motion for particles. They are particularly important since they can be generalized to non-rigid materials such as fluids and elastic bodies.

We have emphasised that the general motion of a rigid body is beyond our task in this book and we shall not pursue Equation (10) further. However, we shall note some results applicable to statics. Suppose the rigid body is at rest, that is, $\dot{\mathbf{r}} = 0$. The linear momentum and moment of momentum must vanish with the result that the equations of *static equilibrium* are, from (11)

$$\mathbf{F} = \mathbf{0} \qquad \mathbf{M} = \mathbf{0}$$

that is, the total force and the total moment of the forces must each balance.

Suppose now that the body is translating but not rotating, which means that every point of the body must have the same velocity as that of the mass-centre, that is $\dot{\mathbf{r}} = \dot{\bar{\mathbf{r}}}$. The angular momentum

$$\mathbf{h} = \int_V \mathbf{r} \times \dot{\mathbf{r}}\rho \ dV = \int_V \mathbf{r} \times \dot{\bar{\mathbf{r}}}\rho \ dV = \int_V \mathbf{r}\rho \ dV \times \dot{\bar{\mathbf{r}}}$$

since $\dot{\bar{\mathbf{r}}}$ is independent of the volume integration. By the definition of mass-centre,

$$\mathbf{h} = M\bar{\mathbf{r}} \times \dot{\bar{\mathbf{r}}}$$

and consequently

$$\frac{d\mathbf{h}}{dt} = M\bar{\mathbf{r}} \times \ddot{\bar{\mathbf{r}}} = \bar{\mathbf{r}} \times \mathbf{F}$$

by (9). Now

$$\mathbf{M} = \frac{d\mathbf{h}}{dt} = \bar{\mathbf{r}} \times \mathbf{F} = \int_V \bar{\mathbf{r}} \times \mathbf{Q}\rho \ dV + \sum_{i=1}^n \bar{\mathbf{r}} \times \mathbf{F}_i - \int_S p\bar{\mathbf{r}} \times \mathbf{n} \ dS$$

and by the definition of $\mathbf{M}$ (see the equation prior to Equation (10)),

$$\int_V (\mathbf{r} - \bar{\mathbf{r}}) \times \mathbf{Q}\rho \ dV + \sum_{i=1}^n (\mathbf{r}_i - \bar{\mathbf{r}}) \times \mathbf{F}_i - \int_S p(\mathbf{r} - \bar{\mathbf{r}}) \times \mathbf{n} \ dS = \mathbf{0}.$$

In other words, for a body simply translating, the moment of the external forces about the *mass-centre* must vanish. The moment about any *fixed* point, of course, will not generally be zero.

Example 8   A uniform bar of mass m and length 2a is supported in a horizontal position at a point distance $\frac{1}{2}a$ from one end at a point distance $\frac{1}{4}a$ from the other end of the bar. How is the weight distributed between the supports?

We are only interested in the vertical forces P and Q acting on the bar at the supports (see Figure 3.12). The rod is in static equilibrium so that the vertical forces must balance:

$$P + Q = mg. \tag{12}$$

The moments of the forces about a fixed point, say the mid-point of the bar, must balance. Thus

$$\tfrac{1}{2}aP = \tfrac{3}{4}aQ \tag{13}$$

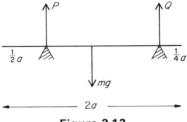

Figure 3.12

since the weight *mg* acts through the mid-point. Solving (12) and (13), we obtain

$$P = \tfrac{3}{5}mg, \qquad Q = \tfrac{2}{5}mg.$$

Example 9   A wagon of length 10 m, width 2 m and depth 3 m can move along a horizontal track on wheels whose axles are 2 m from the ends of the wagon. The wagon contains material of density 10^3 kg/m³. The maximum safe working load of the axles is that occurring when the wagon is full and stationary. If the wagon is drawn by a coupling which exerts a constant horizontal pull of 8×10^4 N on the floor of the wagon, find the maximum safe load assuming that it is evenly distributed.

A sketch of the wagon together with the two vertical reactions *R* and *S* on the axles is shown in Figure 3.13.

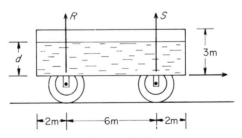

Figure 3.13

When full the wagon contains $10 \times 3 \times 2 \times 10^3 = 6 \times 10^4$ kg. By symmetry each axle can support a load not greater than 3×10^4 kg. Suppose the load reaches a height *d* m above the floor of the wagon when it moves: the mass-centre will be at height $\tfrac{1}{2}d$ above the centre of the floor. The load will now be $10 \times d \times 2 \times 10^3 = 2 \times 10^4 d$ kg. Since the wagon does not rotate and moves horizontally the total vertical force and the moment of the forces about the mass-centre must vanish:

$$R + S = 2 \times 10^4 \, dg$$
$$8 \times 10^4 \times \tfrac{1}{2}d + 3S = 3R$$

From the second of these equations we note that $R > S$ and therefore we need only determine *R*, the greater load. Eliminating *S* between these equations:

$$6 \times 10^4 \, dg + 4 \times 10^4 d - 3R = 3R$$

and, therefore,

$$d = \frac{3R}{(2 + 3g)} \, 10^4 \text{ m.}$$

The maximum value R can take is $3 \times 10^4 \, g$ N with the result that the maximum value of d is 2.8 m. Thus the maximum safe load is

$$5.6 \times 10^4 \text{ kg}$$

approximately.

3.9   IMPULSIVE MOTION

A golf ball struck by a golf club achieves a velocity in a relatively short span of time. When the collision or impact occurs the ball is deformed and large internal stresses are created. These stresses tend to restore the ball to its original shape and it is the combined effect of these stresses which cause the ball to be projected from the face of the club. The precise behaviour of the ball will depend on the *elasticity* of the material of which the ball is composed. If the ball starts from rest, a typical graph of speed v against time t is shown in Figure 3.14. The ball accelerates from zero speed to speed v_1 in a 'small' time interval t_1. The mathematical analysis of large scale deformation of bodies is a very complicated subject and it is theoretically much simpler to imagine that the velocity jumps suddenly at time $t = 0$ from zero to speed v_1. In this way we need not consider the details of the impact. The collision between the club and ball creates linear momentum mv_1 in the ball where m is the mass of the ball. The momentum so created is called the *impulse* which we shall denote by I. Thus

$$I = mv_1.$$

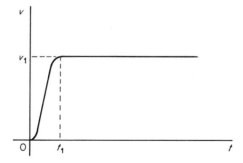

Figure 3.14 Typical velocity response of a body subjected to an impulse.

We can interpret I in a different sense if we return to the real situation. Suppose F is the resultant force which acts on the ball so that

$$F = m\frac{\mathrm{d}v}{\mathrm{d}t}.$$

Integrate with respect to time over the time interval t_1:

$$\int_{t=0}^{t=t_1} F\,\mathrm{d}t = m\int_{t=0}^{t=t_1} \frac{\mathrm{d}v}{\mathrm{d}t}\,\mathrm{d}t = m\int_{v=0}^{v=v_1}\mathrm{d}v = mv_1.$$

We observe that the impulse is the time integral of the force; in a sense, it is the total effect of the elastic force on the ball. Note also that impulse is not a force and if mass is in kg and speed is in m/s, impulse will be measured in kg m/s.

In general, if a particle of mass m moving with velocity $\mathbf{v}_1$ experiences an impulse $\mathbf{I}$ which produces a final velocity $\mathbf{v}_2$ in the particle, then

$$\mathbf{I} = m(\mathbf{v}_2 - \mathbf{v}_1). \tag{14}$$

That is, the impulse is the vector difference of the final and initial linear momenta. We note that in any direction in which the component of $\mathbf{I}$ vanishes, the linear momentum in that direction is conserved.

Since it is assumed that the impulse is created instantaneously, the integrals of other forces which are acting on the body do not enter into Equation (14). It is often assumed, however, that impulses are transmitted along strings joining two bodies, see Example 13 which follows, but again this really represents the integral of a very large force over a very short time. Such impulses are often referred to as impulsive tensions, although strictly they are not tensions which have the dimensions of force.

The form of the balance of momentum change and impulse follows naturally for a rigid body. Suppose a rigid body is subject to impulses $\mathbf{I}_1, \mathbf{I}_2, \ldots, \mathbf{I}_n$ acting at the same instant of time (see Figure 3.15). As explained in the previous section the linear momentum

$$\mathbf{p} = \int_V \dot{\mathbf{r}}\rho\,\mathrm{d}V = \frac{\mathrm{d}}{\mathrm{d}t}\int_V \mathbf{r}\rho\,\mathrm{d}V = M\dot{\bar{\mathbf{r}}}$$

where M is the mass of the body. For rigid bodies the total impulse balances the change in linear momentum. Thus

$$\sum_{i=1}^{n}\mathbf{I}_i = \mathbf{p}_1 - \mathbf{p}_2 = M(\dot{\bar{\mathbf{r}}}_1 - \dot{\bar{\mathbf{r}}}_2)$$

where suffixes 1 and 2 refer respectively to the state of the body before and after the impulses are applied. In other words, the mass-centre behaves as though it is a particle of mass M with all the impulses acting on it.

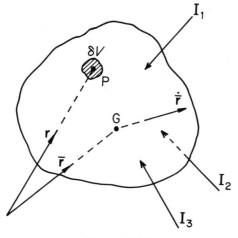

Figure 3.15

If we wish to examine the angular velocity created by the application of impulses we equate the moment of the impulses to the jump in angular momentum.

When two bodies collide, as in Figure 3.16, each body will experience an impulsive reaction due to the impact of the other, $\mathbf{I}$ and $-\mathbf{I}$ in the figure. If friction is negligible, we assume that the impulsive reactions act along the normals to the surfaces at the point of impact (if friction must be taken into account there will, in addition, be an impulsive tangential reaction on the bodies). For each of the bodies the balance of impulse and jump of linear momentum still hold, whilst for the system as a whole, since no external impulse acts, the impulsive reactions must be equal and opposite. For this reason also the linear momentum of the whole system must be conserved.

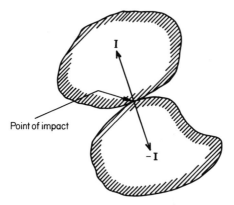

Figure 3.16

Returning to our illustration of the golf ball, we noted that the total propulsive force exerted by the internal stresses will depend very much on the material of which a golf ball is made. We must specify the elastic law empirically. The simplest law is *Newton's law of restitution*. This states that when two bodies collide their relative parting velocity in the direction of the common normal at the point of impact is $-e$ times their relative approach velocity in this direction. We shall assume that for any two materials, the *coefficient of restitution e* is a positive constant. The reason for the minus sign prefixing e in the statement of the law is that the two relative velocities are taken in the same direction and that bodies usually rebound on impact.

Example 10   A steel ball is released from rest 1 m above a horizontal steel plate. On rebound it is found to reach a height of 81 cm. Find e for steel against steel.

Since the ball falls with constant acceleration g it hits the plate with speed $\sqrt{200g}$ cm/s. Since the ball reaches a height of 81 cm, it must rebound with speed $(2g \times 81)^{1/2}$ cm/s upwards. By Newton's law

$$\sqrt{162g} = e\sqrt{200g}.$$

Therefore

$$e = \left(\frac{81}{100}\right)^{1/2} = 0.90.$$

There are no units for e since it is a ratio of two speeds.

This example provides a simple experimental method of determining e.

Example 11   Two spheres of masses m and M moving in the same straight line with velocities u_1 and v_1 collide. Find their subsequent speeds if the coefficient of restitution is e.

In any impact between spheres the impulsive reactions will act along the line joining the centres of the spheres. Assume that sphere m overtakes sphere M, that is, $u_1 > v_1$. Let u_2 and v_2 be the corresponding final velocities of the spheres in the same sense. For the system as a whole the linear momentum will be conserved, that is the initial momentum balances the final momentum:

$$mu_1 + Mv_1 = mu_2 + Mv_2. \tag{15}$$

Newton's law gives (taking care with the signs)

$$v_2 - u_2 = -e(v_1 - u_1). \tag{16}$$

Solving equations (15) and (16), we find that

$$u_2 = \frac{1}{m + M} [Mv_1(1 + e) + u_1(m - eM)]$$

$$v_2 = \frac{1}{m + M} [mu_1(1 + e) + v_1(M - em)]$$

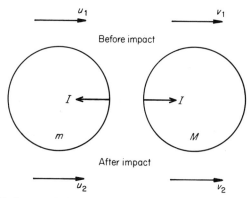

Figure 3.17 Before and after velocities for the direct impact of two spheres.

the final velocities of the two spheres. All velocities which appear may be positive or negative: if a velocity is negative it will be in the opposite sense to that drawn in Figure 3.17.

If $e = 0$ for the two spheres, $u_2 = v_2$ and the two bodies adhere after impact: we say that they are *inelastic*. If $e = 1$, the relative velocity before impact equals in *magnitude* the relative velocity after impact. In this ideal case we describe the bodies are *perfectly elastic*. For real substances $0 < e < 1$. Steel against steel has a high value for e (Example 10) whereas dough, for instance, is almost inelastic.

Example 12   A sphere sliding on a smooth table (that is, frictionless) with speed u hits a smooth vertical wall at angle θ to the wall. If e is the coefficient of restitution between the sphere and wall, find the velocity of the sphere after impact.

Let the sphere rebound from the wall with speed v in a direction ϕ to the wall (Figure 3.18). Since the wall is smooth the impulsive reaction on the sphere will be perpendicular to the wall. Thus the linear momentum of the sphere parallel to the wall will be conserved:

$$u \cos \theta = v \cos \phi. \tag{17}$$

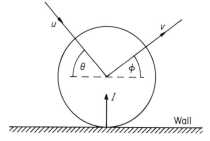

Figure 3.18 Oblique impact of a sphere with a vertical wall.

Newton's law gives

$$v \sin \phi = eu \sin \theta. \tag{18}$$

From (17) and (18):

$$v = u(e^2 \sin^2 \theta + \cos^2 \theta)^{1/2} \quad \text{and} \quad \tan \phi = e \tan \theta.$$

The ball rebounds from the wall with speed $u(e^2 \sin^2 \theta + \cos^2 \theta)^{1/2}$ in a direction $\tan^{-1}(e \tan \theta)$ to the wall.

Example 13   A tug of mass m is attached to a ship of mass M by a cable of negligible mass. Initially the cable is slack. The tug moves off and attains a speed u_1 before the cable becomes taut and the ship is jerked into motion. If $e = \frac{1}{2}$ for the cable, find, neglecting any impulsive resistance of the water, the speed imparted to the ship. If $m = 5 \times 10^5 \, kg$, $M = 1.5 \times 10^7 \, kg$ and $u_1 = 1 \, m/s$, find the mean tension in the cable supposing the jerk takes 2s.

Let u_2 and v_2 be respectively the final speeds of the tug and ship. The momentum of the whole system is conserved. Thus

$$mu_1 = mu_2 + Mv_2. \tag{19}$$

Newton's law gives

$$u_2 - v_2 = -eu_1 = -\tfrac{1}{2}u_1. \tag{20}$$

Solving (19) and (20) for v_2, we obtain

$$v_2 = \frac{3mu_1}{2(m + M)}$$

the final speed of the ship. Note that, after the jerk in the cable,

$$u_2 = v_2 - \tfrac{1}{2}u_1 < v_2$$

which implies that the cable becomes slack again since the ship moves off more quickly than the tug.

The above analysis assumes that the jerk takes place instantaneously. In order to investigate the tension in the cable we must look at the integral of the equation of motion for the system. Let T be the tension in the cable. Whilst the cable is taut the equation of motion for the tug is

$$-T = m \frac{dv}{dt}.$$

Integrate over the time interval of 2 s, so that

$$-\int_0^2 T \, dt = m \int_{u_1}^{u_2} dv = m(u_2 - u_1).$$

Over a time interval (t_1, t_2) we define the mean value $\bar{T}$ of T by

$$\bar{T} = \frac{1}{t_2 - t_1} \int_{t_1}^{t_2} T \, dt$$

In this case $\bar{T}$ is given by

$$\int_0^2 T \, dt = 2\bar{T}$$

so that

$$\bar{T} = -\tfrac{1}{2}m(u_2 - u_1) = \tfrac{1}{2}Mv_2 \quad \text{[by (19)]}$$

$$= \frac{3mMu_1}{4(m + M)}.$$

Thus

$$\bar{T} = \frac{3 \times 5 \times 10^5 \times 1.5 \times 10^7 \times 1}{4 \times 155 \times 10^5}$$

$$= 3.6 \times 10^5 \ N.$$

Example 14   Two particles A and B of the same mass are connected by a taut string and lie at rest on a horizontal table. Particle A is struck a blow of magnitude I in a direction inclined at 45° to BA as shown in Figure 3.19. Find the initial velocities of the particles.

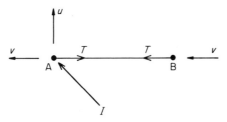

Figure 3.19

When A is struck, an impulsive tension T will be created in the string since this is taut (we assume the string is inextensible). Let A move off with velocity components u perpendicular to the string and v parallel to the string (Figure 3.19). Since A and B are connected by a taut string, B must move off in the direction BA with the same speed v. B will have no velocity component perpendicular to AB since the only impulse acting on B is the impulsive tension acting along the string. If m is the mass of each particle, we have for A

$$\frac{I}{\sqrt{2}} = mu, \qquad \frac{I}{\sqrt{2}} - T = mv$$

and for B

$$T = mv.$$

Eliminating T, the velocity components are given by

$$u = I/m\sqrt{2}, \qquad v = I/2m\sqrt{2}.$$

Remember these are the *initial* velocities; subsequently the system moves as two connected particles.

Exercises

1. What force in newtons is required to give a car of mass 10^3 kg an acceleration of 2m/s²?

2. A particle of unit mass has position vector $\mathbf{r} = t\mathbf{i}$ in one inertial frame of reference. A second frame is translating with respect to the inertial frame such that its origin has a position vector $a\mathbf{i} + b\mathbf{j} + t^2\mathbf{k}$ where a and b are constants. Find the path the particle takes relative to the second frame and show that it will appear to be subject to a force $-2\mathbf{k}$.

3. Atwood's machine consists of two weights fastened to the ends of a light string which passes over a light free pulley. If the two weights have masses 7 kg and 9 kg, find their acceleration.

4. A proton moving in a horizontal path in vacuo at a speed of 3×10^6 m/s enters a uniform electric field which exerts a downward force of 6×10^{-15} N on it. The mass of the proton is 1.66×10^{-27} kg. If the field extends over a region 0.5 m long, through what angle is the particle deflected?

5. A particle is projected vertically upwards under gravity with a speed of 16 m/s. One second later another particle is fired upwards from the same point. Find the initial speed of this particle in order that the two particles will collide when the first particle has reached its highest point.

6. A balloon of total mass M descends with a downward acceleration f_1. Assuming no air resistance, find what mass of ballast should be thrown out in order that the balloon should rise with acceleration f_2.

7. A circular face of a solid circular cylinder of radius a and height a is attached to the circular face of a solid hemisphere of radius a and of the same density. Find the mass-centre of the composite body.

8. An object falls vertically past a window 2 m high in $\frac{1}{12}$ s. Find the height above the bottom of the window from which the object was dropped.

9. A stone is dropped from a balloon rising at 10 m/s and reaches the ground in 8 s. How high was the balloon when the stone was dropped?

10. Find the mass of the Sun given that the gravitational constant $\gamma = 6.67 \times 10^{-11}$ m³/kg s² and that the Earth's orbit is a circle of radius 1.5×10^8 km which it completes in 365 days.

11. A locomotive can keep a train of mass 3×10^6 kg moving up an incline of 1 (vertically) in a 100 (horizontally) at a constant speed. How long will it take to accelerate

the locomotive to 45 km/h from rest on the level? If the drawbar weighs 150 kg, how much harder will the locomotive pull the drawbar than the drawbar pulls the train?

12. A barge whose deck is 4 m below the level of a dock is pulled in by means of a cable attached to a ring on the dock, the cable being hauled in by a windlass on the deck of the barge at the rate of 2 m/min. Find the tension in the cable when the windlass is 5 m horizontally from the dock given that the barge has a mass of 10^5 kg.

13. An aircraft flies along the upper arc of a circle of radius 8000 m in order to simulate weightlessness in the cabin. Obtain the speed of the aircraft for this to be achieved at the highest point of the path.

14. A rectangular gate of mass M and width $2b$ is supported by two hinges symmetrically placed on the side of the gate and distance l apart. The upper hinge provides no vertical support. Find the magnitude of the support at both hinges.

15. A car has a wheelbase of 1.8 m and runs on wheels of radius 0.3 m. The car weighs 1000 kg and its mass-centre is centrally placed with respect to the wheels at a height 0.45 m above the road. The car accelerates at 1 m/s². Assuming that the propulsive force can be interpreted as a horizontal force acting on the rear axle, find the distribution of the weight between the axles.

16. A girl of mass 40 kg moving at 16 km/h jumps on to a toboggan of mass 10 kg, sliding at 32 km/h in the same direction, as it passes her. What is the final speed of the toboggan?

17. A particle of mass m falls freely a distance h and then jerks into motion a mass M ($M > m$) connected to m by a light string which passes over a frictionless pulley. Find the time which elapses before M returns to its original position.

18. A ball is projected vertically upwards from the ground with a speed of 29 m/s. Simultaneously a similar ball is dropped from the highest point that the first ball would reach. If the coefficient of restitution between the balls is 0.6, calculate the time which elapses between the balls hitting the ground.

19. A set of n wagons, each of mass m, standing in a railway siding is set in motion by a locomotive of mass rm starting with speed V. Initially the couplings between the wagons are slack, as is the coupling between the first wagon and the locomotive, and each wagon moves a distance s before it jerks the succeeding wagon into motion. The coefficient of restitution of the coupling is e. Find the time which ensues between the movement of the first and last wagons, neglecting the effect of friction and assuming that each coupling does not remain taut until the last wagon moves.

20. One smooth sphere hits another smooth sphere which is at rest. The two spheres are perfectly elastic. After the collision they move off at right angles to each other. Show that their masses are equal.

21. Four particles A, B, C, D, each of mass M, are connected by equal light strings AB, BC, DC, DA and lie at rest on a smooth horizontal table with the strings taut so that

ABCD is a rhombus and $B\hat{A}D = 2\alpha$ $(\alpha < 45°)$. The particle A is given a velocity U in the direction CA. Find the initial velocities of B, C, D.

22. Two small spheres P and Q of equal mass lie touching each other in a smooth circular horizontal groove of radius a. Sphere P is projected away from Q with speed U. Assuming no resistance to motion in the groove, ascertain

  (i) the speeds of P and Q after the nth collision,
  (ii) the time that elapses before the nth collision

if the coefficient of restitution for each collision is e.

23. A ball of mass 110 g is struck by a club, the contact lasting 1/100th of a second. The ball is given a speed of 10 m/s. Assuming that the force increases from zero linearly with time for 1/200th of a second and decreases linearly to zero again, find the impulse of the blow and the maximum force on the ball.

24. A man weighing 80 kg stands 1 m from the bow of a boat which is just touching the shore of a lake, the line of the boat being perpendicular to the shore. The boat weighs 40 kg. Assuming that the resistance to motion of the boat in the water is negligible, find:

  (i) the distance the man would be from the bank if he moved to the bow of the boat,
  (ii) whether it would be feasible for him to jump to the bank (a man could expect to jump about 2 m from a standing position on firm ground),
  (iii) what impulse he would impart horizontally to the boat if he did jump.

4

Applications in Particle Dynamics

4.1 INTRODUCTION

In this chapter we shall investigate some applications in particle dynamics of the principles of mechanics developed in the previous chapter. These will include projectiles, resisted motion, simple harmonic motion and the pendulum.

Newton's law of motion for a particle of constant mass m subject to a force vector $\mathbf{F}$ is (see Equation (1), Chapter 3) is

$$\mathbf{F} = m\ddot{\mathbf{r}}$$

which is a *differential equation* for the position vector $\mathbf{r}$ of the particle. This vector equation is equivalent to the three scalar equations

$$F_x = m\ddot{x}, \qquad F_y = m\ddot{y}, \qquad F_z = m\ddot{z}$$

where $\mathbf{F} = F_x\mathbf{i} + F_y\mathbf{j} + F_z\mathbf{k}$. Whenever possible our aim is to *solve* these equations for x, y and z in terms of the time t for given force components. A significant feature of particle (and rigid-body) dynamics is that the principles generally lead to *ordinary* differential equations, that is, ordinary derivatives and not partial derivatives. The reason for this is the dependence of coordinates only on the single variable—time. Consequently, we need to be familiar with some of the simpler methods of solving ordinary differential equations. These will be pointed out as they arise in the text, but a summary of the main methods is given in the Appendix.

We start with two examples where the motion is in a straight line. Both equations use the important identity concerning the derivative of the velocity first derived in Section 2.3. Consider a particle moving along a straight line with displacement x and velocity v. We can treat the velocity v as either a function of the time t or the displacement x. In the identity

$$\frac{\mathrm{d}v}{\mathrm{d}t} = \frac{\mathrm{d}x}{\mathrm{d}t}\frac{\mathrm{d}v}{\mathrm{d}x} = \frac{\mathrm{d}v}{\mathrm{d}x}$$

87

v on the left is interpreted as a function of t and v on the right as a function of x. This identity is particularly useful if the force is a function of displacement rather than time. The following examples illustrate the point.

Example 1   A particle of mass m is subject to a force kx. At x = 0 its velocity is v_0. Find its velocity when x = 2.

The equation of motion is

$$kx = m \frac{d^2 x}{dt^2}$$

which is a second-order equation. However its velocity $v = dx/dt$, so that the equation can be written

$$kx = m \frac{dv}{dt}.$$

We now consider velocity as dependent on displacement x rather than time t:

$$kx = m \frac{dx}{dt} \frac{dv}{dx} = mv \frac{dv}{dx}$$

which is a first-order equation of the variables separable type. The solution is

$$\tfrac{1}{2}kx^2 = \tfrac{1}{2}mv^2 + C.$$

Since $v = v_0$ at $x = 0$, $C = -\tfrac{1}{2}mv_0^2$ so that

$$\tfrac{1}{2}kx^2 = \tfrac{1}{2}m(v^2 - v_0^2).$$

At $x = 2$,

$$\tfrac{1}{2}mv^2 = \tfrac{1}{2}mv_0^2 + 2k$$

or

$$v^2 = v_0^2 + \frac{4k}{m}.$$

Example 2   Determine the lowest speed with which a rocket must be fired from the Earth's surface in order that it should escape the Earth's gravitational field. Neglect air resistance and the Earth's spin. Take $\gamma = 6.7 \times 10^{-11}$ m^3/kg s^2, mass of Earth, M = 6.0×10^{24} kg and radius of Earth a = 6400 km.

When at a distance r from the centre of the Earth, the rocket is subject to a force per unit mass of $\gamma M/r^2$ towards the centre of the Earth. The equation of motion is therefore

$$-\frac{\gamma M}{r^2} = \ddot{r}.$$

If V is the projection speed, we see that initially the velocity $v = V$ at $r = a$. The equation of motion can be rewritten as

$$-\frac{\gamma M}{r^2} = \frac{dv}{dt} = \frac{dr}{dt}\frac{dv}{dr} = v\frac{dv}{dr}$$

an equation of the variables separable type. The solution is

$$-\gamma M \int \frac{dr}{r^2} = \int v \, dv + C$$

that is,

$$\frac{\gamma M}{r} = \tfrac{1}{2}v^2 + C.$$

From the initial condition we see that $C = (\gamma M/a) - \tfrac{1}{2}V^2$, resulting in the required particular solution

$$\gamma M \left(\frac{1}{r} - \frac{1}{a}\right) = \tfrac{1}{2}(v^2 - V^2).$$

The rocket just escapes from the Earth's gravitational field if $v \to 0$ as $r \to \infty$: the rocket then just reaches 'infinity' with zero velocity. For this to be the case, we must have

$$\frac{\gamma M}{a} = \tfrac{1}{2}V^2$$

or

$$V = \left(\frac{2\gamma M}{a}\right)^{1/2} = \left(\frac{2 \times 6.7 \times 10^{-11} \times 6.0 \times 10^{24}}{6.4 \times 10^6}\right)^{1/2} \text{m/s}$$

$$= 1.1 \times 10^4 \text{ m/s}$$

$$= 40\,000 \text{ km/h}$$

approximately. This is called the *escape velocity* for the Earth.

4.2 PROJECTILES

The motion of a projectile near the Earth's surface is governed by Equations (7), Section 3.7, and provides a simple illustration of motion in a uniform field of force. Suppose that a particle is fired from a point on the Earth's surface with speed V in a direction making angle α to the horizontal (the particle could be interpreted as a shell fired from a gun whose barrel is inclined at an angle α). Take the origin O at the point of projection, the z-axis vertical and the x-axis horizontal in the vertical plane through the initial velocity vector, with the y-axis completing the triad. The equations of motion are given by

$$\ddot{x} = \ddot{y} = 0, \qquad m\ddot{z} = -mg$$

where m is the mass of the particle.

Initially (at time $t = 0$), $x = y = z = 0$, $\dot{x} = V \cos \alpha$, $\dot{y} = 0$, $\dot{z} = V \sin \alpha$. We first observe that since $\dot{y} = 0$ initially it must remain so throughout the motion since $\ddot{y} = 0$. The particle must therefore move in the xz-plane as indicated in Figure 4.1. Similarly since $\dot{x} = V \cos \alpha$ initially, the horizontal component of the velocity must remain $V \cos \alpha$. Thus our equations of motion reduce to

$$\dot{x} = V \cos \alpha \qquad \ddot{z} = -g$$

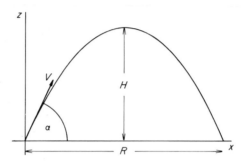

Figure 4.1   Range R and maximum height H for a projectile.

(we can ignore the y-coordinate henceforward). These two elementary differential equations have solutions

$$x = Vt \cos \alpha + A, \qquad z = B + Ct - \tfrac{1}{2}gt^2$$

where A, B and C are constants to be determined from the initial conditions. Note that motion in the z-direction is one of constant acceleration which has been examined in detail in Section 2.3. Using the given initial conditions, we see that $A = 0$, $B = 0$, $C = V \sin \alpha$, so that the horizontal and vertical displacements are given by

$$x = Vt \cos \alpha \tag{1}$$
$$z = Vt \sin \alpha - \tfrac{1}{2}gt^2 \tag{2}$$

in terms of the time t.

From (2), we observe that $z = 0$ when

$$t(V \sin \alpha - \tfrac{1}{2}gt) = 0$$

which has two solutions $t = 0$ and $t = 2V \sin \alpha/g$. The former solution refers to the initial condition whilst the latter gives the time when the particle meets the horizontal through the point of projection, in other words it gives the *time of flight* of the particle over horizontal ground. Denoting this time by T, we see that

$$T = \frac{2V}{g} \sin \alpha.$$

During this time the particle covers a horizontal distance, called the horizontal *range* and denoted by R, of

$$R = VT \cos \alpha = \frac{2V^2}{g} \sin \alpha \cos \alpha = \frac{V^2}{g} \sin 2\alpha \qquad (3)$$

from Equation (1) and the trigonometric identity $\sin 2\alpha = 2 \sin \alpha \cos \alpha$. The *maximum* range for a *fixed* projection speed but variable projection angle occurs where $\sin 2\alpha$ takes its maximum value. For $0 \le \alpha \le \frac{1}{2}\pi$, the maximum value of $\sin 2\alpha$ is 1 when $\alpha = 45°$. Thus

$$R_{max} = \frac{V^2}{g}$$

which is achieved when the particle is fired at $45°$ to the horizontal.

It should be acknowledged that the range represents that of a particle *in vacuo* and ignores the effects, which can be significant, of air resistance.

The particle attains its greatest height where z has its maximum value. This occurs at the time when dz/dt vanishes. From equation (2)

$$\frac{dz}{dt} = V \sin \alpha - gt$$

and $dz/dt = 0$ when $t = V \sin \alpha/g = \frac{1}{2}T$. This is a maximum since $d^2z/dt^2 = -g < 0$. We note that the maximum is reached after half the total time of flight. The greatest height attained, H, is obtained by substituting the time back into Equation (2) so that

$$H = \frac{V^2 \sin^2 \alpha}{g} - \frac{V^2 g \sin^2 \alpha}{2g^2} = \frac{V^2 \sin^2 \alpha}{2g}. \qquad (4)$$

Equations (1) and (2) are parametric equations of the *path* or *trajectory* of the particle in space in terms of the parameter t. The equation of the path is obtained by eliminating t between the equations:

$$z = x \tan \alpha - \frac{gx^2}{2V^2} \sec^2 \alpha. \qquad (5)$$

This path is a parabola with its vertex upward as shown in Figure 4.1. The path is symmetric about a vertical line through the maximum point of the path.

Example 3   A woman finds that she can throw a ball less easily upwards than horizontally and that the speed of projection of the ball varies approximately as 25 cos $\frac{1}{2}\alpha$ m/s, where α is the angle of projection. Find the maximum values of the range and height, and the angles at which the ball should be projected to attain these.

The equations of motion of the ball, (1) and (2), become

$$x = 25t \cos \tfrac{1}{2}\alpha \cos \alpha$$

$$z = 25t \cos \tfrac{1}{2}\alpha \sin \alpha - \tfrac{1}{2}gt^2$$

since $V = 25 \cos \frac{1}{2}\alpha$. The range

$$R = \frac{25^2}{g} \cos^2 \tfrac{1}{2}\alpha \sin 2\alpha \tag{6}$$

from (3). The range is a maximum when $\cos^2 \frac{1}{2}\alpha \sin 2\alpha$ is a maximum, which occurs where $dR/d\alpha$ vanishes. Now

$$\frac{dR}{d\alpha} = \frac{25^2}{g} (2 \cos^2 \tfrac{1}{2}\alpha \cos 2\alpha - \cos \tfrac{1}{2}\alpha \sin \tfrac{1}{2}\alpha \sin 2\alpha)$$

and $dR/d\alpha = 0$ when

$$2 \cos^2 \tfrac{1}{2}\alpha \cos 2\alpha = \cos \tfrac{1}{2}\alpha \sin \tfrac{1}{2}\alpha \sin 2\alpha.$$

Using the identities $2 \cos^2 \beta = 1 + \cos 2\beta$ and $\sin 2\beta = 2 \sin \beta \cos \beta$, this equation can be written

$$(1 + \cos \alpha)(2 \cos^2 \alpha - 1) = \sin \alpha \cdot \sin \alpha \cos \alpha = (1 - \cos^2 \alpha) \cos \alpha$$

or as

$$(1 + \cos \alpha)(3 \cos^2 \alpha - \cos \alpha - 1) = 0$$

of which there are three solutions:

$$\cos \alpha = -1, \qquad \cos \alpha = \tfrac{1}{6}(1 \pm \sqrt{13}).$$

We reject the two negative roots since we are interested only in the range $0 \le \alpha \le \frac{1}{2}\pi$. Thus the required angle is given by $\cos \alpha = \frac{1}{6}(1 + \sqrt{13}) = 0.7676$ so that $\alpha = 39° \, 51'$. The maximum range from (6) works out to be about 55.4 m.

The maximum height for *fixed* α is given by (4):

$$H = \frac{25^2}{2g} \cos^2 \tfrac{1}{2}\alpha \sin^2 \alpha. \tag{7}$$

The maximum possible height as α varies occurs when $dH/d\alpha$ vanishes. Now

$$\frac{dH}{d\alpha} = \frac{25^2}{2g} (2 \cos^2 \tfrac{1}{2}\alpha \sin \alpha \cos \alpha - \cos \tfrac{1}{2}\alpha \sin \tfrac{1}{2}\alpha \sin^2 \alpha)$$

which vanishes when

$$2 \cos^2 \tfrac{1}{2}\alpha \sin \alpha \cos \alpha = \cos \tfrac{1}{2}\alpha \sin \tfrac{1}{2}\alpha \sin^2 \alpha.$$

By using the trigonometric identities again,

$$(1 + \cos \alpha) \sin \alpha \cos \alpha = \tfrac{1}{2} \sin \alpha \sin^2 \alpha = \tfrac{1}{2} \sin \alpha(1 - \cos^2 \alpha)$$

or

$$\sin \alpha(1 + \cos \alpha)(3 \cos \alpha - 1) = 0$$

giving the three roots $\sin \alpha = 0$, $\cos \alpha = -1$, $\cos \alpha = \tfrac{1}{3}$. The first two roots give the minimum heights. The third root is the one we require and gives an angle of $70° 32'$. On substituting this angle back into (7) the maximum height is about 18.9 m, achieved when the ball is thrown at an angle of $70° 32'$ to the horizontal.

Example 4   Find the range of a projectile which is fired with speed V at angle α to an inclined plane which is itself inclined at an angle β to the horizontal. For fixed V, what is the maximum range?

   The particle is assumed to be projected in the vertical plane through the line of greatest slope on the inclined plane. It is convenient in this problem to choose a modified coordinate system. Take the x-axis up the plane and the z-axis perpendicular to it as indicated in Figure 4.2. The acceleration components in the x- and z-directions become respectively $-g \sin \beta$ and $-g \cos \beta$ resulting in the equations of motion

$$\ddot{x} = -g \sin \beta, \qquad \ddot{z} = -g \cos \beta.$$

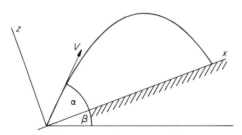

Figure 4.2   Range on an inclined plane.

The initial conditions for the problem are

$$x = z = 0, \qquad \dot{x} = V \cos \alpha, \qquad \dot{z} = V \sin \alpha \quad \text{at } t = 0.$$

Integrating the equations twice and using the initial conditions, the reader may verify that

$$x = Vt \cos \alpha - \tfrac{1}{2}gt^2 \sin \beta \tag{8}$$

$$z = Vt \sin \alpha - \tfrac{1}{2}gt^2 \cos \beta. \tag{9}$$

When the projectile hits the plane, $z = 0$, which, from (9), occurs after a time T where

$$T = \frac{2V \sin \alpha}{g \cos \beta}.$$

The inclined range R at this time is, from (8),

$$R = VT \cos \alpha - \tfrac{1}{2}gT^2 \sin \beta$$

$$= \frac{2V^2}{g} \frac{\sin \alpha}{\cos^2 \beta} (\cos \alpha \cos \beta - \sin \alpha \sin \beta)$$

$$= \frac{2V^2 \sin \alpha \cos (\alpha + \beta)}{g \cos^2 \beta} = \frac{V^2[\sin (2\alpha + \beta) - \sin \beta]}{g \cos^2 \beta}$$

by employing two standard trigonometric identities. For fixed V and β but variable α, the range will be a maximum where $\sin(2\alpha + \beta)$ has its maximum. The greatest value of the sine function is 1, which corresponds to $2\alpha + \beta = \tfrac{1}{2}\pi$, that is to $\alpha = \tfrac{1}{4}\pi - \tfrac{1}{2}\beta$. The maximum range

$$R_{\text{max}} = \frac{V^2(1 - \sin \beta)}{g \cos \beta} = \frac{V^2(1 - \sin \beta)}{g(1 - \sin^2 \beta)} = \frac{V^2}{g(1 + \sin \beta)}, \quad (\beta \neq \tfrac{1}{2}\pi).$$

Example 5   Mud is thrown off the rim of a wheel, of radius a, on a car travelling with speed v. If $v^2 \geq ga$ show that no mud can be thrown higher than

$$a + \frac{v^2}{2g} + \frac{ga^2}{2v^2}$$

above the ground.

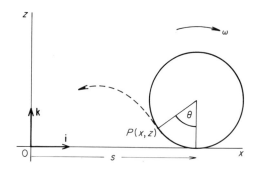

Figure 4.3   Coordinate scheme for Example 5.

Figure 4.3 shows the wheel rolling along a straight line Ox with angular speed $\omega = v/a$. Suppose at time t the point of contact between the wheel and road is distance s from O and that mud is being thrown from the wheel at P through which the radius makes an angle θ with the downward vertical. If $\mathbf{r} = x\mathbf{i} + z\mathbf{k}$ is the position vector of P then, in terms of s and θ,

$$\mathbf{r} = (s - a \sin \theta)\mathbf{i} + a(1 - \cos \theta)\mathbf{k}.$$

The velocity of P

$$\dot{\mathbf{r}} = (\dot{s} - a \cos \theta \dot{\theta})\mathbf{i} + a \sin \theta \dot{\theta}\mathbf{k}$$
$$= v(1 - \cos \theta)\mathbf{i} + v \sin \theta \mathbf{k}$$

since $\dot{s} = v$ and $\dot{\theta} = \omega = v/a$. This is the velocity with which the mud is thrown from the wheel.

The vertical equation of motion of a particle of mud is

$$\ddot{z} = -g$$

which has a solution

$$z = A + Bt - \tfrac{1}{2}gt^2.$$

Initially for fixed θ, $\dot{z} = v \sin \theta$, $z = a(1 - \cos \theta)$. Thus $A = a(1 - \cos \theta)$ and $B = v \sin \theta$ and

$$z = a(1 - \cos \theta) + vt \sin \theta - \tfrac{1}{2}gt^2.$$

The maximum height reached occurs when $\dot{z} = 0$, that is when $t = v \sin \theta/g$. Thus the maximum height

$$H = a(1 - \cos \theta) + \frac{v^2 \sin^2 \theta}{2g}.$$

For varying θ this height has an overall maximum where $dH/d\theta = 0$.

$$\frac{dH}{d\theta} = a \sin \theta + \frac{v^2}{g} \sin \theta \cos \theta$$

which vanishes where $\sin \theta = 0$ or $\cos \theta = -ag/v^2$. If $v^2 < ag$, there is one real root: $\sin \theta = 0$ giving $\theta = 0$ and π. The first of these corresponds to the point of contact which is at rest and is clearly a minimum whilst the second corresponds to the mud being thrown off at the highest point of the wheel which will be the highest point of its path. If $v^2 \geq ag$, the second root, $\cos \theta = -ag/v^2$, gives an overall maximum height of

$$H_{max} = a\left(1 + \frac{ag}{v^2}\right) + \frac{v^2}{2g}\left(1 - \frac{a^2g^2}{v^4}\right)$$

$$= a + \frac{a^2g}{2v^2} + \frac{v^2}{2g}$$

as required. To achieve this maximum, mud is thrown off in the range $\tfrac{1}{2}\pi \leq \theta \leq \pi$.

For a wheel of radius 0.3 m travelling at 45 km/h, $H_{max} = 8.27$ m approximately. Remember however that the mud is more likely to be thrown off in the range $0 \leq \theta \leq \tfrac{1}{2}\pi$ since adhesion will generally not hold the mud to the wheel beyond this range.

Example 6   *A gun can fire a shell with speed V in any direction. Show that a shell can reach any target within the surface*

$$g^2r^2 = V^4 - 2gV^2z$$

where z is the height of the target and r is the horizontal distance of the target from the gun.

The shell can be fired in any vertical plane through the gun and in such a plane the trajectory of the shell is given by Equation (5) with r replacing x:

$$z = r \tan \alpha - \frac{gr^2}{2V^2} \sec^2 \alpha.$$

Using the identity $\sec^2 \alpha = 1 + \tan^2 \alpha$, this equation can be rewritten in the form

$$\frac{gr^2}{2V^2} \tan^2 \alpha - r \tan \alpha + z + \frac{gr^2}{2V^2} = 0. \tag{10}$$

Interpret (r, z) now as the coordinates of a target. Equation (10) is a quadratic equation in $\tan \alpha$ with the two roots

$$\tan \alpha = \frac{1}{gr} [V^2 \pm (V^4 - 2gzV^2 - g^2r^2)^{1/2}].$$

These two roots will be real if the discriminant

$$V^4 - 2gzV^2 - g^2r^2 \geq 0$$

and a shell will reach a target with coordinates (r, z) only if this inequality is satisfied. The *critical* surface is therefore given by

$$V^4 - 2gzV^2 - g^2r^2 = 0$$

a *paraboloid of revolution*, sometimes called the paraboloid of safety. Any target on or within this bounding paraboloid can be reached by a shell. Note also that for any target within the paraboloid there are *two* possible directions in which the barrel of the gun can be inclined; for a target on the paraboloid there is just one. Some typical paths and a section of the paraboloid are shown in Figure 4.4.

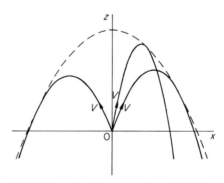

Figure 4.4   The dashed parabola is a section of the bounding paraboloid.

4.3 RESISTED MOTION

Unless the projectiles of the previous section are being fired from the surface of the Moon or some similar satellite without an atmosphere, mathematical models of projectiles should take account of air resistance. In practice the frictional effects of air can make significant changes to the ideal parabolic trajectory. It is found experimentally that the magnitude of the resistance depends largely on three factors: the area of the projectile perpendicular to the direction of motion (its *aspect*), its speed and the density of the air. The resistance seems to be proportional to the aspect and density, varies as the speed for low speeds but nearer the square of the speed for high speeds. In still air the resistance opposes the motion and acts in a direction directly opposite of the velocity. If the projectile is moving in a wind then the resistance will be in the direction of the velocity of the projectile relative to the moving air.

We shall concentrate on the variation of the drag with velocity and absorb the other factors into the constant of proportionality. In still air, suppose that the magnitude of the resistance $\mathbf{R}$ is a function of the speed $|\mathbf{v}|$ of the particle. Since the drag opposes the motion we can express $\mathbf{R}$ in the form

$$\mathbf{R} = -mf|\mathbf{v}|\mathbf{v}.$$

From our earlier remarks, we would expect $f(|\mathbf{v}|)$ to be a positive increasing function of $|\mathbf{v}|$ representing a law of resistance which has to be defined according to data or experimental evidence for a particular application.

For a particle moving under gravity subject to this resistance, its equation of motion will be

$$-mg\mathbf{k} - mf(|\mathbf{v}|)\mathbf{v} = m\dot{\mathbf{v}} \tag{11}$$

where $\mathbf{k}$ is the upward unit vector. Since the resistance opposes the direction of motion, just as in the unresisted case the projectile will move in still air in a vertical plane through the point of projection which contains the initial velocity vector.

Let us first consider the simpler special case of vertical motion under gravity (the general projectile problem will be looked at in Section 4.7). Equation (11) takes the one-dimensional form

$$\dot{w} = -g - f(|w|)w \tag{12}$$

where w is the vertical velocity component, positive being upwards. Further we shall assume a power-law resistance in which

$$f(|w|) = c|w|^{n-1}$$

where c and n are constants. Equation (12) then becomes

$$\dot{w} = -g - c|w|^{n-1}w. \tag{13}$$

As we remarked earlier, n usually lies in the range $1 \leq n \leq 2$. This differential equation does not seem to have simple general solutions except for the cases $n = 1$ and $n = 2$. Suppose that the particle is dropped from rest well clear of the ground. Then the drag will increase with speed until the resistance approaches the gravitational force (remember that $w < 0$ in (13) for a falling particle), so that the acceleration tends to zero. The particle then approaches a constant speed known as the *terminal speed*. On putting $\dot{w} = 0$ in (13), we find that

$$w = -(g/c)^{1/n}. \tag{14}$$

In fact (14) is known as a *singular solution* of (13): these particular solutions are frequently easy to identify even when general solutions are difficult to find. The terminal speed, say w_τ, is the magnitude of w in (14), namely

$$w_\tau = (g/c)^{1/n}. \tag{15}$$

It is easy to compute solutions from (13) both for the velocity and the displacement. To find the displacement, put $w = \dot{z}$ and eliminate c in favour of w_τ using (15) so that (13) can be replaced by the first-order system

$$\dot{z} = w$$

$$\dot{w} = -g\left(1 + \frac{|w|^{n-1}w}{w_\tau^{n-1}}\right).$$

Some remarks and help with the computation of solutions of first-order initial value problems can be found in the Appendix. However, before we plot some solutions numerically (see p. 108) let us consider two examples which illustrate the analytic solutions for $n = 1$ and $n = 2$.

Example 7   A parachutist falling with speed 50 m/s at a height of 200 m opens his parachute. With resistance proportional to speed the terminal speed of the parachutist is 5 m/s. Estimate the time of descent.

With z upwards with origin at ground level, the equation of motion is

$$\frac{d^2z}{dt^2} = -g - c\frac{dz}{dt}$$

with the initial conditions at $t = 0$:

$$z = 200\,\text{m}, \quad \frac{dz}{dt} = -50\,\text{m/s}.$$

The terminal speed must equal g/c so that

$$c = g/5 = 9.8/5 = 1.96\,\text{s}^{-1}.$$

The differential equation

$$\frac{d^2z}{dt^2} + c\frac{dz}{dt} = -g$$

is of the second order with characteristic equation $m(m + c) = 0$. It is easy to guess that $-gt/c$ is a particular integral of this equation. The general solution is therefore

$$z = A + B e^{-ct} - \frac{gt}{c}.$$

The initial conditions give

$$200 = A + B,$$

$$-50 = -Bc - \frac{g}{c}.$$

Thus

$$B = \frac{1}{c}\left(50 - \frac{g}{c}\right) = \frac{1}{1.96}(50 - 5) = \frac{45}{1.96} = 22.96$$

and

$$A = 200 - 22.96 = 177.04.$$

The required solution is

$$z = 177.04 + 22.96\, e^{-1.96t} - 5t.$$

After 3 s we obtain approximately that the value of the second term is

$$23\, e^{-1.96t} = 23\, e^{-5.88} = 23 \times 0.003 \simeq 0.07$$

This term is very small compared with the first term. The parachutist reaches the ground $z = 0$ when

$$177.04 + 22.96\, e^{-1.96t} - 5t = 0$$

or, since the second term is negligible, when

$$177.04 - 5t = 0$$

which gives a time of descent of approximately 35 s.

Example 8   A ball is thrown vertically upwards with speed w_0 from the level $z = 0$. Air resistance exerts a drag on the ball proportional to the square of the speed. Find the maximum height reached by the ball and the time taken.

If $n = 2$ Equation (13) becomes

$$\frac{dw}{dt} = -g - c|w|w$$

where, in terms of the terminal speed w_τ, $c = g/w_\tau^2$ (see Equation (15)). Until the ball reaches its highest point, w is positive. Thus

$$\frac{dw}{dt} = -g - cw^2$$

for this part of its path. Separating the variables and integrating, we find that

$$\int \frac{dw}{g + cw^2} = -\int dt + C$$

where C is a constant. Using the substitution $w = (g/c)^{1/2} \tan \phi$, we find that

$$\frac{1}{(cg)^{1/2}} \int d\phi = -t + C \quad \text{or} \quad \phi = -(cg)^{1/2}(t - C)$$

Hence

$$w = (g/c)^{1/2} \tan [-(cg)^{1/2}(t - C)].$$

Use the expansion formula for $\tan (A + B)$ and let $w = w_0$ when $t = 0$. Thus

$$w = \frac{w_0 - (g/c)^{1/2} \tan [(cg)^{1/2}t]}{(g/c)^{1/2} + w_0 \tan [(cg)^{1/2}t]} \tag{16}$$

since $w_0 = (g/c)^{1/2} \tan [C(cg)^{1/2}]$. At the highest point $w = 0$ so that it is reached at time

$$t_1 = (cg)^{1/2} \tan^{-1} [w_0(c/g)^{1/2}].$$

(Note that $\tan^{-1}$ is defined in the principal range $(0, \frac{1}{2}\pi)$.)

To obtain the relation between velocity and displacement z, we use the identity $dw/dt = w\,dw/dz$. Thus

$$w \frac{dw}{dz} = -g - cw^2$$

for the ascending particle. Separating the variables and integrating, we find that

$$\int \frac{w\,dw}{g + cw^2} = -\int dz + D.$$

Hence

$$\frac{1}{2c} \ln (g + cw^2) = -z + D.$$

From the initial condition $w = w_0$ at $z = 0$, it follows that $D = [\ln (g + cw_0^2)]/2c$ and that

$$z = \frac{1}{2c} \ln \left(\frac{g + cw_0^2}{g + cw^2} \right). \tag{17}$$

At the highest point, $w = 0$ so that this is reached at height H where

$$H = \frac{1}{2c} \ln \left(\frac{g + cw_0^2}{g} \right).$$

On the ascent the relation between z and t can be obtained by eliminating w between (16) and (17).

For the subsequent descent $w < 0$ for the particle so that its equation of motion becomes

$$\frac{dw}{dt} = -g + cw^2.$$

We can now solve this equation subject to the new initial conditions $w = 0$ at $t = t_1$ where $z = H$ (see Exercise 13). Note that we have to solve a different differential equation.

Figure 4.5 shows some $z - t$ graphs for the cases $n = 1$, $n = 2$ and various terminal speeds.

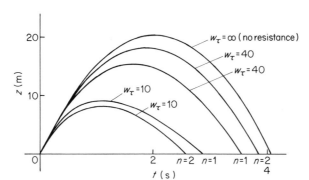

Figure 4.5 The maximum height versus time for cases $n = 1$, $n = 2$ and various values of the terminal speed w_τ. The initial speed is $w_0 = 20$ m/s in all cases, and w_τ is also given in m/s.

4.4 MOTION IN A CIRCLE

Consider a particle or bob of mass m attached by a piece of string of length a with the other end attached to a point on the table. Project the bob horizontally along the table perpendicular to the string with the string taut. The constraining effect of the string causes the bob to move in a circle. We shall assume that there is friction between the table and the particle, and that the string is light and inextensible.

The particle experiences three forces, the vertical reaction of the table which does not affect the horizontal motion, the frictional force F which opposes the motion and the tension T in the string which acts along the string. Let the centre of the circle be O, the origin of coordinates. Since the particle is constrained to move in a circle it is convenient to locate its position by the polar coordinates (r, θ) (see Figure 4.6). The radial and transverse components of acceleration are given by (see Section 1.10) $(\ddot{r} - r\dot{\theta}^2, 2\dot{r}\dot{\theta} + r\ddot{\theta})$. In this case $r = a$ so that the acceleration vector is $(a\dot{\theta}^2, a\ddot{\theta})$. The equations of motion become, taking account of the directions of F and T,

$$T = ma\dot{\theta}^2, \quad -F = ma\ddot{\theta}. \tag{18}$$

Let us assume that the frictional force is proportional to the speed according to the law $F = c|\dot{\theta}|a$, where c is a constant. In the sense in which θ is increasing

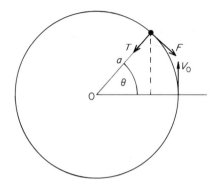

Figure 4.6

in Figure 4.6, the second equation in (18) becomes

$$-c\dot{\theta} = m\ddot{\theta}.$$

We can treat this as a second-order constant-coefficient equation with characteristic equation

$$mp^2 + cp = 0.$$

The roots are $p_1 = 0$ and $p_2 = -c/m$. Thus the angle θ is given by

$$\theta = C + D\exp(-ct/m).$$

Suppose that the particle starts at $\theta = 0$ and is projected with transverse speed V_0 (Figure 4.6). The initial conditions are $\theta = 0$, and $\dot{\theta} = V_0/a$ when $t = 0$. Hence

$$0 = C + D, \quad V_0/a = -Dc/m$$

implying that

$$D = -mV_0/ac = -C.$$

Finally the required solution is

$$\theta = \frac{mV_0}{ac}[1 - \exp(-ct/m)].$$

By letting $t \to \infty$ we can see that the string will ultimately turn through an angle mV_0/ka before coming to rest.

If the table is smooth then $F = 0$ and the equation of motion simplifies to $\ddot{\theta} = 0$, which has the solution V_0t/a satisfying the same initial conditions.

4.5 SIMPLE HARMONIC MOTION

Consider the one-dimensional motion of a particle P of mass m in which it experiences an opposing force T which is proportional to its displacement x (Figure 4.7). Thus $T = kx$ and the equation of motion becomes

$$-kx = m\ddot{x}$$

or

$$\ddot{x} + \omega^2 x = 0, \quad \omega^2 = k/m. \tag{19}$$

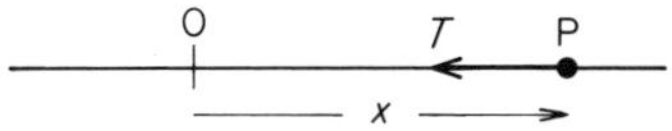

Figure 4.7

Any system which is governed by Equation (19) is said to exhibit simple harmonic motion (SHM). We shall say more about the generation of SHM by springs in Section 5.5.

Equation (19) is a second-order homogeneous differential equation with constant coefficients. Its characteristic equation is

$$p^2 + \omega^2 = 0$$

which has the imaginary roots $p_1 = i\omega$, $p_2 = -i\omega$. Hence we can write the solution in the exponent form

$$x = A\,e^{i\omega t} + B\,e^{-i\omega t} \tag{20}$$

where A and B are constants to be determined by initial conditions on the displacement and velocity. An alternative form which displays the oscillatory character of the solutions can be deduced by using the identities

$$e^{\pm i\omega t} = \cos \omega t \pm i \sin \omega t.$$

Thus (20) becomes

$$x = A(\cos \omega t + i \sin \omega t) + B(\cos \omega t - i \sin \omega t)$$
$$= (A + B) \cos \omega t + i(A - B) \sin \omega t.$$

The constants $A + B$ and $i(A - B)$ can be replaced by two further constants C and D respectively. Hence the solution, in real form, can be written as

$$x = C \cos \omega t + D \sin \omega t.$$

Finally we can also express the solution in a third phase-amplitude version by substituting $C = K \sin \varepsilon$ and $D = K \cos \varepsilon$, where the angle ε is chosen to make $K > 0$, so that

$$x = K \sin \varepsilon \cos \omega t + K \cos \varepsilon \sin \omega t$$
$$= K \sin (\omega t + \varepsilon). \tag{21}$$

The solution (21) clearly shows the oscillatory character of the output (see Figure 4.8). With the exception of $x = 0$, all solutions of (19) are periodic with

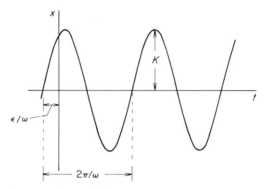

Figure 4.8   Displacement-time graph for simple harmonic motion.

minimal period $2\pi/\omega$. In mathematical terms a function $x(t)$ is said to be periodic with period τ or τ-periodic if $x(t) = x(t + \tau)$ for all t. The minimal period is the smallest value of τ for which this holds. Generally we use the word period to mean the minimal period.

In (21), K is called the *amplitude* of the oscillation, ω the (angular) *frequency* in radians per second, and $\omega t + \varepsilon$ describes the *phase* at time t with ε the initial phase. The reciprocal of the period defines the frequency in Hertz (Hz) or cycles per second.

We shall investigate the theory and application of oscillatory systems in more detail in Chapters 7 and 9.

4.6 THE SIMPLE PENDULUM

A simple pendulum is a bob (idealised as a particle) of mass suspended from a fixed point by a light rod (that is, of negligible mass) of length a which is allowed to move in a fixed vertical plane through the point of suspension O (the rod could be replaced by a string: as long as the string remains taut the problem is analogous). If the rod is freely hinged (freely means that the hinge is smooth or frictionless) and the pendulum oscillates in a vacuum so that there is no air resistance, then two forces act on the bob, namely its own weight mg and the tension T in the rod which acts towards O (see Figure 4.9).

Let θ be the angle between the rod and the downward vertical. We can construct the equation in two ways either by taking moments about O or by equating the transverse component of the weight to the product of the mass and the transverse acceleration (see Section 1.10). In both cases the result is

$$-mg \sin \theta = ma\ddot{\theta}. \tag{22}$$

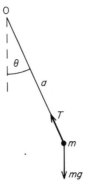

Figure 4.9 The simple pendulum.

A first integral of this equation can be obtained if we use the standard identity $\ddot{\theta} = \dot{\theta}\,d\dot{\theta}/d\theta$. Hence (22) becomes the separable equation

$$g \sin \theta = -a\dot{\theta}\,\frac{d\dot{\theta}}{d\theta}$$

which can be separated and integrated to give

$$\int g \sin \theta \, d\theta = -a \int \dot{\theta} \, d\dot{\theta} + C$$

or

$$-g \cos \theta = -\tfrac{1}{2}a\dot{\theta}^2 + C$$

where C is the constant of integration. This equation can be integrated again but not in terms of simple functions: this requires some knowledge of certain special functions known as elliptic functions but this is beyond the scope of this book.

However, we can solve the equation approximately by assuming that the deviation from the downward vertical is small (say, for θ, less than 40°). The function $\sin \theta$ in (22) can be expanded by its MacLaurin series:

$$\sin \theta = \theta - \tfrac{1}{6}\theta^3 + \cdots .$$

We then argue that, for small θ, θ is a good approximation to $\sin \theta$ (compare θ with $\sin \theta$ for small θ using a calculator). With this approximation, the equation of motion becomes

$$g\theta = -a\ddot{\theta} \quad \text{or} \quad \ddot{\theta} + (g/a)\theta = 0$$

which is the equation for simple harmonic motion (see previous section). As we have seen its solution is

$$\theta = K \sin\left[\left(\frac{g}{a}\right)^{1/2} t + \varepsilon\right].$$

Note that the period of oscillation of the simple pendulum is approximately $2\pi(a/g)^{1/2}$, and that it is independent of the amplitude. This is not the case with the true Equation (22). Figure 4.10 shows the relation between period and amplitude for the pendulum equation: this has been computed from numerical solutions of Equation (22) in the special case $a = g$. The rising curve indicates that the period increases with amplitude but this increase is very gradual for $|\theta|$ less than one radian.

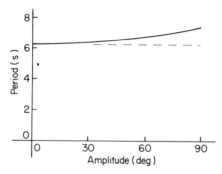

Figure 4.10 The curve shows how period varies with amplitude. The results were computed for the dimensionless equation $\theta'' + \sin\theta = 0$ where $'$ refers to the derivative with respect to τ and $\tau = t(g/a)^{1/2}$. The broken line indicates the SHM approximation.

4.7 SOME FURTHER PLANE PROBLEMS

As we have seen a dynamical problem can be considered as plane if one component of the force **F** in the equation of motion

$$\mathbf{F} = m\ddot{\mathbf{r}}$$

vanishes. We can always arrange that this plane coincides with a pair of coordinate planes, say the x- and y-planes. In this case $F_z = 0$ and $\ddot{z} = 0$ which implies that $\dot{z} = $ constant. For plane motion this constant must also be zero since otherwise the particle would have a constant velocity component in the z direction. Further integration then implies that $z = $ constant which we can choose to be zero also. The vector equation of motion now effectively reduces to the pair of equations

$$F_x = m\ddot{x}, \qquad F_y = m\ddot{y}.$$

Each force component F_x and F_y could depend on x, y, $\dot{x}$, $\dot{y}$ and t which means that the dependent variables x and y satisfy *simultaneous* differential equations in x and y. There are some general methods of solution for such systems of equations (mainly for sets of *linear* equations) but usually we have to search for solutions on a case-by-case approach, or abandon analytic methods and solve the equations numerically. The following example leads to a pair of linear differential equations.

Example 9   A particle moves in the force field $k^2(y\mathbf{i} + x\mathbf{j})$ per unit mass, where k is a constant. If it starts from the origin with velocity $V\mathbf{i}$, find the subsequent path of the particle.

If **r** is the position vector of the particle at time t, the equation of motion is

$$k^2(y\mathbf{i} + x\mathbf{j}) = \ddot{\mathbf{r}}$$

or, in scalar form,

$$\frac{d^2x}{dt^2} = k^2 y, \qquad \frac{d^2y}{dt^2} = k^2 x, \qquad \frac{d^2z}{dt^2} = 0. \qquad (23)$$

The initial conditions at $t = 0$ are

$$x = y = z = 0, \qquad \frac{dy}{dt} = \frac{dz}{dt} = 0, \qquad \frac{dx}{dt} = V.$$

From the third of equations (23),

$$z = A_1 t + A_2$$

and, if $z = dz/dt = 0$, at $t = 0$, we must have $A_1 = A_2 = 0$. Throughout the motion $z = 0$ and the particle moves in the (x,y)-plane.

The first two members of (23) are simultaneous equations in x and y. Eliminating y, we find that x satisfies

$$\frac{d^4x}{dt^4} - k^4 x = 0.$$

The general solution of this is,

$$x = A\,e^{kt} + B\,e^{-kt} + C\cos kt + D\,\sin kt$$

whilst

$$y = \frac{1}{k^2}\frac{d^2x}{dt^2} = A\,e^{-kt} + B\,e^{-kt} - C\cos kt - D\sin kt.$$

The initial conditions give the following four equations for A, B, C and D:

$$A + B + C = 0 \qquad (24)$$

$$A - B + D = V/k \qquad (25)$$

$$A + B - C = 0 \qquad (26)$$

$$A - B - D = 0. \qquad (27)$$

Subtracting (26) from (24), $C = 0$. From (25) and (27), $D = V/2k$. Equations (24) and (27) become

$$A + B = 0 \qquad A - B - V/2K = 0.$$

Therefore $A = V/4k$, $B = -V/4k$. The required solution is

$$x = V(e^{kt} - e^{-kt} + 2 \sin kt)/4k$$
$$= V(\sinh kt + \sin kt)/2k$$

and

$$y = V(e^{kt} - e^{-kt} - 2 \sin kt)/4k$$
$$= V(\sinh kt - \sin kt)/2k.$$

These are the parametric equations of the path in the (x,y)-plane.

Let us reconsider the projectile problems discussed in Section 4.2 when some account is now taken of air resistance. The general equation of motion was derived in Section 4.3, and the relevant equation we now require is (11), namely

$$-mg\mathbf{k} - mf(|\mathbf{v}|)\mathbf{v} = m\dot{\mathbf{v}} \tag{28}$$

in which the resistance is assumed to be a function of the speed v. If we assume a power-law dependence

$$f(v) = c|\mathbf{v}|^{n-1}$$

as in Section 4.3, then Equation (28) has components (assuming that the motion takes place in the plane $y = 0$)

$$\ddot{x} = -c\dot{x}(\dot{x}^2 + \dot{z}^2)^{(n-1)/2} \tag{29}$$
$$\ddot{z} = -g - c\dot{z}(\dot{x}^2 + \dot{z}^2)^{(n-1)/2}. \tag{30}$$

These are simultaneous differential equations in x and z. However, they do uncouple in the case $n = 1$ and this is the subject of the next example.

Example 10   A projectile with a terminal speed of w_t is fired with an initial speed V at an inclination of α to the horizontal. If the resistance is proportional to velocity find the path taken by the projectile.

The equations of motion are

$$\frac{d^2x}{dt^2} = -c\frac{dx}{dt}$$

$$\frac{d^2z}{dt^2} = -g - c\frac{dz}{dt}.$$

These are two standard second-order equations with solutions

$$x = A + B e^{-ct}$$

$$z = C + D e^{-ct} - \frac{g}{c} t$$

where A, B, C, and D are constants. With the origin at the point of projection the initial conditions are

$$x = 0, \qquad \dot{x} = V \cos \alpha$$

$$z = 0, \qquad \dot{z} = V \sin \alpha$$

at $t = 0$. The constants consequently take the values

$$A = -B = \frac{V}{c} \cos \alpha, \qquad C = -D = \frac{1}{c^2} (Vc \sin \alpha + g)$$

giving the required solution

$$x = \frac{V}{c} (1 - e^{-ct}) \cos \alpha \tag{31}$$

$$z = \frac{1}{c^2} (Vc \sin \alpha + g)(1 - e^{-ct}) - \frac{g}{c} t. \tag{32}$$

These are the parametric equations of the path of the projectile. Since the terminal speed is w_τ, it follows that

$$c = \frac{g}{w_\tau}.$$

To find the horizontal range (assuming $0 < \alpha < \frac{1}{2}\pi$) through the point of projection we require first the non-zero solution of $z = 0$ from (32) for the time of flight T. Once T is known the range can be found by substituting $t = T$ in (31). Unfortunately, no explicit formula for T can be found: T appears in a mixture of exponential and linear terms in what is known as a *transcendental* equation. Hence it is difficult to construct an analytic solution for the range. However, it is a straightforward matter to compute the graph of x against z for small time steps: the range is then approximately the value of x for which z next becomes negative.

Before actually applying a programming routine to (31) and (32) we need to express the solutions in suitably reduced form so that they have the widest applicability. Essentially the model contains two parameters which remain unchanged for the same particle. They are $w_\tau = g/c$ which has the dimensions of (length)/(time) and g which has the dimensions of (length)/(time)^2. Thus a typical length for the model is w_τ^2/g, and a typical time w_τ/g. Hence we introduce so-called dimensionless variables

$$X = gx/w_\tau^2, \quad Z = gz/w_\tau^2, \quad s = gt/w_\tau,$$

into (31) and (32) with the result that the answers

$$X = V_0 (1 - e^{-s}) \cos \alpha$$

$$Z = (V_0 \sin \alpha + 1)(1 + e^{-s}) - s$$

now contain the single parameter $V_0 = V/w_\tau$.

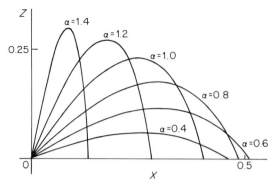

Figure 4.11 Trajectories computed from the solution in Example 10 for $V_0 = 1$ and various angles of projection.

Figure 4.11 shows some typical trajectories for $V_0 = 1$. The paths are no longer symmetric about their highest points but show the characteristic steepening on the descent in resisted motion.

Other features are also affected by air resistance. Figure 4.11 suggests that the maximum range is no longer at $\alpha = \frac{1}{2}\pi$. We can compute the horizontal range for small increments of the angle of projection α. For the same $V_0 = 1$, Figure 4.12 displays the horizontal range R plotted against the angle α. In this particular case the maximum horizontal range occurs at approximately $\alpha = 35.5°$ compared with $\alpha = 45°$ for the unresisted projectile.

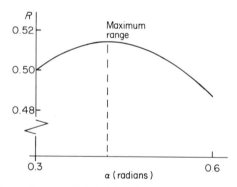

Figure 4.12 Horizontal range R plotted against angle α with $V_0 = 1$ in Example 10. The maximum value occurs at about an angle of projection, $\alpha = 35.5°$.

For other values of n in Equation (29) and (30) we have to resort to numerical methods of solution. For example, if $n = 2$, then the equations become

$$\ddot{x} = -c\dot{x}(\dot{x}^2 + \dot{z}^2)^{1/2}, \quad \ddot{z} = -g - c\dot{z}(\dot{x}^2 + \dot{z}^2)^{1/2}. \tag{33}$$

From the second of these equations we can infer that the terminal speed $w_\tau = (g/c)^{1/2}$. On this occasion we will look at the qualitative effect of decreasing terminal speed on the trajectories and range for a projectile fired with the same initial speed V_0 and angle of projection. Before we compute solutions it is advisable again to express (33) in dimensionless form but this time, since we wish to vary w_τ we use the length V_0^2/g and time V_0/g. Thus if $x = V_0^2 X/g$, $z = V_0^2 Z/g$ and $t = V_0 s/g$, then X and Z satisfy

$$X'' = -\gamma X'(X'^2 + Z'^2)^{1/2}, \quad Z'' = -1 - \gamma Z'(X'^2 + Z'^2)^{1/2}$$

where $X' = dX/ds$, etc, and $\gamma = (V_0/w_\tau)^2$. To apply standard numerical packages it is usual to express the equations as the first-order system

$$X' = U, \quad U' = -\gamma U(U^2 + W^2)^{1/2} \tag{34}$$

$$Z' = W, \quad W' = -1 - \gamma W(U^2 + W^2)^{1/2} \tag{35}$$

Figure 4.13 gives a qualitative picture of the variation of paths with the terminal speed w_τ. As we might expect both the range and maximum height reached decrease as w_τ decreases, and also the descent part of the path becomes steeper than the ascent.

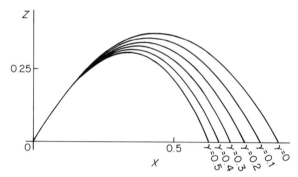

Figure 4.13 Typical trajectories for projectiles with $n = 2$ computed from equations (34) and (35). In the computations $V_0 = 1$ and $\alpha = \frac{1}{3}\pi$, and $\gamma = (V_0/w_\tau)^2$ takes the values indicated.

Exercises

1. A particle moves in the force-field

$$\mathbf{F} = y^2\mathbf{i} - \mathbf{j}$$

per unit mass. If the particle starts from rest at the origin, obtain the equation of its path.

2. A particle moving in space is subject to a time-varying force

$$\mathbf{F} = e^{-t}\mathbf{i} + \sin t\mathbf{j} + \cos t\mathbf{k}$$

per unit mass. If the particle starts from rest, find its velocity at any subsequent time.

3. A falling raindrop increases its mass at a rate proportional to the product of its surface area and speed v. Show that its radius is proportional to its distance s below a fixed level.

Assuming that the rate of change of linear momentum is equal to the force applied, show that the acceleration is $g - 3v^2/s$. Show also that the acceleration eventually approaches $\frac{1}{7}g$.

4. A particle is projected with speed 20 m/s in a direction inclined at 30° to the horizontal. Calculate the horizontal range of the particle and the maximum height it reaches.

5. A gun emplacement is on the edge of a cliff of height h. Prove that the greatest horizontal distance at which a shell from the emplacement can hit the ship is

$$2[k(k + h)]^{1/2},$$

and that the greatest horizontal distance at which a gun in the ship can hit the emplacement is

$$2[k(k - h)]^{1/2}(k > h),$$

if $(2gk)^{1/2}$ is the firing speed of a shell in each case.

6. A particle is thrown from the highest point of a hemispherical mound of radius 100 m. Find the minimum throwing speed necessary for it to clear the mound.

7. A particle is fired at an inclination $\alpha + \beta$ to the horizontal with a speed u. Show that the particle will strike, at right angles, a plane through the point of projection inclined at angle β to the horizontal if $\cot \beta = 2 \tan \alpha$. If the particle strikes the plane with speed $\frac{1}{2}u$, show that the range is $u^2(21)^{1/2}/8g$.

8. It is required to hit a squash ball from a given point with a given speed v so as to strike a vertical wall above a horizontal line on the wall. It is found that when the ball is projected in the vertical plane perpendicular to the wall, the ball just hits the line if its *initial projected direction* is inclined at θ_1 or θ_2 ($\theta_1 > \theta_2$) to the horizontal. Show that the ball can reach any point on the wall within a circle of radius

$$v^2 \sin (\theta_1 - \theta_2)/g \sin (\theta_1 + \theta_2).$$

9. The line joining the net and the point of projection of a netball makes an angle α with the horizontal. If the net is at height h above the point of projection, show that to score the initial speed of the ball must be not less than

$$[gh(1 + \operatorname{cosec} \alpha)]^{1/2}.$$

10. A shell bursting on the ground throws fragments in all directions with speeds up to 30 m/s. Find for what period of time a man standing 30 m from the explosion is in danger.

11. A baseball is hit from 1 m above the ground and attains a maximum height of 21 m and a horizontal range of 100 m. A fielder can catch the ball between the ground and 3 m. Over what horizontal distance would a fielder be capable of catching the ball?

12. Two parallel vertical walls have height h_0 and h_1 and the distance between these top edges is l. A particle is projected from the ground in a plane perpendicular to both walls. Show that the minimum speed of projection necessary for the particle to clear both walls is $[g(h_0 + h_1 + l)]^{1/2}$.

13. A body is projected vertically upwards in a medium for which the resistance is $k \times (\text{speed})^2$ per unit mass. If the initial speed is v_0 show that the body returns to the point of projection with speed v_1, where

$$v_1^2 = \frac{gv_0^2}{g + kv_0^2}.$$

14. A particle of mass m moves in the plane $z = 0$ under the attractive force $2m\pi^2 r$ towards the origin, r being its distance from the origin. In addition there is a force of magnitude $m\pi v$ in the direction $\mathbf{v} \times \mathbf{k}$, where $\mathbf{v}$ is the velocity of the particle and $\mathbf{k}$ a unit vector perpendicular to $z = 0$. If the particle is projected from the origin at $t = 0$ with velocity $3\mathbf{i}$, find its position at $t = 3$.

15. A cyclist exerts a constant propulsive force and is subject to air resistance which is proportional to the square of his speed. He can ride at a maximum speed of 36 km/h on the level and of 18 km/h up a slope of 1 in 20. Show that he can attain a speed of 20 km/h on the level, when starting from rest, in about 31 m.

16. A body having a mass of 45×10^5 kg is acted on by a force of $10^4(6 - v)$ N where v is its speed in m/s and experiences a resistance proportional to the square of its speed. If its maximum speed is 2 m/s show that it attains a speed of 1.5 m/s from rest in approximately 161 s and find the distance it has travelled to the nearest metre.

17. A body of mass 150 kg is dropped by parachute with negligible initial velocity. While the parachute is opening the motion is resisted by a force of $30v$ N, where v m/s is the velocity of the body. If the parachute is fully open at the end of 5 s, prove that the velocity will then be

$$5g(1 - e^{-1}) \text{ m/s.}$$

The subsequent motion is resisted by a force of $10v^2$ N. Find the minimum safe dropping height (subject to the parachute opening fully), given that the impact velocity must not exceed $(0.2g)^{1/2}$ m/s.

18. The force exerted on a charged particle by the constant magnetic field $\mathbf{H}$ is $Q\mathbf{H} \times \dot{\mathbf{r}}$ where Q is a constant. If $\dot{\mathbf{r}}$ is perpendicular to $\mathbf{H}$ at some instant show that the particle describes a circle with constant speed. (Neglect gravity.)

19. A projectile is fired with initial velocity $V_0 \cos \alpha \mathbf{i} + V_0 \sin \alpha \mathbf{k}$ into a wind of velocity $-V\mathbf{i}$. The resistance is proportional to the vector difference of projectile velocity and wind velocity and opposes the motion. If the terminal speed of the projectile is 60 m/s, $V_0 = 30$ m/s, $\alpha = 30°$ and $V = 3$ m/s, find the horizontal range of the particle. What wind speed would cause the projectile to return to its firing point?

20. A hydrofoil starts a journey with its hull submerged in the water, called stage (a), and when an appropriate speed is reached its hull is raised out of the water, called stage (b). In stage (a) the resistance to motion is of magnitude $3mkv$, where m is the mass of the boat, v its speed and k a constant, and the boat is subject to a constant force which would enable it to reach a maximum speed V. In stage (b) the resistance to motion has a magnitude mkv and it is subject to a different constant force which would enable it to achieve a speed $2V$ in the limit. Find at what speed the hull of the hydrofoil should be raised in order that a minimum time should be used in attaining a speed V_1, where $2V > V_1 \geq V$, and show that to attain a speed V the minimum time would be $(1/3k)\ln(27/4)$.

21. A particle is attached by a string of length a to a fixed point on a table. The particle is fired, with the string taut, with speed V perpendicular to the string. The friction of the table resists the motion according to the formula $\alpha v + \beta v^2 (\alpha > 0, \beta > 0)$, where v is the speed of the particle. Find the angle through which the string turns before the particle comes to rest.

22. A projectile with a terminal speed w_τ is fired with initial speed V at an angle $\alpha(0 < \alpha < \frac{1}{2}\pi)$ to the horizontal. Show that the projectile reaches its highest point after a time

$$\frac{w_\tau}{g} \ln \left(\frac{V \sin \alpha}{w_\tau} + 1 \right)$$

assuming that the air resistance is proportional to its speed. Find the horizontal and vertical displacements of this point.

23. A block of mass m can slide on a rough horizontal plane. Consider the two problems in which (i) the resistance is cv, (ii) the resistance behaves as cv^2, where v is the speed of the block and c is a constant. If the block is fired with initial speed V, find, in both cases, how far the block travels before it comes to rest.

24. A simple pendulum of length 1 m is suspended at rest from a fixed point. Its bob is given a speed 2 m/s horizontally. Find the subsequent amplitude of the pendulum assuming no resistance.

25. (Conical pendulum) A pendulum of length a is set in motion so that its bob describes a horizontal circle. If the string is inclined at an angle α to the downward vertical, find the period of the pendulum.

26. A bend in a road is an arc of a circle of radius a. Determine the banking of the road, given by the angle α to the horizontal, in order that vehicles travelling round the bend at speed v should experience no radial force parallel to the road surface. Find the angle of banking required for a bend of radius 100 m for vehicles moving with speed 50 km/h.

5

Work and Energy

5.1 WORK

The *work* done on a particle by a constant force F when the particle is moved a distance s in the direction of the force is defined to be Fs—the product of force and distance. For example, if a man carries a 30 kg load up a vertical ladder through a height of 10 m, the work done by the man in lifting the load will be $30 \times 10 \times g = 300\,g$N m since he lifts the load against the gravitational attraction of the Earth. As its name suggests, work is a measure of the effort expended by the man: it does not describe how or at what rate the load is lifted.

This simple description of work must now be generalised to include variable forces and situations in which the particle does not move in a straight line. Suppose a particle has a position vector $\mathbf{r}$ and moves in a field of force $\mathbf{F}$. The work $\mathcal{W}$ done by the force-field $\mathbf{F}$ on the particle as it moves from point A to point B on a specified curve C (Figure 5.1) is defined by the line integral

$$\mathcal{W} = \int_C \mathbf{F} \cdot d\mathbf{r}. \tag{1}$$

We can show that this agrees with simpler ideas of of work given above. Suppose A is at the origin and B is the point (1, 0, 0), and C is the straight line

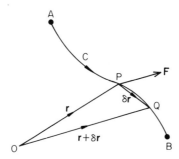

Figure 5.1   Path of integration C of the line integral defining work.

joining A and B. If $\mathbf{F} = F_0\mathbf{i}$, a constant force,

$$\mathscr{W} = \int_C F_0\mathbf{i} \cdot d\mathbf{r} = \int_0^1 F_0 \, dx = F_0 \times 1$$

that is, it is the product of force and distance moved.

Alternatively we can think of

$$\delta\mathscr{W} = \mathbf{F} \cdot \delta\mathbf{r}$$

as approximately the increment of work done as the point P moves to the neighbouring point Q in Figure 5.1. If θ is the angle between $\mathbf{F}$ and $\overline{PQ}$,

$$\delta\mathscr{W} = |\mathbf{F}||\delta\mathbf{r}| \cos \theta$$

by the definition of the scalar product; it is the product of the component of $\mathbf{F}$ in the direction $\overline{PQ}$ and the distance moved.

Remember that integral (1) is a line integral and must be evaluated as outlined in Section 1.12.

The S.I. unit of work is the joule (J) which represents 1 N acting through a distance of 1 m.

Example 1   Find the total work done in moving a particle in a force field given by $\mathbf{F} = 3xy\mathbf{i} - 2yz\mathbf{j} + y\mathbf{k}$ N along the curve $x = t$, $y = 2t^2$, $z = 1 + t$ from $t = 1$ to $t = 2$ where x, y, z are measured in metres.

The work done

$$\mathscr{W} = \int_C \mathbf{F} \cdot d\mathbf{r} = \int_C (3xy\mathbf{i} - 2yz\mathbf{j} + y\mathbf{k}) \cdot (\mathbf{i} \, dx + \mathbf{j} \, dy + \mathbf{k} \, dz)$$

$$= \int_C (3xy \, dx - 2yz \, dy + y \, dz)$$

$$= \int_{t=1}^{t=2} [3(t)(2t^2) \, dt - 2(2t^2)(1 + t) \, d(2t^2) + 2t^2 \, d(1 + t)]$$

$$= \int_1^2 (6t^3 - 16t^3 - 16t^4 + 2t^2) \, dt$$

$$= \int_1^2 (-16t^4 - 10t^3 + 2t^2) \, dt$$

$$= \left[-\frac{16t^5}{5} - \frac{5t^4}{2} + \frac{2t^3}{3} \right]_1^2$$

$$= -\frac{3961}{30} \, \text{J}.$$

The symbol t in this example is not necessarily time but merely a parameter which describes the path C.

5.2 CONSERVATIVE AND NON-CONSERVATIVE FORCES

If the man of the preceding section carries the 30 kg load up an inclined ladder to a height of 10 m, he still performs work of $300g$ N m. For, if the ladder is inclined at an angle θ to the horizontal, the work done is $30g \cos \theta \times (10/\cos \theta) = 300g$ N m: the first term is the component of weight along the ladder and the second term is the length of the ladder. In fact, whatever path the man climbs in reaching 10 m the work done is unchanged because

$$\mathscr{W} = \int_C 30g\mathbf{k} \cdot d\mathbf{r}$$

where C is any path joining a point at zero height and a point at a height of 10 m. Thus

$$\mathscr{W} = \int_0^{10} 30g \, dz = 300g \text{ N m}$$

the work done is *independent* of the path taken. We say that the uniform gravitational field is a *conservative* field of force.

More generally, if the work done by a force on a particle which is moved from A to B is independent of the path joining the two points, the force is said to be conservative. A force which is not conservative is called a *non-conservative* force.

The contents of Section 1.13 are relevant here. We showed there that if a line integral was independent of the path joining the two points, then curl $\mathbf{F} = \mathbf{0}$, provided certain conditions were satisfied. This implies that $\mathbf{F}$ can be expressed as the gradient of a scalar potential function i.e. $\mathbf{F} = -\text{grad } \phi$. Thus, under the action of a conservative field of force, the work done depends only on the values of ϕ at the end points of the path of integration, and if a particle moves round a closed curve the total work done is zero. A consequence of this is that the work done is recoverable.

The definition of work so far refers only to points or particles and not to bodies with finite dimensions. Suppose a rigid body occupying position 1 in a field of force $\mathbf{F}$ per unit mass in Figure 5.2 is moved into position 2 which involves a translation and a rotation. Consider a small element $\rho \delta V$ of the body which moves from position 1 to 2 along the path C. We adopt the following plausible reasoning. The work done in moving the element $\rho \delta V$ is

$$\int_C \mathbf{F} \cdot d\mathbf{r} \rho \delta V$$

since $\mathbf{F}\rho \delta V$ is the force on this element. The total work done will be the sum of these elements through the volume:

$$\mathscr{W} = \int_V \int_C \mathbf{F} \cdot d\mathbf{r} \rho \, dV \tag{2}$$

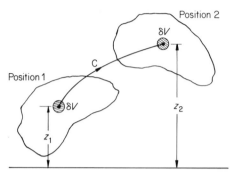

Figure 5.2 Work done by a rigid body translated and rotated in a uniform gravitational field.

which is a multiple integral involving a line integral and a volume integral. In general, this integral cannot be simplified greatly. Suppose the field of force is now a uniform gravitational one, $\mathbf{F} = -g\mathbf{k}$. Then

$$\mathcal{W} = \int_V \int_C (-g)\, \mathrm{d}z\rho\, \mathrm{d}V$$

$$= \int_V (z_1 - z_2)\rho\, \mathrm{d}V$$

$$= g\int_V z_1\rho\, \mathrm{d}V - g\int_V z_2\rho\, \mathrm{d}V$$

$$= Mg\bar{z}_1 - Mg\bar{z}_2$$

$$= Mg(\bar{z}_1 - \bar{z}_2) \tag{3}$$

where $\bar{z}_1$ and $\bar{z}_2$ are the initial and final heights of the mass-centre of the body. The work done is therefore the product of the weight of the body and height through which its mass-centre rises. Note that any rotation of the body does not affect the work done.

Let us consider now a scalar non-conservative force. Suppose a block slides along a rough horizontal table. Any motion of the block will be opposed by the *friction* between the block and the table and we assume that the frictional force F acts in a direction opposite to the direction of motion of the block. Thus if, in Figure 5.3, the block is moving in a straight line A to B, the frictional force F on the block acts from B to A. The mathematical description of F is a matter for definition based, in part, on experimental evidence. It may depend on such properties as the nature of the surfaces in contact, the weight of the block and the speed of the block. In one of its simplest forms F is a constant multiple of the reaction (in this case the weight of the block), the constant depending on the two surfaces in contact.

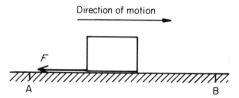

Figure 5.3 A block sliding on a rough horizontal plane.

In the example under consideration F will be constant so that the work done on the block by the frictional force will be $-F \cdot AB$ (negative sign because F opposes the motion). If now the block is slid back to A the total work done will be $-2F \cdot AB$ since F will act in the opposite direction in the return motion of the block. The work done is not recoverable because it is transformed principally into heat by the friction.

Conservative systems are important for the good theoretical reason that they are usually simpler to handle mathematically. This explains why so many problems in dynamics assume smooth bodies, frictionless pulleys, etc. For example, as we saw in Section 4.2, the motion of projectiles under a constant gravitational field without air resistance is reasonably easy to analyse. The same problem with air resistance (which is a non-conservative force acting on the projectile) considered in Section 4.7 is much more complicated.

A body may experience simultaneously conservative and non-conservative forces in which case only part of the work done will be recoverable.

Example 2   A particle of mass m at rest at the origin generates a gravitational field force

$$\mathbf{F} = -\frac{\gamma m \mathbf{r}}{r^3}$$

(this is the force of attraction which a particle of unit mass placed at $\mathbf{r}$ *would experience).*
Show that $\mathbf{F}$ *is conservative.*

By the definition of curl (Section 1.13):

$$\mathrm{curl}\left(-\frac{\gamma m \mathbf{r}}{r^3}\right) = -\gamma m \left\{ \mathbf{i}\left[\frac{\partial}{\partial y}\left(\frac{z}{r^3}\right) - \frac{\partial}{\partial z}\left(\frac{y}{r^3}\right)\right] + \mathbf{j}\left[\frac{\partial}{\partial z}\left(\frac{x}{r^3}\right) - \frac{\partial}{\partial x}\left(\frac{z}{r^3}\right)\right] \right.$$
$$\left. + \mathbf{k}\left[\frac{\partial}{\partial x}\left(\frac{y}{r^3}\right) - \frac{\partial}{\partial y}\left(\frac{x}{r^2}\right)\right] \right\}$$

$$= 3\gamma m \left[\mathbf{i}\left(\frac{yz}{r^5} - \frac{zy}{r^5}\right) + \mathbf{j}\left(\frac{zx}{r^5} - \frac{xz}{r^5}\right) + \mathbf{k}\left(\frac{yx}{r^5} - \frac{xy}{r^5}\right)\right]$$

$$= \mathbf{0}$$

where $r = (x^2 + y^2 + z^2)^{1/2}$. Thus curl $\mathbf{F} = \mathbf{0}$, $\mathbf{F}$ is conservative and there exists a scalar function ϕ such that $\mathbf{F} = -\text{grad } \phi$. The reader should verify that the *gravitational potential* is

$$\phi = -\frac{\gamma m}{r}.$$

Example 3　Consider a model of a hurricane in which the wind field is given by $\mathbf{v} = g(r)\mathbf{e}_\theta$, *where* $\mathbf{e}_\theta$ *is a unit vector in the direction of* θ *increasing and* (r, θ) *are polar coordinates with the centre of the hurricane as origin. Suppose that a ship in such a hurricane is subjected to a force* $\alpha\mathbf{v}$ *due to the action of the wind with* α *a constant. Is such a field of force conservative?*

Expressed in Cartesian coordinates, the force on the ship,

$$\alpha\mathbf{v} = \alpha g(r)\mathbf{e}_\theta = \alpha[-g(r) \sin \theta\, \mathbf{i} + g(r) \cos \theta\, \mathbf{j}]$$

$$= \alpha\left(-\frac{g(r)}{r} y\mathbf{i} + \frac{g(r)}{r} x\mathbf{j}\right)$$

where $r^2 = x^2 + y^2$. Consequently

$$\text{curl }(\alpha\mathbf{v}) = \alpha\mathbf{k}\left[\frac{\partial}{\partial x}\left(\frac{xg(r)}{r}\right) + \frac{\partial}{\partial y}\left(\frac{yg(r)}{r}\right)\right]$$

$$= \alpha\mathbf{k}\left[\frac{g(r)}{r} + \frac{x^2}{r}\frac{d}{dr}\left(\frac{g(r)}{r}\right) + \frac{g(r)}{r} + \frac{y^2}{r}\frac{d}{dr}\left(\frac{g(r)}{r}\right)\right]$$

noting that $\dfrac{\partial}{\partial x}(r) = \dfrac{x}{r}$ and $\dfrac{\partial}{\partial y}(r) = \dfrac{y}{r}$. Thus

$$\text{curl }(\alpha\mathbf{v}) = \alpha\mathbf{k}\left[\frac{2g(r)}{r} + r\frac{d}{dr}\left(\frac{g(r)}{r}\right)\right]$$

which will vanish only if

$$\frac{2g(r)}{r} + r\frac{d}{dr}\left(\frac{g(r)}{r}\right) = 0.$$

Putting $z = g(r)/r$, we find that the equation is of separable type

$$2z + r\frac{dz}{dr} = 0$$

with solution

$$z = \frac{A}{r^2} \quad \text{or} \quad g(r) = \frac{A}{r}$$

where A is a constant. Thus, if $g(r) \neq A/r$, the wind field will certainly be non-conservative. In the particular case $g(r) = A/r$, which is not much different from the wind field exterior to the central core of a hurricane, then the comment made in Section 1.13 becomes applicable. We have curl $\mathbf{F} = \mathbf{0}$ but for a path enclosing the centre of the hurricane one of the two conditions for the theorem

$$\text{`} \int_C \mathbf{F} \cdot \mathbf{dr} = 0 \text{ implies curl } \mathbf{F} = \mathbf{0} \text{ and conversely'}$$

is not satisfied. If the origin is included in the region then $\mathbf{F}$ is not well-behaved there; if the origin is excluded by surrounding it with an interior contour, then the region is not simply-connected (see Section 1.13 for this definition). If we integrate the line integral around a circle of constant radius r_0 we get

$$\int_C \mathbf{F} \cdot \mathbf{dr} = \int_C \frac{A}{r} \mathbf{e}_\theta \cdot (dr\mathbf{e}_r + r\, d\theta \mathbf{e}_\theta)$$

$$= \int_0^{2\pi} A\, d\theta = 2\pi A$$

confirming the point just made. If however the path does *not* include the origin, the conditions of the theorem are satisfied and the value of any such line integral is zero.

5.3 POTENTIAL ENERGY

We have seen that a force $\mathbf{F}$ is conservative if curl $\mathbf{F} = \mathbf{0}$ throughout an appropriate region and that, if this is so, there must exist a scalar function of position ϕ such that $\mathbf{F} = -\text{grad } \phi$. Suppose that a particle of mass m is moved from position A to B in a conservative force field $\mathbf{F} = -\text{grad } \phi$ (remember that $\mathbf{F}$ is the force which a particle of unit mass would experience) along a curve C. The negative of the work done is called the *potential energy* $\mathscr{V}$ relative to the first position, A. Thus

$$\mathscr{V} = -\mathscr{W} = -m \int_A^B \mathbf{F} \cdot \mathbf{dr} = m \int_A^B \text{grad } \phi \cdot \mathbf{dr}$$

$$= m(\phi_B - \phi_A)$$

which is proportional to the difference in the potentials at the two points. Energy can be described as the capacity for doing work and is distinguished in conservative systems from work in general since the work is mechanically recoverable. Much confusion often arises between the meaning of potential and potential energy. The potential of a force may be thought of as an abstraction: if a particle of unit mass is placed at a point it experiences a force which is the negative gradient of the potential. The potential energy of an actual particle is a quantity associated with the position in space of that particle.

Whilst potential and potential energy are simply related for particles, they are not so for bodies. The potential energy of a body in moving it from position 1 to 2 (Figure 5.2) is, from Equation (2)

$$\mathscr{V} = -\mathscr{W} = -\int_V \int_C \mathbf{F} \cdot \mathrm{d}\mathbf{r}\rho \; \mathrm{d}V$$

$$= \int_V \int_C \mathrm{grad}\; \phi \cdot \mathrm{d}\mathbf{r}\rho \; \mathrm{d}V$$

$$= \int_V (\phi_2 - \phi_1)\rho \; \mathrm{d}V$$

where ϕ_1 and ϕ_2 are the values of the potential at the end-points of the path C. For bodies, the potential energy involves the volume integral of the potential initially and finally. Such general treatments are beyond the scope of this book and we restrict our attention to bodies in uniform gravitational fields.

From Equation (3) we see that if the mass-centre of a body is raised through a height h in a uniform gravitational field $-g\mathbf{k}$, the body acquires potential energy

$$\mathscr{V} = Mgh$$

where M is the mass of the body. In any specific problem all potential energies must be referred to the *same* reference level.

5.4 POWER AND KINETIC ENERGY: CONSERVATION OF ENERGY

Suppose a particle is subject to a force $\mathbf{F}$. The work done by the force in moving the particle from $\mathbf{r}$ to the neighbouring point $\mathbf{r} + \delta\mathbf{r}$ is given by

$$\delta\mathscr{W} = \mathbf{F} \cdot \delta\mathbf{r}$$

approximately. The *rate of working* or *work-rate* $\dot{\mathscr{W}}$ is given by

$$\dot{\mathscr{W}} = \lim_{\delta t \to 0} \frac{\delta\mathscr{W}}{\delta t} = \lim_{\delta t \to 0} \frac{\mathbf{F} \cdot \delta\mathbf{r}}{\delta t} = \mathbf{F} \cdot \dot{\mathbf{r}}$$

which is the product of force and the velocity of the particle.

The work-rate is usually known by the more familiar name of *power*. The S.I. unit of power is the watt (W) which is 1 J/s. The unit of 1 kWh (kilowatt-hour) $= 3.6 \times 10^6$ J is still an acceptable unit of energy and work. One British unit of power, which is gradually being superseded by the kilowatt, is the horse-power (hp) which is defined as 17 700 foot/poundals/s. In terms of S.I. units

$$1 \text{ hp} = 0.7457 \text{ kW}.$$

If the particle has mass m, $\mathbf{F} = m\ddot{\mathbf{r}}$ and the work-rate

$$\dot{\mathscr{W}} = m\ddot{\mathbf{r}} \cdot \dot{\mathbf{r}} = \frac{\mathrm{d}}{\mathrm{d}t}(\tfrac{1}{2}m\dot{\mathbf{r}} \cdot \dot{\mathbf{r}})$$

$$= \frac{\mathrm{d}}{\mathrm{d}t}(\tfrac{1}{2}mv^2)$$

where v is the speed of the particle. The term $\tfrac{1}{2}mv^2$ is called the *kinetic energy* of the particle and denoted by the symbol $\mathscr{T}$. Suppose that the force $\mathbf{F}$ is the sum of conservative and non-conservative forces, the conservative part having potential ϕ. We can write

$$\mathbf{F} = \mathbf{F}' - m \text{ grad } \phi$$

where $\mathbf{F}'$ is the non-conservative force. Thus

$$\frac{\mathrm{d}\mathscr{T}}{\mathrm{d}t} = \mathbf{F} \cdot \dot{\mathbf{r}} = \mathbf{F}' \cdot \dot{\mathbf{r}} - m\dot{\mathbf{r}} \cdot \text{grad } \phi.$$

The force $\mathbf{F}'$ will *not* do work during the motion of the particle only if $\mathbf{F}' \cdot \dot{\mathbf{r}} = 0$, that is, if this force always remains perpendicular to the motion. Let us assume that this is the case and also that ϕ is a function of position only. Then

$$\frac{\mathrm{d}\mathscr{T}}{\mathrm{d}t} = -m\dot{\mathbf{r}} \cdot \text{grad } \phi = -m\left(\frac{\mathrm{d}x}{\mathrm{d}t}\frac{\partial\phi}{\partial x} + \frac{\mathrm{d}y}{\mathrm{d}t}\frac{\partial\phi}{\partial y} + \frac{\mathrm{d}z}{\mathrm{d}t}\frac{\partial\phi}{\partial z}\right)$$

$$= -m\frac{\mathrm{d}\phi}{\mathrm{d}t}.$$

For a particle, however, $m\phi = \mathscr{V}$ where $\mathscr{V}$ is the potential energy of the particle. Thus

$$\frac{\mathrm{d}\mathscr{T}}{\mathrm{d}t} = -\frac{\mathrm{d}\mathscr{V}}{\mathrm{d}t} \quad \text{or} \quad \frac{\mathrm{d}}{\mathrm{d}t}(\mathscr{T} + \mathscr{V}) = 0$$

which implies that

$$\mathscr{T} + \mathscr{V} = \text{constant.}$$

This is the energy principle for a conservative system, namely that the sum of the kinetic and potential energies remains constant throughout the motion. The meaning of conservative has been extended slightly to include systems in which there may occur non-conservative forces which do no work. The conservation of energy means that energy can be transferred continuously between its kinetic and potential states: if the particle slows down there must be a corresponding increase in its potential energy and vice-versa.

For a system of particles subject to *external* conservative forces and non-conservative forces which do no work the total kinetic energy and potential energy is conserved.

The units of energy are the same as those of work.

Example 4   A bead can slide on a smooth circular wire of radius a which is fixed in a vertical plane. The bead is displaced slightly from the highest point of the wire. Find the subsequent speed of the bead.

The bead experiences two forces—its weight and the reaction due to the wire. The weight mg acts vertically downwards and the reaction R acts radially outwards since there is no friction between the bead and the wire (Figure 5.4). Since the bead is always moving tangentially, no work is done by R. The system is conservative and the energy is conserved. At time t let the radius to the bead make an angle θ to the vertical and let the speed of the bead be v. Let the lowest point of the wire represent zero potential energy. The kinetic and potential energies are

$$\mathcal{T} = \tfrac{1}{2}mv^2, \qquad \mathcal{V} = mga(1 + \cos\theta).$$

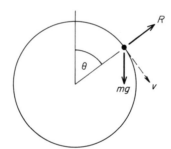

Figure 5.4   A bead sliding on a vertical wire.

By the energy principle, $\mathcal{T} + \mathcal{V} = \text{constant}$,

$$\tfrac{1}{2}mv^2 + mga(1 + \cos\theta) = \text{constant} = \text{initial value} = 2mga$$

since $v = 0$ and $\theta = 0$ initially. The speed at any angle θ is therefore

$$v = [2ga(1 - \cos\theta)]^{1/2}.$$

Note that the use of the energy principle avoids one integration of the equations of motion. Thus the tangential equation of motion is (see p. 101)

$$g\sin\theta = a\ddot{\theta} = a\dot{\theta}\,\frac{d\dot{\theta}}{d\theta}$$

and integration of this equation will give the energy equation and therefore the speed. Note also that if friction were present the energy equation would no longer hold and the equation of motion would then have to be solved to find the speed.

Following the usual extension of our postulates and definitions from particles to bodies, the kinetic energy of a body is given by the volume integral

$$\mathscr{T} = \tfrac{1}{2} \int_V \dot{\mathbf{r}} \cdot \dot{\mathbf{r}} \rho \, \mathrm{d}V.$$

If the body is rigid and translates without rotation every point of the body will have the same velocity $\mathbf{v} = \dot{\mathbf{r}}$. Thus, since $\mathbf{v}$ can only depend on time,

$$\mathscr{T} = \tfrac{1}{2} v^2 \int_V \rho \, \mathrm{d}V = \tfrac{1}{2} M v^2$$

where M is the mass of the body. If in addition the body rotates, then a further term will be included in the kinetic energy due to the rotational motion which the body has. We shall consider rotational effects in Chapter 10.

Suppose a rigid body is subject to point forces $\mathbf{F}_1, \mathbf{F}_2, \ldots, \mathbf{F}_n$ at $\mathbf{r}_1, \ldots, \mathbf{r}_n$ and moves without rotation in a uniform gravitational field $-g\mathbf{k}$ (since the body does not rotate the moment of the forces $\mathbf{F}_1, \ldots, \mathbf{F}_n$ about the mass-centre must vanish). We know that the mass-centre $\bar{\mathbf{r}}$ moves as a particle with all the forces acting there. The rate of working of the forces is given by

$$\dot{\mathscr{W}} = \mathbf{F}_1 \cdot \dot{\mathbf{r}}_1 + \mathbf{F}_2 \cdot \dot{\mathbf{r}}_2 + \cdots + \mathbf{F}_n \cdot \dot{\mathbf{r}}_n - M g \mathbf{k} \cdot \dot{\bar{\mathbf{r}}}$$
$$= (\mathbf{F}_1 + \mathbf{F}_2 + \cdots + \mathbf{F}_n - M g \mathbf{k}) \cdot \dot{\bar{\mathbf{r}}}$$

since $\dot{\mathbf{r}}_i = \dot{\bar{\mathbf{r}}}$ $(i = 1, 2, \ldots, n)$. Therefore, since

$$M\ddot{\bar{\mathbf{r}}} = \mathbf{F}_1 + \mathbf{F}_2 + \cdots + \mathbf{F}_n - M g \mathbf{k}$$

then

$$\dot{\mathscr{W}} = M\ddot{\bar{\mathbf{r}}} \cdot \dot{\bar{\mathbf{r}}} = \frac{\mathrm{d}}{\mathrm{d}t} (\tfrac{1}{2} M \dot{\bar{\mathbf{r}}} \cdot \dot{\bar{\mathbf{r}}}) = \frac{\mathrm{d}\mathscr{T}}{\mathrm{d}t}$$

and

$$\frac{\mathrm{d}\mathscr{T}}{\mathrm{d}t} = \mathbf{F}_1 \cdot \dot{\mathbf{r}}_1 + \mathbf{F}_2 \cdot \dot{\mathbf{r}}_2 + \cdots + \mathbf{F}_n \cdot \dot{\mathbf{r}}_n - M g \mathbf{k} \cdot \dot{\bar{\mathbf{r}}}.$$

Suppose now that the point forces do not work so that

$$\mathbf{F}_1 \cdot \dot{\mathbf{r}}_1 = \mathbf{F}_2 \cdot \dot{\mathbf{r}}_2 = \cdots = \mathbf{F}_n \cdot \dot{\mathbf{r}}_n = 0$$

which implies that

$$\frac{\mathrm{d}\mathscr{T}}{\mathrm{d}t} = -M g \mathbf{k} \cdot \dot{\bar{\mathbf{r}}} = -M g \frac{\mathrm{d}\bar{z}}{\mathrm{d}t}.$$

We conclude that

$$\frac{\mathrm{d}}{\mathrm{d}t}(\mathscr{T} + Mg\bar{z}) = 0$$

or

$$\mathscr{T} + Mg\bar{z} = \text{constant}$$

which is the energy principle again since $Mg\bar{z}$ is the potential energy $\mathscr{V}$ of the body.

We must emphasise that the principle asserted above has been derived for non-rotating bodies; it is still true for rotating bodies if the rotational energy of the body is included in the kinetic energy. Rigid bodies *translating* in uniform force-fields are equivalent to particles.

Example 5   A rocket is fired from the Earth's surface with speed v at an angle α to the radius through the point of projection. Show that the rocket's subsequent greatest distance from the Earth is the larger root of

$$\left(v^2 - \frac{2\gamma M}{a}\right)r^2 + 2\gamma Mr - a^2 v^2 \sin^2 \alpha = 0$$

if $v^2 < 2\gamma M/a$, where a is the radius and M the mass of the Earth, and γ is the gravitational constant. Deduce that the escape velocity is independent of α. Assume that the Earth is stationary and that air resistance is ignored, and that the rocket fuel is burnt instantaneously.

Let the position of the rocket have polar coordinates (r, θ) referred to the centre of the Earth as depicted in Figure 5.5. If we treat the rocket as a particle of mass m, the

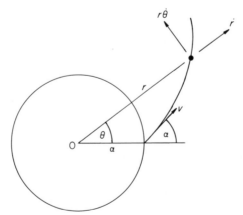

Figure 5.5   Coordinate scheme for the path of the rocket.

total energy of the rocket is conserved. The gravitational potential energy of a particle of mass m distance r from the centre of the Earth is

$$\mathscr{V} = -\frac{\gamma Mm}{r}$$

and the kinetic energy is

$$\mathscr{T} = \tfrac{1}{2}m[\dot{r}^2 + (r\dot{\theta})^2].$$

The energy principle gives

$$\tfrac{1}{2}m[\dot{r}^2 + (r\dot{\theta})^2] - \frac{\gamma Mm}{r} = \tfrac{1}{2}mv^2 - \frac{\gamma Mm}{a} \qquad (4)$$

the initial value of the energy.

Secondly, since the only force on the rocket acts towards the centre of the Earth, the transverse component of acceleration must vanish (Section 1.10):

$$\frac{1}{r}\frac{d}{dr}(r^2\dot{\theta}) = 0$$

or

$$r^2\dot{\theta} = \text{constant} = av \sin \alpha$$

since $r = a$ and $r\dot{\theta} = v_0 \sin \alpha$ initially. Substituting for $\dot{\theta}$ in (4):

$$\dot{r}^2 + \frac{a^2v^2 \sin^2 \alpha}{r^2} - \frac{2\gamma M}{r} = v^2 - \frac{2\gamma M}{a}. \qquad (5)$$

At the highest point of the rocket's path, $\dot{r} = 0$ so that the required height is a solution of the quadratic equation

$$\left(v^2 - \frac{2\gamma M}{a}\right)r^2 + 2\gamma Mr - a^2v^2 \sin^2 \alpha = 0.$$

If $v^2 < 2\gamma M/a$ the equation has two positive roots which correspond to furthest and nearest points of the orbit of the rocket from the centre of the Earth. If $v = (2\gamma M/a)^{1/2}$, Equation (5) becomes

$$\dot{r}^2 = \frac{1}{r^2}(2\gamma Mr - a^2v^2 \sin^2 \alpha)$$

$$= \frac{2\gamma M}{r^2}(r - a \sin^2 \alpha)$$

and $\dot{r}$ has real value for any value of $r \geq a$. In other words the rocket will just escape the Earth's gravity since $\dot{r} \to 0$ as $r \to \infty$. The value of this initial speed is independent of α.

Example 6   Find the power required to pump 6 m³ of water per minute to a height of 20 m through a pipe of cross-sectional area 0.004 m². (The mass of 1 m³ of water is 10³ kg.)

The water must be raised through a height of 20 m. Volume ejected per second $= 6/60 = 0.1$ m^3. The speed of the water in the pipe

$$= \frac{\text{Volume ejected per second}}{\text{Cross-sectional area of pipe}}$$

$$= \frac{0.1}{0.004} = 25 \text{ m/s}.$$

Mass of water ejected per second $= 100$ kg. Hence kinetic energy is produced at a rate per second of

$$\tfrac{1}{2} \times 100 \times 25^2.$$

Each minute 6000 kg of water must be lifted through 20 m. The rate of production of potential energy per second is therefore

$$\frac{6000 \times 20 \times 9.81}{60} = 19\,620 \text{ J}$$

Therefore each second the total energy created

$$= 31\,250 + 19\,620$$
$$= 50\,870 \text{ J}$$

The pump is working at 50 870 J/s or approximately 51 kW.

5.5   SPRINGS AND ELASTIC STRINGS

We may describe an elastic body as one which deforms when subjected to a load and resumes its original shape when the load is removed. We have discussed in Section 3.9 the overall elastic behaviour of impact in very simple terms without examining in detail the actual nature of the elasticity within the body. We shall adopt the same procedure with springs.

Springs serve many purposes in engineering and are designed to have particular characteristics to suit what is required of them. They are used to store energy as in a clock or to absorb energy and so protect a mechanism which might otherwise be damaged by a sudden applied force as is required of the shock absorbers of a car.

We shall think principally in terms of the coil spring and the elastic string where the former can be extended and compressed along its length whilst the latter cannot sustain compression. The elastic string may become slack during the motion. The *ideal* spring or string is assumed to have no mass and also a *natural length* which it takes up when subject to no load. (The natural length of a heavy spring would be the length it took up if placed horizontally on a smooth horizontal table: a suspended heavy spring will be extended by its own weight.)

Consider an ideal spring suspended from a fixed point. By attaching loads to the free end we can measure the resulting extensions of the spring and plot

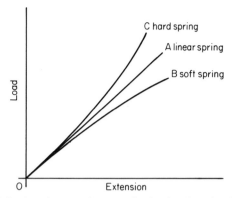

Figure 5.6   Load-extension graphs for hard and soft springs.

graphically load against extension. Some examples of load/extension graphs are shown in Figure 5.6. The straight line OA implies a linear relation between load and extension. Such an idealized spring is called a *linear* spring and is said to obey *Hooke's law*. It is a reasonable approximation for loads which do not produce excessive extensions of the spring. Springs whose extension increases at a faster rate than the load as on OB are called *soft* springs. On the other hand, *hard* springs, as on OC, tend to resist extension with increasing load.

Our simple model for a linear spring can be summarized:

$$\text{load} \propto \text{extension}$$

or

$$F = \frac{\lambda x}{a} \tag{6}$$

where F is the force or load applied, x is the extension, a is the natural length and λ is the *modulus of elasticity* of the spring, that is, that force which produces unit extension in a spring of unit length. The ratio $k = \lambda/a$ is called the *stiffness* of the spring and is used as an alternative measure of the spring's strength. We shall assume that the spring under compression has similar properties to the extended spring so that relation (6) still holds with x negative for compression. On the other hand for an elastic string we must apply the restriction $x \geq 0$.

These remarks refer only to the static behaviour of springs but we assume that law (6) still applies when the spring is in motion.

Example 7   A bob of mass m is attached to the free end of a spring of natural length a and modulus λ, the other end of the spring being fixed. The bob is displaced from its position of equilibrium and moves in the vertical line of the spring. Determine the subsequent motion of the bob.

Let b be the extension of the spring when in equilibrium so that

$$mg = \frac{\lambda b}{a}. \tag{7}$$

Suppose at time t the spring has an *additional* extension x and that the tension is then T (Figure 5.7). The equation of motion of the bob is

$$mg - T = m\ddot{x} \tag{8}$$

where

$$T = \frac{\lambda(x + b)}{a}. \tag{9}$$

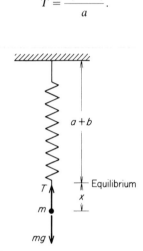

Figure 5.7   Spring supporting a bob in vertical motion.

From (7), (8) and (9), we can easily see that the displacement x satisfies

$$\ddot{x} + \frac{\lambda}{am} x = 0. \tag{10}$$

This is the equation for simple harmonic motion (see Section 4.5). The solution is

$$x = K \cos(\omega t + \varepsilon), \qquad \omega^2 = \lambda/am$$

where the constants $K > 0$ and ε are to be determined by the initial displacement and velocity of the bob. The period of oscillation of the bob is $2\pi(am/\lambda)^{1/2}$.

   If the spring is replaced by an elastic string the motion will be simple harmonic provided the string remains extended, that is provided the amplitude of the oscillation $K \leq b$. If $K > b$ the string will become slack for part of the motion and the bob will move as a particle vertically under gravity.

   We shall have more to say about oscillations in general in Chapters 7 and 9.

Obviously a spring stores energy when extended or compressed. The potential energy of a spring with extension x is the work done in extending the spring from its natural length. Thus

$$\mathcal{V} = \int_0^x T \, dy$$

where T is the tension at extension y. Since $T = (\lambda/a)y$ for the linear spring,

$$\mathcal{V} = \int_0^x \frac{\lambda y}{a} \, dy = \tfrac{1}{2} \frac{\lambda}{a} x^2.$$

Since there is no energy dissipation in the ideal spring the energy principle must hold: $\mathcal{T} + \mathcal{V} = $ constant. In Example 7, we must have

$$\tfrac{1}{2} m \dot{x}^2 + \tfrac{1}{2} \frac{\lambda}{a} x^2 = \text{constant} \tag{11}$$

which is, of course, a first integral of Equation (10). Put another way, Equation (10) is the time-derivative of (11).

5.6 EQUILIBRIUM AND STABILITY

Imagine a ball placed at the highest point of a fixed sphere and at the lowest point of a fixed hollow sphere (Figure 5.8). We can balance the ball in both positions and we say that the ball is *in equilibrium*. The *stability* of the ball in both these positions is concerned with what happens to the ball when it is given a small push. Clearly in the first case the ball rolls off the sphere whilst in the second case the ball rolls to and fro across the equilibrium position. The former we would call *unstable* and the latter *stable* equilibrium. We now attempt to formulate mathematical definitions of stability.

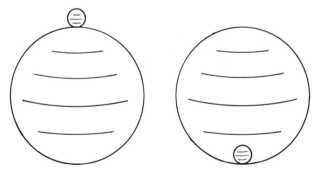

Figure 5.8

We have already encountered equilibrium in Section 3.8. A particle is in equilibrium if the forces acting on it balance and if it is at rest. A rigid body is in equilibrium if the forces balance and the moment of forces about any fixed point balance.

Stability can be defined in a variety of ways. For example, a position of stable equilibrium can be defined as one in which, in any disturbance from the position of equilibrium, the forces acting tend to restore the system to its position of equilibrium or as one in which any sufficiently small disturbance produces a bounded motion about the equilibrium position. It is largely a matter of translating into mathematical language what we think of intuitively as stable equilibrium.

Any investigation of the state of a mechanical system requires a recognition of the positions which the system can take up consistent with the *constraints* imposed upon it. For example, suppose the rod in Figure 5.9(a) is hinged at O to a fixed point and moves in a fixed vertical plane. We say that the rod is *constrained* so that one end point is fixed and the rod moves in a plane. Every possible position of the rod can be described by the angle θ. This example requires one parameter or coordinate θ to describe every *configuration* of the system and for this reason we say that the body has one *degree of freedom*. If we remove the vertical plane restriction then the rod will have two degrees of freedom and we shall require a second spherical polar angle to cover the second degree of freedom.

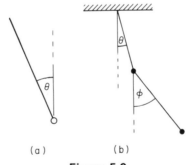

(a)   (b)

Figure 5.9

The double pendulum shown in Figure 5.9(b) has two degrees of freedom if constrained in a fixed vertical plane. The angles θ and ϕ can describe every configuration. A particle moving in space has three degrees of freedom since three independent coordinates are required to fix its position. A rigid body moving freely has six degrees of freedom; three coordinates are needed to determine the translation and three the orientation of the body.

A simple method of testing whether sufficient coordinates have been selected is to fix them: if parts of the system can still be moved, more coordinates are

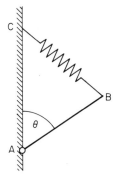

Figure 5.10

needed. Figure 5.10 shows a hinged rod AB supported by a spring CB. Once the inclination of AB is fixed by θ the inclination of the spring is determined, and the system has one degree of freedom.

We can view stability in general terms as follows. Suppose that the system is placed in an equilibrium state, that is, at rest with zero velocity. Assign (small) positive constants as upper bounds for the displacements and velocities of the system. The question is: can we find non-zero initial values for the displacements and velocities such that subsequently their values are always bounded by the assigned constants *no matter how small they may be*? If the answer is yes then the equilibrium is said to be *stable*: if it is no then the equilibrium is unstable. The formalising of these definitions into precise mathematical language is more technical but the intuitive idea above gives the sense of the definitions. Figure 5.11 shows a length of wire formed into various curves in a vertical plane. The equilibrium positions of a bead which can slide on the wire are shown as S (stable) or U (unstable). The equilibrium in (a) is stable since we can always create a bounded motion of any amplitude no matter how small. In (b) there are two unstable and one stable positions of equilibrium: the inflexion must be unstable since once disturbed the bead cannot remain within an arbitrarily small distance of U. The equilibrium in (c) must be unstable.

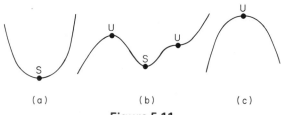

(a)   (b)   (c)

Figure 5.11

5.7 CONSERVATIVE SYSTEMS WITH ONE DEGREE OF FREEDOM

For a system with one degree of freedom any point of the system can be expressed in terms of a single coordinate q. Then $\mathbf{r}(q)$ is the position vector of a point of the system. The velocity and acceleration of this point are given by

$$\dot{\mathbf{r}} = \frac{d\mathbf{r}}{dq}\frac{dq}{dt}$$

$$\ddot{\mathbf{r}} = \frac{d^2\mathbf{r}}{dq^2}\left(\frac{dq}{dt}\right)^2 + \frac{d\mathbf{r}}{dq}\frac{d^2q}{dt^2}.$$

For an equilibrium configuration $\dot{\mathbf{r}} = \ddot{\mathbf{r}} = \mathbf{0}$ which imply that $\dot{q} = \ddot{q} = 0$.

The kinetic energy $\mathcal{T}$ will depend on q and $\dot{q}$ since $\mathcal{T}$ is formed by summing or integrating terms involving the square of the speed $|\dot{\mathbf{r}}|$. Thus the kinetic energy must take the form $\mathcal{T} = f(q)\dot{q}^2$ where $f(q) > 0$. Let us suppose that the potential energy $\mathcal{V}$ depends only on the coordinate q. For a conservative system

$$\mathcal{T} + \mathcal{V} = \text{constant}.$$

Take the total derivative with respect to q of this equation:

$$\frac{d(\mathcal{T} + \mathcal{V})}{dq} = 0$$

or

$$\frac{\partial \mathcal{T}}{\partial q} + \frac{\partial \mathcal{T}}{\partial \dot{q}}\frac{d\dot{q}}{dq} + \frac{d\mathcal{V}}{dq} = 0,$$

or

$$f'(q)\dot{q}^2 + 2f(q)\ddot{q} + \mathcal{V}'(q) = 0. \tag{12}$$

In an equilibrium configuration $\dot{q} = \ddot{q} = 0$ and the last equation implies that $d\mathcal{V}/dq = 0$ in such a configuration. In order to find the equilibrium configuration we simply determine the turning points of the potential energy.

This theory assumes that the potential energy is a smooth function of q; the potential energy may also have minima or maxima which is not revealed by considering the derivative of the potential energy. A ball placed in a cone is self-evidently in stable equilibrium but the potential energy does not have an analytic minimum, that is a minimum which can be found by setting the derivative equal to zero.

Let the system represented by the equation of motion (12) be in equilibrium at $q = q_0$, so that $\mathcal{V}'(q_0) = 0$. Let $q(t) = q_0 + q_1(t)$ and assume that $|\dot{q}_1(t)|$ and $|q_1(t)|$ are small. In (12) we shall retain only first degree terms in these quantities, expanding the functions $f(q)$, $f'(q)$ and $\mathcal{V}'(q)$ as Taylor series about $q = q_0$ as required. In fact the only approximations needed are $f(q) \approx f(q_0)$ and $\mathcal{V}'(q) \approx \mathcal{V}'(q_0) + (q - q_0)\mathcal{V}''(q_0) = q_1 \mathcal{V}''(q_0)$. Thus q_1 satisfies

$$2f(q_0)\ddot{q}_1 + \mathcal{V}''(q_0)q_1 = 0. \tag{13}$$

Hence the perturbed motion will be SHM if $\mathcal{V}''(q_0) > 0$ since $f(q_0)$ must be non-negative. Thus close to the equilibrium the motion must be bounded within any small amplitude we care to name, and thus this equilibrium must be stable. If $\mathcal{V}''(q_0) < 0$ then one solution must grow exponentially with time from every neighbourhood of $q = q_0$ no matter how small.

The condition $\mathcal{V}''(q_0) > 0$ for stability is a sufficient condition for $\mathcal{V}(q)$ to have a minimum at $q = q_0$. Hence we can say that if $\mathcal{V}'(q_0) = 0$ and $\mathcal{V}''(q_0) > 0$ then the equilibrium is stable, but not asymptotically stable since oscillations persist in SHM

It can happen that $\mathcal{V}''(q_0) = 0$ but the potential energy still has a minimum. The equilibrium is still stable but the justification becomes more difficult to establish since we cannot usefully linearise the equation of motion about the equilibrium state. We shall say more about equilibrium and stability of non-linear systems in Chapter 9.

Example 8 Suppose in the system depicted in Figure 5.10 the rod **AB** *has mass m and length a,* **AC** $= a$ *and the spring* **CB** *has natural length* $\frac{1}{2}a$ *and modulus* $\frac{1}{2}mg$. *Find the equilibrium configurations.*

Let θ be the inclination of the rod to the vertical. The potential energy has two terms, one for the rod and one for the stretched spring:

$$\mathcal{V} = \tfrac{1}{2}mga \cos \theta + \tfrac{1}{2} \cdot \frac{2\lambda}{a}(d - \tfrac{1}{2}a)^2$$

where

$$d = BC = 2a \sin \tfrac{1}{2}\theta.$$

Putting $\lambda = \tfrac{1}{2}mg$, we have

$$\mathcal{V} = \tfrac{1}{2}mga \cos \theta + \tfrac{1}{2}mga(2 \sin \tfrac{1}{2}\theta - \tfrac{1}{2})^2.$$

In the equilibrium configurations

$$\frac{d\mathcal{V}}{d\theta} = 0 = -\tfrac{1}{2}mga \sin \theta + mga \cos \tfrac{1}{2}\theta(2 \sin \tfrac{1}{2}\theta - \tfrac{1}{2}).$$

Expressing $\sin \theta$ in terms of half-angles the condition reduces to

$$\cos \tfrac{1}{2}\theta \cdot (\sin \tfrac{1}{2}\theta - \tfrac{1}{2}) = 0$$

or

$$\cos \tfrac{1}{2}\theta = 0 \quad \text{and} \quad \sin \tfrac{1}{2}\theta = \tfrac{1}{2}.$$

The equilibrium configurations are essentially $\theta = 180°$ and $\theta = 60°$.

To investigate the stability we obtain the second derivative of $\mathscr{V}$. It is easy to verify that

$$\mathscr{V}''(\theta) = \tfrac{1}{4}mga(2 \cos \theta + \sin \tfrac{1}{2}\theta)$$

whence

$$\mathscr{V}''(\pi) = \tfrac{1}{4}mga(-2 + 1) < 0, \qquad \mathscr{V}''(\tfrac{1}{3}\pi) = \tfrac{1}{4}mga(1 + \tfrac{1}{2}) > 0.$$

The configuration in which ABC is an equilateral triangle is in stable equilibrium, and the one with B vertically below A is unstable.

Example 9   A simple pendulum consists of a bob of mass m suspended by a light rod of length a from a free pivot. A spring of stiffness k and natural length b is attached to the bob and to a wheel which runs in a horizontal groove passing through the support (Figure 5.12). The spring remains vertical.

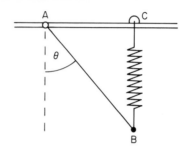

Figure 5.12

Let θ be the inclination of the pendulum to the downward vertical.

The potential energy of the bob $= -mga \cos \theta$.

The potential of the spring $= \tfrac{1}{2}k(a \cos \theta - b)^2$.

The total potential energy, $\mathscr{V} = -mga \cos \theta + \tfrac{1}{2}k(a \cos \theta - b).^2$.

The turning points of $\mathscr{V}$ are given by

$$\mathscr{V}'(\theta) = 0 = mga \sin \theta - ak \sin \theta(a \cos \theta - b)$$

which has solutions

$$\sin \theta = 0 \quad \text{and} \quad \cos \theta = (mg + bk)/ak.$$

The second solution is real if $mg + bk \le ak$. There are two solutions $\theta = 0$, $\theta = \pi$ if $mg + bk \ge ak$ and essentially three solutions (taking account of symmetry about $\theta = 0$) if $mg + bk < ak$; the additional equilibrium position is $\cos^{-1}[(mg + bk)/ak]$.
    Now

$$\mathscr{V}''(\theta) = \cos \theta \cdot (mga - a^2k \cos \theta + akb) + a^2k \sin^2 \theta.$$

For the equilibrium position $\theta = 0$:

$$\mathscr{V}''(0) = a(mg + kb - ak) > 0 \quad \text{if} \quad mg + kb > ak$$
$$\le 0 \quad \text{if} \quad mg + kb \le ak.$$

The situation in which B is below A is stable if only two equilibrium positions exist, and unstable otherwise.

For $\theta = \pi$:

$$\mathcal{V}''(\pi) = -a(mg + kb + ak) < 0$$

and the position with B above A is always unstable.

The third position exists only if $mg + bk < ak$ in which case (call the angle θ_1):

$$\mathcal{V}''(\theta_1) = a^2 k \sin^2 \theta_1 > 0$$

and this position is always stable.

Exercises

1. Find the work required to lift a satellite of mass 200 kg to a height of 1000 km above the Earth's surface. (Radius of Earth = 6400 km.)

2. Calculate the work done in sliding a block of weight 22 kg up a plane inclined at $30°$ to the horizontal through a distance of 15 m against a frictional force of 30 N.

3. Water flows over a waterfall of height 100 m at a rate of 250 m^3/s. Estimate the total power of the waterfall.

4. Determine the efficiency of a pump which is driven by a 1.6 kW motor if it takes 60 hours to raise $10^4 \, m^3$ of water to a height of 3 m. (The mechanical efficiency of a machine is the ratio of output work to input work and must always be less than unity since work is lost in overcoming internal resistance in the machine.)

5. With the steam shut off, a train of mass 3×10^5 kg descends an incline of 1 in 100 at a constant speed of 50 km/h. Find the resistance to motion. At the foot of the incline the steam is turned on and the train moves on a level track. If the engine develops 500 kW and the resistance to motion is unchanged, find the initial acceleration of the train, and its maximum speed on the level.

6. A train of mass 10^5 kg starts from rest along a level track. It reaches a speed of 50 km/h in 2 min and the resistance to motion is 4000 N. Calculate the power at which the engine must be capable of working assuming it accelerates uniformly.

7. A sphere sliding along a horizontal plane with velocity v_1 collides with a second sphere moving in the same straight line with velocity v_2 in the same direction. If the coefficient of restitution is e and the spheres both have the same mass m, find the kinetic energy lost in the impact.

8. The power output of the engine of a car is given in kW by

$$P = An \, e^{-nk}$$

where n is the engine speed in r.p.m. and A and k are constants. The maximum output of 50 kW occurs at 4000 r.p.m. In top gear 5000 r.p.m. corresponds to 100 km/h. Obtain the shortest time for the car to accelerate from 50 km/h to 120 km/h in top gear if its weight is 1000 kg.

9. Express the kinetic energy of a particle moving in a plane in terms of polar coordinates.

Two particles of masses M and m are connected by an inextensible string of negligible mass which passes through a small smooth ring on a smooth horizontal table. The particles are at rest with the string taut and straight and M at a distance a from the ring. M is now projected at right angles to the string. Prove that its path until m reaches the ring is

$$r = a \sec \{\theta[M/(M + m)]^{1/2}\}$$

in polar coordinates.

10. A light elastic string passing over a smooth peg has masses M and m attached to its ends. The system is released from rest with the string just slack. Show that after time t, the tension (assumed to be the same throughout the string) is $2mMg(1 - \cos nt)/(m + M)$ where $n^2 = \lambda(M + m)/aMm$, λ and a being the modulus of elasticity and natural length of the string respectively.

11. An elastic string of natural length 50 cm has an extension of 5 cm when a mass of 10 g is suspended by it. The mass is pulled down a further 10 cm and released. Find the maximum height reached by the mass subsequently.

12. The overhead door shown in Figure 5.13 weighs 100 kg and is released from rest in the position shown. The two springs each have an unstrained length of 0.4 m and are designed so that the door just comes to rest when closed, and is held in that position by a catch. Determine the elastic properties of the springs. Ignore friction and any energy losses occurring at the impact of the door and springs.

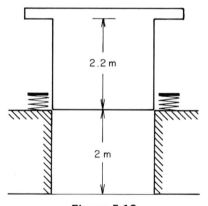

Figure 5.13

13. Figure 5.14 shows (schematically) a 'block and tackle'. What force F must be applied at the free end of the rope in order to lift a load of weight W? The weight of the lower block is w and friction is negligible.

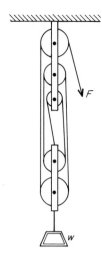

Figure 5.14

14. A heavy uniform rod AB of length $2a$ rests with A in contact with a smooth vertical wall and a point of it against a smooth peg distance $b\,(<a)$ from the wall. Find the equilibrium position and examine its stability.

15. On the smooth surface of a fixed horizontal circular cylinder lie two masses m and M connected by a light cord, the cord subtending an angle ϕ at the centre of the circular cross-section in which it lies. Find the position of equilibrium and show that it is always unstable.

16. A uniform plank of thickness $2h$ can roll without slipping on a fixed horizontal circular cylinder of radius a. Show that the equilibrium position in which the plank lies perpendicular to the cylinder is stable or unstable according as $a \gtrless h$.

17. A heavy uniform rod AB of length $2a$ and mass $2M$ is hinged at a fixed point A. End B is tied to a light string which passes over a smooth peg fixed vertically above A and distance c above A. A particle of mass M is attached to the other end of the string. Find the positions of equilibrium and show that the position in which AB is vertical is stable if $a < c$.

18. A pulley of radius a, free to rotate in a vertical plane, has its centre fixed at a distance $2a$ from a vertical rod. A mass m hangs from a light cord which is attached to the circumference of the pulley at a point R on the upper semicircle of the pulley. A spring of natural length $2a$ and modulus $8mg$ extends from R to a slide B which runs freely on the vertical rod such that RB always remains horizontal. Find the positions of equilibrium of the system and determine their stability.

19. A bead of mass m can slide freely on a vertical circular wire of radius a. A spring of natural length $3a/2$ and modulus λ joins the bead to the highest point of the wire. Investigate the stability of the equilibrium positions in the two cases $\lambda = 12mg$, $\lambda = mg$.

20. An inverted pendulum consists of a light rigid rod of length l which carries a mass m at its upper end and is smoothly pivoted at its lower end. At a distance d from the pivot, two springs each of modulus λ with the same natural length are attached to the rod and join the rod to two vertical walls on either side of the rod and equidistant from it when the rod is vertical. In this position the springs are horizontal. Find the condition that this position of equilibrium is stable in the vertical plane perpendicular to the walls.

21. A square frame, consisting of four equal uniform rods of length $2a$ joined rigidly together, hangs at rest in a vertical plane on two smooth pegs P and Q at the same level. If PQ $= c$ and the pegs are not both in contact with the same rod, show that there are three positions of equilibrium if $c < a < \sqrt{2}c$ and one otherwise. Investigate their stability.

22. Using the MacLaurin expansion for $\sin x$, explain why the pendulum behaves as a soft spring.

6

Variable Mass Problems: Rocket Motion

6.1 THE EQUATION OF MOTION

So far in this book we have dealt exclusively with the dynamics of particles and bodies whose masses remain constant during any motion. In certain applications we cannot make this assumption. A rocket is propelled by ejecting burnt fuel which causes the mass of the rocket to decrease substantially as the rocket accelerates. A raindrop falling through a damp atmosphere coalesces with smaller droplets which increase its mass. In both of these illustrations the mass of the body may be thought of as varying with time: the term 'variable mass' is slightly misleading since we do not intend to mean that mass is being created or being destroyed, but that it is being removed or added to the body.

Suppose that a body having variable mass $m(t)$ is moving with velocity $v(t)$. At time $t + \delta t$ let its main mass be $m(t + \delta t)$ and its velocity be $v(t + \delta t)$. The body has either gained or lost incrementally mass $-m(t + \delta t) + m(t)$ depending on the sign of this difference. For the sake of discussion let us suppose that an increment of mass has broken from the main body with absolute velocity $u(t)$ (see Figure 6.1). At time t this mass $m(t) - m(t + \delta t)$ (which will be positive in

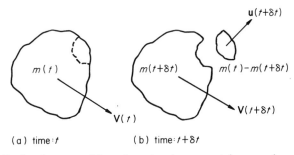

(a) time: t (b) time: $t + \delta t$

Figure 6.1 Body of mass $m(t)$ undergoing incremental mass change. (a) shows the body immediately before the mass $m(t) - m(t + \delta t)$ is ejected with absolute velocity $u(t)$. (b) shows the situation at time $t + \delta t$ with the main mass now $m(t + \delta t)$.

this case) experiences a sudden velocity change from $\mathbf{v}(t)$ to $\mathbf{u}(t)$. For the whole system the momentum at time $t + \delta t$ is

$$m(t + \delta t)\mathbf{v}(t + \delta t) + [m(t) - m(t + \delta t)]\mathbf{u}(t + \delta t)$$

whilst at time t it was $m(t)\mathbf{v}(t)$. We are only concerned with the translation of the body: any rotational effects on the body will not be considered here. The change of momentum is

$$m(t + \delta t)\mathbf{v}(t + \delta t) + [m(t) - m(t + \delta t)]\mathbf{u}(t + \delta t) - m(t)\mathbf{v}(t)$$
$$= [m(t + \delta t) - m(t)]\mathbf{v}(t) + m(t)[\mathbf{v}(t + \delta t) - \mathbf{v}(t)]$$
$$- [m(t + \delta t) - m(t)]\mathbf{u}(t + \delta t)$$

where the previous line has been prepared for division by the time increment δt. We now divide the right-hand side above by δt so that

$$\frac{[m(t + \delta t) - m(t)]}{\delta t}\mathbf{v}(t) + m(t)\frac{[\mathbf{v}(t + \delta t) - \mathbf{v}(t)]}{\delta t} - \frac{[m(t + \delta t) - m(t)]}{\delta t}\mathbf{u}(t + \delta t).$$

From the definition of the derivative

$$\lim_{\delta t \to 0} \frac{m(t + \delta t) - m(t)}{\delta t} = \frac{dm(t)}{dt}$$

etc. Consequently the rate of change of the linear momentum of the body becomes

$$\frac{dm(t)}{dt}\mathbf{v}(t) + m(t)\frac{d\mathbf{v}(t)}{dt} - \frac{dm(t)}{dt}\mathbf{u}(t) = [\mathbf{v}(t) - \mathbf{u}(t)]\frac{dm(t)}{dt} + m(t)\frac{d\mathbf{v}}{dt}(t).$$

Suppose that the body is also subject to an external force $\mathbf{F}$. Then Newton's second law of motion is now interpreted in the form, force equals the rate of change of the linear momentum of the body which is the same as the previous hypothesis (Section 3.2) if the mass of the body remains constant. Hence it follows that the equation of motion assumes the form

$$\mathbf{F} = (\mathbf{v} - \mathbf{u})\frac{dm}{dt} + m\frac{d\mathbf{v}}{dt} \tag{1}$$

where we have now dropped the time arguments of m, $\mathbf{u}$ and $\mathbf{v}$.

It may appear at first sight that (1) is inconsistent with the expression

$$\frac{d}{dt}(m\mathbf{v}) = \mathbf{v}\frac{dm}{dt} + m\frac{d\mathbf{v}}{dt}$$

which may be thought of as the rate of change of momentum $m\mathbf{v}$. However, any disposal or accretion of mass which involves a velocity difference will have a continuous impulsive effect on the remaining mass. Thus the disposal of the increment $m(t) - m(t + \delta t)$ involves a velocity difference $\mathbf{u} - \mathbf{v}$. Hence over the time interval δt this means that the remaining mass experiences an impulse

$$[m(t) - m(t + \delta t)][\mathbf{v}(t) - \mathbf{u}(t)].$$

The corresponding force $\mathbf{F}_I$, say, as a continuous function of time is given by

$$\mathbf{F}_I = \lim_{\delta t \to 0} \frac{[m(t) - m(t + \delta t)]}{\delta t} [\mathbf{v}(t) - \mathbf{u}(t)].$$

Thus we could interpret (1) also as

$$\mathbf{F} + \mathbf{F}_I = m \frac{d\mathbf{v}}{dt}$$

or 'force = mass × acceleration'.

Notice that equation (1) becomes

$$\mathbf{F} = m \frac{d\mathbf{v}}{dt}$$

if $\mathbf{u} = \mathbf{v}$. In this case mass is being lost or acquired but at zero relative velocity. If $\mathbf{u} = \mathbf{0}$, then

$$\mathbf{F} = \mathbf{v} \frac{dm}{dt} + m \frac{d\mathbf{v}}{dt} = \frac{d}{dt}(m\mathbf{v}).$$

This corresponds, for example, to the case of the raindrop falling through a stationary cloud of droplets.

Equation (1) is the fundamental relation for motion with variable mass. However, in order to be able to analyse a problem we shall still need to specify the rate of mass change and its velocity in addition to the external force.

Example 1   A balloon of mass M contains a bag of sand of mass m_0, and the balloon is in equilibrium. The sand is released at a constant rate and is disposed of in a time t_0. Find the height of the balloon and its velocity when all the sand has been released. Assume that the balloon experiences a constant upthrust and neglect air resistance.

In equilibrium the upthrust F must balance the weight of the balloon and sand:

$$F = (M + m_0)g.$$

Let m be the mass of sand at time t where $0 \le t \le t_0$. Then

$$m = m_0\left(1 - \frac{t}{t_0}\right) \tag{2}$$

since the sand is released at a constant rate. The velocity of the sand relative to the balloon is zero on release with the result that $\mathbf{v} = \mathbf{u}$ in Equation (1). Let x be the subsequent displacement of the balloon. Its equation of motion becomes

$$(M + m_0)g - (M + m)g = (M + m)\frac{dv}{dt} \tag{3}$$

where $v = \dot{x}$. On substituting for m from (2) into (3):

$$\frac{dv}{dt} = \frac{m_0 gt}{(M + m_0)t_0 - m_0 t} = -g + \frac{(M + m_0)gt_0}{(M + m_0)t_0 - m_0 t}.$$

This is a variables separable equation with solution

$$v = -gt - \frac{(M + m_0)gt_0}{m_0} \ln\left(1 - \frac{m_0 t}{(M + m_0)t_0}\right) \tag{4}$$

where the initial condition $v = 0$ when $t = 0$ has been used.

The differential equation for the displacement, .

$$\frac{dx}{dt} = -gt - \frac{(M + m_0)gt_0}{m_0} \ln\left(1 - \frac{m_0 t}{(M + m_0)t_0}\right)$$

is again of separable type with solution

$$x = C - \int\left(gt + \frac{g}{k}\ln(1 - kt)\right)dt, \qquad k = m_0/t_0(M + m_0)$$

$$= C - \tfrac{1}{2}gt^2 - \frac{gt}{k}\ln(1 - kt) - g\int\frac{t\,dt}{1 - kt}, \quad \text{integrating by parts}$$

$$= C - \tfrac{1}{2}gt^2 - \frac{gt}{k}\ln(1 - kt) - \frac{g}{k}\int\left(-1 + \frac{1}{1 - kt}\right)dt$$

$$= C - \tfrac{1}{2}gt^2 - \frac{gt}{k}\ln(1 - kt) + \frac{gt}{k} + \frac{g}{k^2}\ln(1 - kt)$$

$$= C - \tfrac{1}{2}gt^2 + \frac{g}{k^2}(1 - kt)\ln(1 - kt) + \frac{gt}{k}.$$

Taking the initial condition to be $x = 0$ when $t = 0$, we see that $C = 0$. Thus

$$x = \frac{gt}{k} - \tfrac{1}{2}gt^2 + \frac{g}{k^2}(1 - kt)\ln(1 - kt). \tag{5}$$

All equations and solutions hold only during the time interval $0 \le t \le t_0$.

At time $t = t_0$ the balloon has reached a height

$$x_0 = \frac{gt_0^2}{2m_0^2}\left[(2M + m_0)m_0 + 2M(M + m_0)\ln\left(\frac{M}{M + m_0}\right)\right]$$

and is moving with speed

$$v_0 = \frac{gt_0}{m_0}\left[(M + m_0)\ln\left(\frac{M + m_0}{M}\right) - m_0\right].$$

In practice the quantity of sand carried is small so that the ratio m_0/M is small: denote this ratio by ε. Then, for small ε,

$$x_0 = \frac{gt_0^2}{2\varepsilon^2} \left[(2 + \varepsilon)\varepsilon - 2(1 + \varepsilon) \ln (1 + \varepsilon) \right]$$

$$= \frac{gt_0^2}{2\varepsilon^2} \left[2\varepsilon + \varepsilon^2 - 2(1 + \varepsilon)\left(\varepsilon - \frac{\varepsilon^2}{2} + \frac{\varepsilon^2}{3} - \cdots \right) \right]$$

$$\simeq \frac{gt_0^2\varepsilon}{6}$$

by using the MacLaurin expansion for $\ln (1 + \varepsilon)$. Similarly

$$v_0 = \frac{gt_0}{\varepsilon} \left[(1 + \varepsilon) \ln (1 + \varepsilon) - \varepsilon \right]$$

$$= \frac{gt_0}{\varepsilon} \left[(1 + \varepsilon)\left(\varepsilon - \frac{\varepsilon^2}{2} + \cdots \right) - \varepsilon \right]$$

$$\simeq \tfrac{1}{2} gt_0 \varepsilon.$$

A simple calculation shows that a balloon of mass 500 kg will rise through a height of about 330 m from equilibrium if 10 kg of sand is released over a period of 100 s.

6.2   ROCKET MOTION

The rocket motor is an important application of the variable mass equation. It can be thought of in very simple terms as a cylinder closed at one end in which fuel is burnt and ejected through the open end. The analysis of rocket motion can be extremely complicated when such factors as gravitational effects and rocket orientation, and the construction of the rocket are taken into account. We shall look at some simple models which are capable of relatively simple analysis. Two parameters are assumed known—the rate at which the propellant is ejected and its exhaust velocity. The exhaust velocity c is the velocity of the burnt fuel relative to the rocket casing; that is, $\mathbf{c} = \mathbf{u} - \mathbf{v}$. Thus the equation of motion of a rocket of mass m moving with velocity $\mathbf{v}$ subject to an external force $\mathbf{F}$ is

$$\mathbf{F} = m \frac{d\mathbf{v}}{dt} - \mathbf{c} \frac{dm}{dt}. \tag{6}$$

Remember that dm/dt is negative in rocket problems.

We will now turn our attention to rocket dynamics with one direction of flight (see Figure 6.2) chosen so that $\mathbf{v} = w\mathbf{k}$, $\mathbf{c} = -c\mathbf{k}$ and $\mathbf{F} = F\mathbf{k}$ which reduces the equation of motion to the single scalar relation

$$F = m \frac{dw}{dt} + c \frac{dm}{dt}.$$

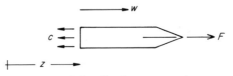

Figure 6.2   Single-stage rocket.

Let us consider first the case in which the rocket is remote from external effects, perhaps in deep space, with force $F = 0$. Thus

$$m \frac{dw}{dt} = -c \frac{dm}{dt} \tag{7}$$

in which we can eliminate time through the ratio

$$\frac{dw}{dm} = \frac{dw/dt}{dm/dt} = -\frac{c}{m}.$$

If c is a constant, this equation can be integrated to obtain the relation between the rocket's velocity and its mass. Suppose that the rocket starts from rest with total mass, including fuel and payload, M and assume that the fuel has mass εM ($0 < \varepsilon < 1$). Let W be the final velocity of the rocket when all the fuel has been consumed. Equation (7) is a separable first-order equation with solution given by

$$\int_0^W dw = -c \int_M^{M(1-\varepsilon)} \frac{dm}{m}$$

or

$$W = -c[\ln m]_M^{M(1-\varepsilon)} = -c \ln (1 - \varepsilon). \tag{8}$$

Interestingly, in a force-free situation the final speed imparted to the rocket depends on the exhaust speed of the burning fuel and the proportion of fuel burnt but not on the rate of combustion. However, this latter factor is required in order to proceed further to find the displacement of the rocket during the powered flight.

A typical maximum value for ε is about 0.85 (that is, 85% of the total mass of the rocket is fuel, the remaining 15% being the mass of the rocket casing, payload, guidance mechanisms, etc.). Such a rocket could achieve an added approximate speed of $-c \ln (0.15) \approx 1.9c$, that is, $1.9 \times$ exhaust speed of the burning propellant. Since high speeds are necessary in space travel, both high exhaust speeds and high proportions of fuel are required features of rocket design.

Example 2   A rocket of total mass M contains fuel of mass εM (0 < ε < 1). When ignited the fuel burns at a constant mass-rate k, ejecting exhaust gases with constant speed c. In a force-free environment, find the distance travelled at burn-out assuming the rocket starts from 'rest'.

To find the displacement we require the velocity as a function of time. Since the fuel is burnt at a constant mass-rate,

$$\frac{dm}{dt} = -k$$

which implies that

$$m = M - kt$$

assuming $m = M$ when $t = 0$. Burn-out occurs at time $T = εM/k$. Equation (7) becomes

$$\frac{dw}{dt} = \frac{ck}{M - kt}.$$

Thus

$$w = ck \int \frac{dt}{M - kt} + A = -c \ln (M - kt) + A$$

where A is a constant. Assume that $w = 0$ at time $t = 0$. Then $A = c \ln M$. Hence

$$w = -c \ln [(M - kt)/M], \quad 0 \le t \le εM/k.$$

At time $t = T$, $w = W = -c \ln (1 - ε)$ which agrees with (8).
   Let the displacement be z at time t so that $w = dz/dt$. Hence

$$z = -c \int \ln [(M - kt)/M] \, dt + B$$

where B is a further constant. Integrating by parts, we deduce

$$z = -ct \ln [(M - kt)/M] - ck \int \frac{t \, dt}{M - kt} + B$$

$$= -ct \ln [(M - kt)/M] + c \int \left(1 - \frac{M}{M - kt}\right) dt + B$$

$$= -ct \ln [(M - kt/M] + ct + \frac{cM}{k} \ln (M - kt) + B.$$

Let $z = 0$ when $t = 0$: thus $B = -(cM \ln M)/k$. Finally the displacement at time is given by

$$z = ct + \frac{c}{k} (M - kt) \ln [M - kt)/M]. \tag{9}$$

At burn-out, $t = T = \varepsilon M/k$ and the distance travelled at this time is

$$z = H = \frac{cM}{k} [\varepsilon + (1 - \varepsilon) \ln (1 - \varepsilon)].$$

In Equation (9) introduce dimensionless variables τ and Z through $t = M\tau/k$ and $z = cMZ/k$. Thus solution (9) can be transformed into

$$Z = \tau + (1 - \tau) \ln (1 - \tau).$$

The solid curve in Figure 6.3 shows a graph of Z against τ on the interval $0 \le \tau \le 1$. As $\tau \to 1, (1 - \tau) \ln (1 - \tau) \to 0$ (the ln function tends to minus infinity 'more slowly' than $1 - \tau \to 0$) which means that Z has an upper bound of 1. In practice the graph will terminate at $\tau = \varepsilon$: the value $\tau = 1$ will not be technically attainable. During the burning of the fuel the rocket can never move further than cM/k from rest. Note that, even though the displacement has an upper bound, the final velocity is theoretically unbounded since $W \to \infty$ as $\tau \to 1$.

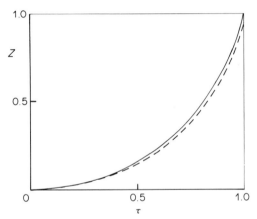

Figure 6.3 Comparison of displacement-time graphs for a single-stage rocket with (---) and without (——) gravity.

Example 3   A rocket engine of mass 3×10^5 kg can eject exhaust gases at 3000 m/s at a rate of 10^4 kg/s. The rocket is fired vertically from the Earth's surface. Show that such a rocket would not be capable, in practice, of escaping from the Earth's gravitational field. Assume that the gravitational force is constant for the period during which the fuel burns.

Let **k** be the upward unit vector at the Earth's surface. In the notation of Equation (6), $\mathbf{F} = -mg\mathbf{k}$, $\mathbf{c} = -c\mathbf{k}$ and $m = M - kt$ where $M = 3 \times 10^5$ kg, $c = 3000$ m/s and $k = 10^4$ kg/s. The equation of motion becomes

$$-(M - kt)g = (M - kt)\frac{dw}{dt} - ck$$

or

$$\frac{dw}{dt} = -g + \frac{ck}{M - kt}.$$

Integrating this equation, we find that

$$w = -gt - c \ln (M - kt) + c \ln M$$

after putting $w = 0$ when $t = 0$. Writing $w = dz/dt$ in the preceding equation and integrating a second time, we obtain

$$z = -\tfrac{1}{2}gt^2 + ct + \frac{c}{k} (M - kt) \ln [(M - kt)/M]$$

using the condition $z = 0$ when $t = 0$.

For a fuel-ratio of 0.85, the fuel will be exhausted at time 25.5 s and at a height of approximately 48 km which, by the way, justifies our uniform approximation for the gravitational force on the rocket. At this height the rocket is moving at about 5400 m/s. The escape velocity from the surface of the Earth is about 11 000 m/s (Example 2, Section 4.1), which differs little from the escape velocity at height 48 km. A *single-stage* rocket with the characteristics given will therefore not escape from the Earth's gravitational field. The effect of gravity on the height reached is small. For the data of this example the height reached using the dimensionless variables of the previous example is given by

$$Z = -\mu\tau^2 + \tau + (1 - \tau) \ln (1 - \tau)$$

where $\mu = gM/(2kc) = 0.049$ approximately. The broken line in Figure 6.3 represents this (Z, τ) graph for comparison with the force-free case.

6.3   THE MULTI-STAGE ROCKET

The overall performance of a rocket can be improved by designing the rocket in stages which in the two-stage rocket means that one rocket is placed on top of a second rocket. When the fuel in the first stage is exhausted its rocket casing is detached and the second stage ignited. The shedding of the surplus mass contained in the casing of the expended fuel considerably improves a rocket's performance.

Consider a rocket consisting of two equal stages each of mass M of which the proportion εM $(0 < \varepsilon < 1)$ in each stage is propellant. The rocket carries a payload of mass m. Suppose we wish to find the final velocity given to the payload. The exhaust speed is c throughout. The problem has two parts. The total mass of the rocket is initially $2M + m$ and during the first stage εM mass of fuel is burnt. This is essentially the same problem as that solved in Example 2 above. Modifying the result slightly, we can deduce that the rocket's velocity at the end of the first stage is

$$-c \ln \left(1 - \frac{\varepsilon M}{2M + m} \right).$$

Mass $(1 - \varepsilon)M$ is now detached from the rocket, and we have exactly the same problem as in Example 2 again with fuel of mass εM and a rocket of mass $M + m$. The rocket receives an additional boost in velocity of

$$-c \ln \left(1 - \frac{\varepsilon M}{M + m} \right)$$

during the second stage. The total velocity imparted to the payload is therefore

$$-c \ln \left(1 - \frac{\varepsilon M}{2M + m} \right) - c \ln \left(1 - \frac{\varepsilon M}{M + m} \right).$$

If $\varepsilon = 0.85$ and $M = 100m$, the total added velocity becomes

$$-c \ln \left(1 - \frac{85}{201} \right) - c \ln \left(1 - \frac{85}{101} \right) = (0.55 + 1.84)c = 2.39c$$

compared with a single-stage rocket with $\varepsilon = 0.85$ and $M = 200m$ (to keep the payload in the same proportion to the total mass) in which the added velocity is

$$-c \ln \left(1 - \frac{170}{201} \right) = 1.87c.$$

The reader should verify that the formula for the final velocity of the payload of a two-stage rocket generalises to a rocket of n equal stages each containing the same proportion ε of fuel. The final velocity of such a payload is

$$-c \ln \prod_{r=1}^{n} \left(1 - \frac{\varepsilon M}{rM + m} \right).$$

In fact, the rocket of equal stages is not generally the optimum construction. We shall obtain now under rather simple conditions the best two-stage rocket of mass M in order that it should give the maximum velocity to a satellite of mass m. Suppose that a proportion ε of each stage is fuel. Let M_1 and M_2 be the masses of the two stages (Figure 6.4) so that

$$M_1 + M_2 = M.$$

Figure 6.4   Two-stage rocket with payload mass m, first stage M_1 and second stage M_2.

Using the technique above again, we see that the velocity achieved by the satellite is given by

$$w = -c \ln \left(1 - \frac{\varepsilon M_1}{M + m}\right) - c \ln \left(1 - \frac{\varepsilon M_2}{M_2 + m}\right)$$

$$= -c \ln \left(1 - \frac{\varepsilon(M - M_2)}{M + m}\right) - c \ln \left(1 - \frac{eM_2}{M_2 + m}\right) \tag{10}$$

expressed in terms of the unknown M_2. We wish to find M_2 such that v takes its maximum value. The turning points of v are given by the equation $dw/dM_2 = 0$, or

$$\frac{\varepsilon}{M + m - \varepsilon(M - M_2)} = \frac{\varepsilon m}{(M_2 + m - \varepsilon M_2)(M_2 + m)}.$$

Simplifying this equation, we have

$$M_2^2 + 2mM_2 - mM = 0.$$

Note that the relation is *independent* of ε. Obviously M_2 must be the *positive* root of this quadratic equation:

$$M_2 = -m + (m^2 + mM)^{1/2}.$$

In practice m/M will be a small number α, say. Thus

$$\frac{M_2}{M} = -\alpha + \alpha^{1/2}(1 + \alpha)^{1/2} \simeq \alpha^{1/2}$$

discarding α compared with $\alpha^{1/2}$. We find from (9) that

$$w_{max} \simeq -c \ln \left(1 - \frac{\varepsilon(1 - \alpha^{1/2})}{1 + \alpha}\right) - c \ln \left(1 - \frac{\varepsilon \alpha^{1/2}}{\alpha + \alpha^{1/2}}\right)$$

$$\simeq -c \ln [1 - \varepsilon(1 - \alpha^{1/2})] - c \ln [1 - \varepsilon(1 - \alpha^{1/2})]$$

$$= -2c \ln [1 - \varepsilon(1 - \alpha^{1/2})]$$

if we neglect powers of α of degree higher than $\alpha^{1/2}$.
   If $\alpha = 1/100$, $M_2 = M/10$, and

$$w_{max} = 2c \ln 4 = 2.9c$$

by using $\varepsilon = 0.85$ again. We conclude that the first stage should be made much larger than the second in order to obtain a high final speed for the satellite.

Exercises

1. A rocket of total mass M contains a proportion εM $(0 < \varepsilon < 1)$ as fuel. If the exhaust speed c is constant show that the final speed of the rocket is independent of the rate at which the fuel is burnt.

2. A rocket of mass M ejects fuel at a constant rate k with exhaust speed c. Show that the rocket will not rise initially from the Earth's surface unless $k > Mg/c$.

3. A rocket of total mass $M + m_0$ contains fuel of mass εM $(\varepsilon < 1)$. The payload is of mass m_0 and $(1 - \varepsilon)M$ is the mass of the rocket casing. Suppose it is technically possible to discard the casing continuously at a constant rate whilst the fuel is burning so that no casing remains when the fuel is burnt. If the fuel is burnt at the constant rate k show that the casing must be discarded at the rate $(1 - \varepsilon)k/\varepsilon$. Verify that, if $\varepsilon = 0.83$ and $m_0 = M/100$, the rocket's final speed will be approximately $3.8c$.

4. A liquid oxygen rocket has an exhaust speed of 2440 m/s. How far will a single-stage rocket burning liquid oxygen travel from the Earth if its fuel/total mass ratio is $\frac{2}{3}$ and the fuel is burnt in 150 s? Assume g to be constant.

5. A balloon of mass 400 kg has suspended from it a rope of mass 100 kg and length 100 m. The buoyancy force of the balloon is sufficient to support a mass of 450 kg. Initially it is falling at its terminal speed of 10 m/s due to air resistance which is proportional to the square of its speed. Show that if m is the total mass of the balloon and rope t s after the rope has first touched the ground, then the equation of motion can be written as

$$m \frac{d^2 m}{dt^2} + g(m - 450) = 0$$

By writing $\dfrac{d^2 m}{dt^2}$ as $\dfrac{dm}{dt} \dfrac{d}{dm} \left(\dfrac{dm}{dt} \right)$, solve the differential equation and find the speed of the balloon:

(i) when 50 m of rope lies on the ground,
(ii) when the balloon hits the ground.

Give a physical explanation for the speeds you obtain.

6. A rocket consists of a payload of mass m propelled by two stages of masses M_1 (first stage) and M_2 (second stage). Each stage has the same exhaust speed c and contains the same proportion $\varepsilon(<1)$ of fuel. Show that the final speed of the rocket is given by

$$v = -c \ln \left(1 - \frac{\varepsilon M_1}{M_1 + M_2 + m} \right) - c \ln \left(1 - \frac{\varepsilon M_2}{M_2 + m} \right).$$

If $\varepsilon = 0.83$ and $M_1 = 9M_2$, show that the maximum payload which can be given a final velocity of $2.5c$ is $0.019 \, (M_1 + M_2)$.

7. A rocket is fired from an aircraft flying horizontally with speed V. The fuel is burnt at a constant rate k and ejected at a constant speed c. The attitude control of the rocket always maintains it in a horizontal position. If the total mass of the rocket is M find the path of the rocket during its powered flight. Assume that g is constant.

8. A uniform layer of snow whose surface is a rectangle with two sides horizontal, rests on a mountain-side of inclination α to the horizontal. The adhesion is just sufficient to hold the snow while at rest. At a certain instant an avalanche starts by the uppermost line of snow moving downwards and collecting with it the snow it meets on the way down. Assuming that there is no friction between the snow and the slope, show that if v is its speed when a distance x of the slope has been uncovered, then

$$\frac{d}{dx}(x^2v^2) = 2gx^2 \sin \alpha.$$

Deduce that the avalanche has a constant acceleration $\frac{1}{3}g \sin \alpha$.

9. Show that it is technically impossible with present rockets to use a single-stage rocket to put a payload on the Moon. Use the following data:

$$\text{exhaust speed of the rocket} = 2440 \text{ m/s}$$
$$\text{mass of the Earth} = 6.0 \times 10^{24} \text{ kg}$$
$$\text{mass of the Moon} = 1/81 \text{ mass of the Earth}$$
$$\text{radius of the Moon's orbit} = 3.9 \times 10^5 \text{ km}$$
$$\gamma = 6.7 \times 10^{-11} \text{ m}^3/\text{kg s}^2$$

The fuel is burnt in 300 s.
(Hint: assume that the take-off mass of the rocket is M of which εM is fuel. Find the burn-out speed required for the rocket just to reach the equilibrium point between the Moon and the Earth so that the rocket will fall to the Moon's surface under the action of the Moon's gravity. Show that this speed leads to an unrealistic value for ε. Make any further assumptions which you feel are justifiable.)

10. A rocket of mass 10 000 kg contains a satellite of mass 100 kg and fuel of mass 7500 kg. The rocket can be designed in two stages each containing the same proportion of fuel which in both cases can be burnt at a rate of 500 kg/s giving an exhaust speed of 2500 m/s. Design the optimum two-stage rocket which will give the maximum final speed to the rocket, assuming that it is fired vertically under constant gravity. What is the maximum final speed?

11. A rocket of initial mass M of which εM $(0 < \varepsilon < 1)$ is fuel, burns the fuel at a constant rate k and ejects the exhaust gases with speed c. The rocket takes off from rest and rises vertically under (constant) gravity. If the air resistance is assumed to be $\alpha \times$ (speed of the rocket), find the speed of the rocket as a function of time whilst the fuel is burning.

12. Two buckets of water each of total mass M are attached to the ends of a light cord which passes over a smooth pulley of negligible mass. The buckets are at rest.

Water starts to leak from one bucket at a constant rate k. Find the equations of motion of each bucket. If v and m are velocity and mass of the leaky bucket deduce that

$$\frac{dv}{dm} = \frac{g(m - M)}{k(m + M)}.$$

If εM ($0 < \varepsilon < 1$) is the mass of water in this bucket initially, determine its speed at the instant when the bucket becomes empty.

7

Mechanical Oscillations: Linear Theory

7.1 INTRODUCTION

Most machinery contains rotating or reciprocating parts which cause *periodic* or continuously repeated forces to be applied to the structure of the machine. These forces cause the machinery to *oscillate* or *vibrate*. Excessive vibrations may damage the support on which the machine rests, may cause undue wear in the moving parts or may be responsible for the emission of noise or seemingly random behaviour.

Periodic motion occurs widely in nature ranging over such examples as the daily rotation of the Earth, the annual change of the seasons, the tides, the vibration of an insect's wings and the heartbeat. The engineer's attitude to vibrations is determined by the purpose for which the mechanism under consideration is designed. In a clock the vibrations of the pendulum or the balance wheel need to be maintained in order to produce an oscillation of fixed period: the energy put into the system must be just sufficient to overcome the energy dissipated through friction. On the other hand the car spring is needed to protect the car and its occupants from sudden jolts, but the initial oscillation set up must be damped as quickly as possible by shock absorbers which dissipate the energy of the spring.

We shall examine in detail some simple models of systems, starting with those having a single degree of freedom. Later in this chapter we shall investigate some simple mechanical systems which have several modes of oscillation. All models and applications in this chapter lead to linear behaviour, although this may require some element of approximation. Large amplitude oscillations of non-linear systems are deferred until Chapter 9.

7.2 OSCILLATIONS OF CONSERVATIVE SYSTEMS

In Section 5.7 we derived the equation of motion for a conservative system with one degree of freedom in which the kinetic energy and potential energy are given respectively by $\mathcal{T} = f(q)\dot{q}^2$ and $\mathcal{V}(q)$:

$$f'(q)\dot{q}^2 + 2f(q)\ddot{q} + \mathcal{V}'(q) = 0 \tag{1}$$

where q specifies the single coordinate required. We showed further that $q = \alpha$ is a position of equilibrium if $\mathcal{V}'(\alpha) = 0$ and that the equilibrium is stable if $\mathcal{V}''(\alpha) > 0$. Let us now consider the motion in the immediate neighbourhood of such a position of stable equilibrium.

Let $q = \alpha + x$ where x is a small quantity and let us assume that $\dot{q} = \dot{x}$ and $\ddot{q} = \ddot{x}$ are also small quantities. Employing Taylor expansions about $q = \alpha$ (assuming the functions to be sufficiently well behaved), we may write

$$f(q) = f(\alpha + x) = f(\alpha) + xf'(\alpha) + \frac{x^2}{2!} f''(\alpha) + \cdots$$

$$f'(q) = f'(\alpha + x) = f'(\alpha) + xf''(\alpha) + \cdots$$

$$\mathcal{V}'(q) = xV''(\alpha) + \frac{x^2}{2!} \mathcal{V}'''(\alpha) + \cdots$$

where, in the last expansion, we have noted that $\mathcal{V}'(\alpha) = 0$. Substituting these series into (1), we obtain

$$\dot{x}^2[f'(\alpha) + \cdots] + 2\ddot{x}[f(\alpha) + \cdots] + [x\mathcal{V}''(\alpha) + \cdots] = 0.$$

If all terms of the second degree and higher in x, $\dot{x}$ and $\ddot{x}$ are neglected, this equation reduces to the second-order linear equation

$$2\ddot{x}f(\alpha) + \mathcal{V}''(\alpha)x = 0$$

or

$$\ddot{x} + \Omega^2 x = 0 \tag{2}$$

where $\Omega^2 = \mathcal{V}''(\alpha)/2f(\alpha)$. For stable equilibrium we already have $\mathcal{V}''(\alpha) > 0$ and the kinetic energy must be non-negative which implies $f(\alpha) > 0$. Thus Ω must be a real number for small *oscillations* about a position of stable equilibrium.

Details of the solution of (2) can be found in Section 4.5. This is SHM with general solutions which can be written as either $x = A \cos \Omega t + B \sin \Omega t$ or $x = K \sin (\Omega t + \varepsilon)$ where A, B, $K > 0$ and ε are constants. From (2) it follows that the period of the SHM is $2\pi[2f(\alpha)/\mathcal{V}''(\alpha)]^{1/2}$.

Example 1   A bead slides on a smooth parabolic wire with equation $y = 4x^2$ fixed in a vertical plane. Show that the bead makes small oscillations with period $\pi/(2g)^{1/2}$ about the position of equilibrium.

The kinetic energy of the bead,

$$\mathcal{T} = \tfrac{1}{2}m(\dot{x}^2 + \dot{y}^2)$$
$$= \tfrac{1}{2}m\dot{x}^2(1 + 64x^2)$$

where m is the mass of the bead. The potential energy

$$\mathscr{V} = mgy = 4mgx^2.$$

Clearly $x = 0$ is the position of equilibrium. In the earlier notation $\mathscr{T} = f(x)\dot{x}^2$ where $f(x) = \frac{1}{2}m(1 + 64x^2)$.

For oscillations of small amplitude, the horizontal displacement x satisfies Equation (2):

$$\ddot{x} + \Omega^2 x = 0$$

with $\Omega^2 = \mathscr{V}''(0)/2f(0) = 8mg/m = 8g$. The period of the oscillations is therefore given by $2\pi/\Omega = \pi/(2g)^{1/2}$.

Example 2   A bob of mass m is suspended by a light spring of natural length a and stiffness k. Find the subsequent displacement of the body if initially

> *(a)   the bob is pulled down a distance $\frac{1}{4}a$ from its position of equilibrium and released,*
> *(b)   the bob is given a downward speed v_0 from its position of equilibrium.*

For an extension y of the spring the restoring force is $-ky$. When the bob is in equilibrium $y = mg/k$. For a further extension x, the equation of motion of the bob is (referring back to Example 7, Section 5.5)

$$m\ddot{x} = mg - k\left(\frac{mg}{k} + x\right) = -kx.$$

Thus

$$\ddot{x} + \Omega^2 x = 0, \qquad \Omega^2 = k/m$$

which has a general solution $x = K \sin(\Omega t + \varepsilon)$.

Case (a). At $t = 0$, $x = \frac{1}{4}a$ and $\dot{x} = 0$. Therefore

$$\tfrac{1}{4}a = K \sin \varepsilon, \qquad 0 = \Omega K \cos \varepsilon.$$

Since $\Omega \neq 0$, $K \neq 0$, the second condition implies that $\varepsilon = \frac{1}{2}\pi$ and the first that $K = \frac{1}{4}a$. The required solution is

$$x = \tfrac{1}{4}a \sin(\Omega t + \tfrac{1}{2}\pi) = \tfrac{1}{4}a \cos \Omega t.$$

Case (b). At $t = 0$, $x = 0$ and $\dot{x} = v_0$. Therefore

$$0 = K \sin \varepsilon, \qquad v_0 = \Omega K \cos \varepsilon$$

which give $\varepsilon = 0$ and $K = v_0/\Omega$. The oscillation is given by

$$x = \frac{v_0}{\Omega} \sin \Omega t.$$

The body oscillates about the equilibrium position with amplitude v_0/Ω. Note that, whatever the initial conditions, the period of the oscillations is the same; only the amplitude and phase vary.

Both problems could have been approached using energy considerations rather than the equation of motion. The motion is truly simple harmonic for the linear spring, no approximation being necessary.

Example 3   Transverse oscillations. A particle of mass m is attached by two elastic strings of the same modulus of elasticity λ and natural lengths a_1 and a_2 to two fixed points on a smooth horizontal plane. The points are distance $k(a_1 + a_2)$ $(k > 1)$ apart. The particle is displaced slightly from its equilibrium position in a direction perpendicular to the strings. Discuss the motion.

In equilibrium, let b_1 and b_2 be lengths of the elastic strings. The tensions in the two strings must balance and it is a simple matter to verify that $b_1 = ka_1$ and $b_2 = ka_2$.

Consider now the situation when the particle has a transverse displacement x. Let T_1 and T_2 be the tensions in the strings and θ_1 and θ_2 the inclinations of the strings as shown in Figure 7.1. The transverse equation of motion becomes

$$T_1 \sin \theta_1 + T_2 \sin \theta_2 = -m\ddot{x}.$$

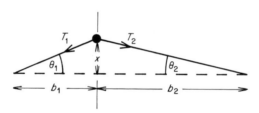

Figure 7.1   Transverse oscillations of a particle on a stretched elastic string.

For oscillations of small amplitude

$$\sin \theta_1 \simeq \theta_1 \simeq x/b_1 = x/ka_1 \qquad \sin \theta_2 \simeq \theta_2 \simeq x/b_2 = x/ka_2$$

so that x satisfies

$$\frac{T_1 x}{ka_1} + \frac{T_2 x}{ka_2} = -m\ddot{x}.$$

We are retaining only terms not greater than the first degree in x, which implies that we may use the equilibrium tension as a sufficiently good approximation to T_1 and T_2. Hence

$$T_1 = T_2 \simeq \lambda(k - 1).$$

The equation of motion becomes

$$\frac{\lambda(k - 1)(a_1 + a_2)x}{ka_1 a_2} = -m\ddot{x} \tag{3}$$

which is simple harmonic motion of period

$$2\pi\left(\frac{mka_1a_2}{\lambda(k-1)(a_1+a_2)}\right)^{1/2}.$$

7.3 DAMPED AND FORCED OSCILLATIONS

In the last section we examined the motion of a conservative system about an equilibrium configuration, this motion taking place only under the action of forces contained within the system. In any real situation two additional factors must be taken into account. The first of these is *friction*. Friction introduced into an otherwise conservative system will oppose the motion and we should expect the amplitude of the oscillations to be progressively reduced or *damped* since energy is being steadily dissipated. In the case of the simple pendulum friction will be present in the air drag on the bob of the pendulum and in the resistance at the point of suspension. Both these effects are usually very small but they do ultimately produce a significant reduction in the amplitude of the pendulum.

The application of an external disturbing force to a system is the second factor which we must include. For example, the motion of an engine on a spring mounting is affected by the periodic motion of the rotating internal machinery of the engine and this leads to a *self-maintained* or *self-excited* oscillation of the engine. However, the general effects of disturbing forces are not easy to predict as we shall show in Chapter 9.

The general system with one degree of freedom may be represented by Figure 7.2. The block B in the figure slides on a smooth table subject to a time-dependent disturbing force $F(t)$ and attached to a fixed wall by a spring S and *dashpot* D, set in parallel. A dashpot may be thought of as a piston sliding in a pot of oil. It provides resistance to the motion of the block and, in its simplest form, this resistance is proportional to the velocity of the piston *relative* to that of the pot. For a linear spring S, a further force is applied to the block which is proportional to the extension of the spring. Measuring x to the right in Figure

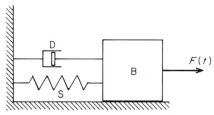

Figure 7.2 Schematic diagram for a block B attached by a spring and a dashpot D to a fixed wall.

7.2, and letting zero displacement of the block occur where the spring is unstrained, we can write the equation of motion of the block as

$$F(t) - kx - c\dot{x} = m\ddot{x}$$

or

$$m\ddot{x} + c\dot{x} + kx = F(t) \tag{4}$$

where m is the mass of the block, k is the stiffness of the spring and c is a constant determined by the damping action of the dashpot.

Equation (4) is a second-order linear differential equation with constant coefficients. The solution is the sum of two terms, one being the complementary function which is the solution of the homogeneous equation

$$m\ddot{x} + c\dot{x} + kx = 0 \tag{5}$$

and represents the unforced motion of the system, and the other being the particular integral or forcing term giving the effect of the disturbing force $F(t)$ on the system.

Let us first consider the unforced system since it is relevant to the forced system also. The characteristic equation corresponding to (5) is

$$mp^2 + cp + k = 0$$

with roots

$$p_1 = [-b + (b^2 - 1)^{1/2}]\Omega, \qquad p_2 = [-b - (b^2 - 1)^{1/2}]\Omega \tag{6}$$

where $\Omega^2 = k/m$ and $b = c/2(mk)^{1/2}$; Ω is the angular frequency of the corresponding *undamped* system (that is, the equation with $c = 0$). The required solution is therefore

$$x = A \exp\{[-b + (b^2 - 1)^{1/2}]\Omega t\} + B \exp\{[-b - (b^2 - 1)^{1/2}]\Omega t\}. \tag{7}$$

The precise effect of the dashpot will depend on the relative magnitudes of m, c and k. We shall now look at the motion of the system in the three cases $b > 1$, $b < 1$ and $b = 1$.

Strong damping ($b > 1$ or $c^2 > 4mk$). The roots in (6) are both real and negative with the result that the displacement will be the sum of two exponentially decreasing terms:

$$x = A e^{-\alpha_1 t} + B e^{-\alpha_2 t} \tag{8}$$

where α_1 and α_2 are positive. For given initial conditions the system can pass at most once through its equilibrium position ($x = 0$) before approaching rest exponentially with time. Figure 7.3 shows some typical solutions.

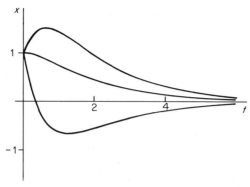

Figure 7.3 Typical solutions for strong damping, computed with $k/m = 1$, $c/m = 2.2$. All have the initial displacement $x(0) = 1$, and from the top, the initial velocities $\dot{x}(0) = 2$, $\dot{x}(0) = 0$, $\dot{x}(0) = -4$, respectively.

Example 4   A block is suspended by a spring and a dashpot with a strong damping action. Show that if the block is displaced downwards and given a downward velocity, it never passes through its equilibrium position again.

With x measured downwards from the equilibrium position,

$$m\ddot{x} + c\dot{x} + kx = 0$$

in the notation used above. The initial conditions are $x = x_0$ and $\dot{x} = v_0$ with both x_0 and v_0 positive. For strong damping the required solution takes the form (8) and without loss of generality we may take $\alpha_2 > \alpha_1$. The constants A and B satisfy

$$x_0 = A + B$$
$$v_0 = -\alpha_1 A - \alpha_2 B$$

so that

$$A = (\alpha_2 x_0 + v_0)/(\alpha_2 - \alpha_1), \qquad B = -(\alpha_1 x_0 + v_0)/(\alpha_2 - \alpha_1).$$

The displacement

$$x = \frac{1}{\alpha_2 - \alpha_1} [(\alpha_2 x_0 + v_0) e^{-\alpha_1 t} - (\alpha_1 x_0 + v_0) e^{-\alpha_2 t}]$$

$$= \frac{1}{\alpha_2 - \alpha_1} [(\alpha_2 e^{-\alpha_1 t} - \alpha_1 e^{-\alpha_2 t})x_0 + (e^{-\alpha_1 t} - e^{-\alpha_2 t})v_0].$$

With $\alpha_2 > \alpha_1$, we must have $e^{-\alpha_1 t} > e^{-\alpha_2 t}$ and $\alpha_2 e^{-\alpha_1 t} > \alpha_1 e^{-\alpha_2 t}$ for $t > 0$ and consequently $x > 0$ for $t > 0$, which means that the block does not pass through the equilibrium position again.

Weak damping ($b < 1$ or $c^2 < 4mk$). In this case the damping action of the dashpot is relatively small. The roots given by (6) are now complex. Put $\beta = \Omega\sqrt{(1 - b^2)}$ in equation (7) so that

$$x = e^{-b\Omega t}(A\ e^{i\beta t} + B\ e^{-i\beta t})$$
$$= e^{-b\Omega t}(A'\ \cos \beta t + B'\ \sin \beta t)$$

for new constants A' and B'. Let $A' = K \sin \varepsilon$ and $B' = K \cos \varepsilon$ ($K > 0$) so that

$$x = K\ e^{-bt\Omega} \sin (\beta t + \varepsilon). \tag{9}$$

The motion can be considered as a distortion of simple harmonic motion in which the amplitude decreases exponentially. Figure 7.4 shows a typical graph of displacement plotted against time. Unlike strong damping, oscillations persist in a weakly damped system. The displacement x vanishes every π/β units of time and we may speak of $2\pi/\beta$ as the period of the oscillation even though the motion never repeats itself as in simple harmonic motion.

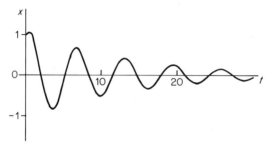

Figure 7.4   Solution for weak damping with parameter values $k/m = 1$, $c/m = 0.15$ and the initial values $x(0) = 1$, $\dot{x}(0) = 0.5$.

Example 5   The angular frequency of harmonic oscillator is 16 rad/sec. With weak damping imposed it is found that the amplitude of two consecutive oscillations are 5 cm and $\frac{1}{4}$ cm. Find the new period of the system.

Since weak damping is present the displacement of the system will be given by (9) above. The amplitude of successive oscillations decreases by a factor $\exp(-2b\Omega\pi/\beta)$. Thus

$$\exp(-2b\Omega\pi/\beta) = \tfrac{1}{4} \times \tfrac{1}{5} = \tfrac{1}{20}$$

or

$$2b\Omega\pi = \beta \log_e 20 = 3.0 \times \Omega(1 - b^2)^{1/2}$$

whence

$$4b^2\pi^2 = 9.0 \times (1 - b^2).$$

From this equation we obtain

$$b^2 = 9/(9 + 4\pi^2).$$

Finally $\beta = \Omega(1 - b^2)^{1/2} = 16 \times 2\pi/(9 + 4\pi^2)^{1/2}$ and the new period

$$\frac{2\pi}{\beta} = \frac{(9 + 4\pi^2)^{1/2}}{16} = 0.44 \text{ s.}$$

Critical damping ($b = 1$ or $c^2 = 4mk$). The roots (6) are now equal:

$$p_1 = p_2 = -\Omega$$

and the corresponding solution is

$$x = e^{-\Omega t}(A + Bt).$$

The behaviour of the system is very similar to that portrayed in Figure 7.3 for strong damping.

The general linear system for forced oscillations is represented by Equation (4) which we repeat:

$$m\ddot{x} + c\dot{x} + kx = F(t).$$

The solution of this equation is the sum of a complementary function (the general solution of the equation with $F(t)$ replaced by 0) and a particular integral (see Appendix for discussion of the solutions of this type of differential equation). The complementary function gives the *free* or *natural* damped motion of the system and is covered by the cases of strong, weak and critical damping described above. However this motion will diminish with time leaving only the particular integral or *forced oscillation*. For this reason the complementary function is known as a *transient* in the theory of oscillations.

The precise effect of a forcing term will naturally depend upon its form. We shall look in detail at the sinusoidal applied force represented by

$$F(t) = F_0 \cos(\omega t + \beta).$$

We proceed to find constants A and B such that

$$A \cos(\omega t + \beta) + B \sin(\omega t + \beta)$$

satisfies the equation

$$m\ddot{x} + c\dot{x} + kx = F_0 \cos(\omega t + \beta).$$

Substituting the expression into the equation and equating the coefficients of $\cos(\omega t + \beta)$ and $\sin(\omega t + \beta)$, we find that

$$-m\omega^2 A + c\omega B + kA = F_0$$
$$-m\omega^2 B - c\omega A + kB = 0.$$

These equations have the solution

$$A = F_0(\Omega^2 - \omega^2)/m[(\Omega^2 - \omega^2)^2 + 4b^2\Omega^2\omega^2]$$
$$B = 2F_0 b\Omega\omega/m[(\Omega^2 - \omega^2)^2 + 4b^2\Omega^2\omega^2]$$

wherein the relations $\Omega^2 = k/m$ and $b = c/2(mk)^{1/2}$ have been used. It is more convenient to put

$$A = (A^2 + B^2)^{1/2} \cos \phi \qquad B = (A^2 + B^2)^{1/2} \sin \phi$$

where

$$(A^2 + B^2)^{1/2} = F_0/m[(\Omega^2 - \omega^2)^2 + 4b^2\Omega^2\omega^2]^{1/2}$$

By using a simple trigonometric identity, the general solution can be expressed as

$$x = \text{transient} + \frac{F_0 \cos(\omega t + \beta - \phi)}{m[(\Omega^2 - \omega^2)^2 + 4b^2\Omega^2\omega^2]^{1/2}}.$$

Usually we are interested in the behaviour of the system after a considerable time has elapsed, in which case the transient becomes negligible and only the forced oscillation persists. The forced oscillation contains certain important general characteristics. Its frequency is the same as that of the applied force but it has suffered a phase shift ϕ. The amplitude K_0 of the oscillation is independent of the initial conditions:

$$K_0 = F_0/m\Omega^2 \left[\left(1 - \frac{\omega^2}{\Omega^2} \right)^2 + \frac{4b^2\omega^2}{\Omega^2} \right]^{1/2}. \tag{10}$$

For any given system the amplitude of the forced oscillation depends on Ω, the undamped natural angular frequency, with b as an additional parameter. The characteristic shape of K_0 is determined by the behaviour of the denominator in (10). Putting $\xi^2 = \omega^2/\Omega^2$, we see that the turning values of

$$u = (1 - \xi^2)^2 + 4b^2\xi^2$$

occur where $du/d\xi = 0$, that is where

$$-4\xi(1 - \xi^2) + 8b^2\xi = 0$$

which has solutions $\xi = 0$ and $\xi^2 = 1 - 2b^2$. The second derivative is given by

$$\frac{d^2u}{d\xi^2} = -4 + 12\xi^2 + 8b^2$$

and at $\xi = 0$, $d^2u/d\xi^2 = 4(2b^2 - 1)$ and at $\xi^2 = 1 - 2b^2$, $d^2u/d\xi^2 = 8(1 - 2b^2)$.

The sign of the second derivative depends on the sign of $1 - 2b^2$: $b = 1/\sqrt{2}$ is a critical value. If $2b^2 > 1$, u has one extreme value—a minimum—at $\xi = 0$. There are two extreme values if $2b^2 < 1$, a maximum at $\xi = 0$ and a minimum at $\xi = (1 - 2b^2)^{1/2}$.

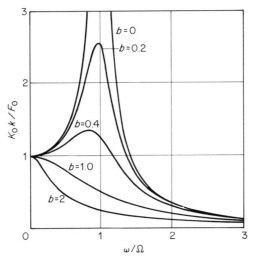

Figure 7.5 Amplitude plotted against frequency ratio of a forced oscillation for selected values of b.

In terms of the amplitude, K_0 has a maximum at $\omega = 0$ if $2b^2 > 1$, and a maximum at $\omega = \Omega(1 - 2b^2)^{1/2}$ and a minimum at $\omega = 0$ if $2b^2 < 1$. The amplitude variation for several values of b is plotted in Figure 7.5 in terms of the angular frequency of the forcing term. The pronounced peak which occurs in the neighbourhood of $\omega = \Omega$ for small values of b displays the phenomenon of *resonance*—large amplitude oscillations which occur when the frequency of the applied force is near the undamped natural frequency of the system. It is these resonant oscillations which can be destructive in mechanisms. This resonant behaviour can always be eliminated by adding a larger damping term, which is obtained by increasing the value of b. For example with $b = 1$ the *response curve* in Figure 7.5 shows no magnification of the amplitude. At the other extreme the amplitude becomes infinite at $\omega = \Omega$ if no damping is present.

Associated with this damped forced system there are three important frequencies:

(i) the undamped natural frequency $\Omega/2\pi$

(ii) the damped frequency $\Omega(1 - b^2)^{1/2}/2\pi$

(iii) the resonant frequency $\Omega(1 - 2b^2)^{1/2}/2\pi$.

The second and third of these frequencies approach the first as $b \to 0$. The angle ϕ is given by the solution of the pair of equations

$$\cos\phi = (\Omega^2 - \omega^2)/[(\Omega^2 - \omega^2)^2 + 4b^2\Omega^2\omega^2]^{1/2}$$
$$\sin\phi = 2b\Omega\omega/[(\Omega^2 - \omega^2)^2 + 4b^2\Omega^2\omega^2]^{1/2}.$$

As $\omega \to \Omega$, $\cos \phi \to 0$ and $\sin \phi \to 1$ which together imply that $\phi \to \frac{1}{2}\pi$ for the resonating system. For low frequency forcing $(\omega \ll \Omega) \sin \phi$ is small and the output is approximately in phase with the applied input. On the other hand for high frequency forcing $(\omega \gg \Omega)$, $\cos \phi$ is negative and $\sin \phi \to 0$ as ω becomes large. Thus ϕ approaches π for the high frequency case, from which we infer that the output is almost exactly out of phase with the input.

Figure 7.6 displays the input oscillation and two output oscillations for $m = 1$, $F_0 = 1$ and $\omega = 1$ (these particular values can always be achieved by suitable scaling of x and t) with $\Omega = 0.5 < 1$ and $\Omega = 1.5 > 1$. The initial transience quickly dies out after a few cycles and both outputs show periodic behaviour: the phase difference for $\Omega = 0.5$ is clearly visible.

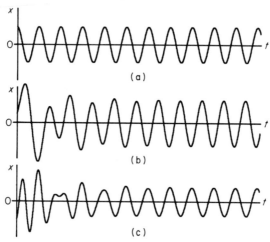

Figure 7.6   (a) shows the input $\cos t$; (b) is the output for $\Omega = 0.5$ and initial conditions $x(0) = 0.5$, $\dot{x}(0) = 0.5$; (c) is the output for $\Omega = 1.5$ and initial conditions $x(0) = -1$, $\dot{x}(0) = 0$. Both outputs approach oscillations of period 2π but are directly out of phase.

Example 6   The mounting of an electric motor of mass 500 kg consists of a spring of stiffness 1.2×10^5 N/m and a dashpot. The motor may take speeds ranging from 0 to 5000 r.p.m. Assuming that the mounting experiences a sinusoidal force having the same frequency as the rotation, what characteristics should the dashpot have in order that the motor should not oscillate with large amplitude?

With x measured upwards, the equation of motion for the system is given by Equation (4),

$$m\ddot{x} + c\dot{x} + kx = F(t)$$

where m is the mass of the motor, c the damping coefficient, k the stiffness of the spring and $F(t) = -mg + F_0 \cos(\omega t + \beta)$, the first term representing the weight of the motor and the second the applied force. The weight merely causes the spring to be depressed and we need not consider it, nor need we consider the transient effects. Let $x = x' - mg/k$ so that x' satisfies

$$m\ddot{x}' + c\dot{x}' + kx' = F_0 \cos{(\omega t + \beta)}.$$

The natural undamped angular frequency

$$\Omega = (k/m)^{1/2} = (120\,000/500)^{1/2} = 4\sqrt{15} \text{ rad/s}.$$

The angular frequency of the applied force ω has a maximum value of

$$5000/60 = 250/3 \text{ rad/s}.$$

Clearly the maximum ω exceeds Ω and consequently resonance could occur if the damping is sufficiently weak. Reference to Figure 7.5 indicates that resonance effects do not appear if we choose $b > 1$ (say). The relative amplitude of the forced oscillations then decreases with increasing ω. Since $b = c/2(mk)^{1/2}$ we conclude that c should exceed $2(500 \times 120000)^{1/2} = 15\,500$ kg/s.

7.4 FORCES TRANSMITTED BY ROTATING MACHINERY

The reduction of vibrations transmitted by rotating machinery is important in engineering. For example, the motor-car engine is usually supported in the frame of a car by rubber mountings which may be thought of as a spring and dashpot system. Their purpose is to insulate the car from uncomfortable vibrations caused by the engine. The system is essentially that shown in Figure 7.2, but with the system turned on its side. The equation of motion is therefore Equation (4) essentially

$$m\ddot{x} + c\dot{x} + kx = -mg + F(t) \tag{11}$$

the weight of the engine now entering the equation on the right-hand side since the motion is taking place vertically. However we can eliminate the weight by the substitution $x = x' - mg/k$ so that x' satisfies

$$m\ddot{x}' + c\dot{x}' + kx' = F(t).$$

We have shown in the previous section that an applied force

$$F(t) = F_0 \cos{(\omega t + \beta)}$$

leads to a displacement $x' = K_0 \cos{(\omega t + \beta - \phi)}$ where K_0 is given by Equation (10). The force on the car or support of the system is (apart from the weight which produces a constant force) ultimately

$$c\dot{x}' + kx' = K_0[k \cos{(\omega t + \beta - \phi)} - c\omega \sin{(\omega t + \beta - \phi)}].$$

The amplitude or maximum value of this force is

$$K_0(k^2 + c^2\omega^2)^{1/2} = K_0 k(1 + 4b^2\omega^2/\Omega^2)^{1/2}$$

since $c = 2(km)^{1/2}$, $b = 2kb/\Omega$. In practice we want to make this quantity small compared with the amplitude of the applied force. It is convenient to overturn the ratio transmitted-force/applied-force so that

$$\frac{F_0}{K_0 k(1 + 4b^2\omega^2/\Omega^2)^{1/2}} = \left(1 + \frac{\omega^2(\omega^2 - 2\Omega^2)}{\Omega^2(\Omega^2 + 4b^2\omega^2)}\right)^{1/2}$$

must be made as large as possible. If $\omega^2 < 2\Omega^2$ the ratio is less than 1 and the transmitted force is greater than the applied force irrespective of the damping. A softer spring will reduce Ω but the critical value $\omega = \sqrt{2}\Omega$ must always occur. If $\omega^2 > 2\Omega^2$, the ratio is always greater than 1 and takes its maximum value for any fixed ω and Ω when $b = 0$, in which case no damping is present. However we must remember that serious resonance of the engine may occur if the damping is too weak and ω is near Ω. The two factors must be balanced in designing the system. A soft spring with modest damping is usually the best combination.

For the motor-car engine ω would be the same as the angular speed of the rotating parts of the engine. The slowest speed of the engine is usually its idling speed (ω_i say) and we would choose $\omega_i^2 > 2\Omega^2$. This explains in part why an engine which is misfiring causes vibrations of large amplitude in the car.

The applied force due to rotating machinery may be periodic but not simple harmonic in form. If the angular speed is ω the period of the applied force can be taken as $2\pi/\omega$ in which case for a periodic force

$$F(t) = F(t + 2\pi/\omega).$$

The equation for x is still given by Equation (11). Introducing the relative displacement $x = x' - mg/k$ again, we have

$$m\ddot{x}' + c\dot{x}' + kx' = F(t).$$

We now write the periodic force $F(t)$ as the sum of cosines and sines:

$$F(t) = \tfrac{1}{2}a_0 + \sum_{n=1}^{\infty}(a_n \cos n\omega t + b_n \sin n\omega t) \tag{12}$$

and determine the coefficients by multiplying both sides successively by 1, $\cos n\omega t$, $\sin n\omega t$ and integrating over the interval $(0, 2\pi/\omega)$. Since (as the reader may easily verify)

$$\int_0^{2\pi/\omega} \cos n\omega t\, dt = \int_0^{2\pi/\omega} \sin \omega t\, dt = 0$$

$$\int_0^{2\pi/\omega} \cos n\omega t \cos r\omega t\, dt = \int_0^{2\pi/\omega} \sin n\omega t \sin r\omega t\, dt = 0 \quad (r \neq n)$$

$$\int_0^{2\pi/\omega} \cos n\omega t \sin r\omega t\, dt = 0$$

$$\int_0^{2\pi/\omega} \cos^2 n\omega t\, dt = \int_0^{2\pi/\omega} \sin^2 n\omega t\, dt = \frac{\pi}{\omega}$$

it follows that

$$a_0 = \frac{\omega}{\pi} \int_0^{2\pi/\omega} F(t)\, dt, \qquad a_n = \frac{\omega}{\pi} \int_0^{2\pi/\omega} F(t) \cos n\omega t\, dt$$

$$b_n = \frac{\omega}{\pi} \int_0^{2\pi/\omega} F(t) \sin n\omega t\, dt. \tag{13}$$

The right-hand side of Equation (12) is known as the *Fourier series* of $F(t)$ and a_n and b_n are called the *Fourier coefficients*.

The significance of this expansion is that it expresses the applied force as an infinite series of harmonic terms each of which we have already discussed separately. The constant term $\frac{1}{2}a_0$ leads to a further constant displacement of x' and is of little interest. We have to solve essentially the differential equations

$$m\ddot{x}' + c\dot{x}' + kx' = a_n \cos n\omega t$$
$$m\ddot{x}' + c\dot{x}' + kx' = b_n \sin n\omega t = b_n \cos (n\omega t - \tfrac{1}{2}\pi).$$

We leave it as an exercise for the reader to show that the relative displacement

$$x' = \frac{a_0}{2k} + \sum_{n=1}^{\infty} Q_n[a_n \cos (n\omega t - \phi_n) + b_n(\sin n\omega t - \phi_n)] \tag{14}$$

where

$$Q_n = 1/m\Omega^2 \left[\left(1 - \frac{n^2\omega^2}{\Omega^2} \right)^2 + \frac{4b^2 n^2 \omega^2}{\Omega^2} \right]^{1/2} \tag{15}$$

and ϕ_n is given by

$$\cos \phi_n = (\Omega^2 - n^2\omega^2)mQ_n, \quad \sin \phi_n = 2b\Omega\omega nmQ_n. \tag{16}$$

We can see that resonance will occur for weak damping ($b \ll 1$) if ω is in the neighbourhood of Ω/n for $n = 1, 2, \ldots$, since the corresponding amplitude Q_n will become relatively large. However this is usually only significant for the first few frequencies Ω/n because the second term in the denominator of Q_n becomes rapidly large as n increases.

Example 7 A *linear damped system is subject to the applied force*

$$F(t) = \begin{cases} \dfrac{F_0 \omega t}{\pi}, & 0 \le t < \dfrac{\pi}{\omega} \\[2ex] -\dfrac{F_0 \omega}{\pi}\left(t - \dfrac{2\pi}{\omega} \right), & \dfrac{\pi}{\omega} \le t < \dfrac{2\pi}{\omega} \end{cases}$$

with $F(t) = F(t + 2\pi/\omega)$. *Obtain the sustained response of the system.*

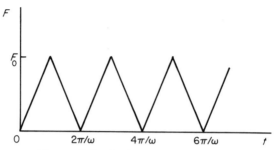

Figure 7.7   A triangular wave input.

The applied force is the triangular wave shown in Figure 7.7. The equation of motion is

$$m\ddot{x} + c\dot{x} + kx = F(t).$$

We express $F(t)$ as a Fourier series whose coefficients are given by Equations (13):

$$a_0 = \frac{2\omega}{\pi} \int_{0}^{\pi/\omega} \frac{F_0 \omega t}{\pi}\, dt = F_0$$

$$a_n = \frac{2\omega}{\pi} \int_{0}^{\pi/\omega} \frac{F_0 \omega}{\pi} t \cos n\omega t\, dt$$

$$= \frac{2F_0 \omega^2}{\pi^2} \left[\left(\frac{1}{n\omega} t \sin n\omega t \right)_{0}^{\pi/\omega} - \frac{1}{n\omega} \int_{0}^{\pi/\omega} \sin n\omega t\, dt \right]$$

$$= \frac{2F_0}{n^2 \pi^2} [\cos n\omega t]_{0}^{\pi/\omega} = \frac{2F_0}{n^2 \pi^2} (\cos n\pi - 1)$$

$$b_n = 0$$

where we have used the symmetry properties of the triangular wave, the sine and the cosine functions. Now $\cos n\pi = (-1)^n$, so that $a_n = 2F_0[(-1)^n - 1]/n^2\pi^2$ which implies that $a_{2n} = 0$ and that $a_{2n-1} = -4F_0/(2n-1)^2\pi^2$ for $n = 1, 2, \ldots$. Thus

$$F(t) = \tfrac{1}{2} F_0 - \frac{4F_0}{\pi^2} \sum_{n=1}^{\infty} \frac{\cos(2n-1)\omega t}{(2n-1)^2}.$$

By Equation (14), the response of the system, apart from the transient, is

$$x' = \frac{F_0}{2k} - \frac{4F_0}{\pi^2} \sum_{n=1}^{\infty} \frac{Q_{2n-1}}{(2n-1)^2} \cos[(2n-1)\omega t - \phi_{2n-1}]$$

where Q_{2n-1} and ϕ_{2n-1} are given by Equations (15) and (16).

7.5   THE SEISMOGRAPH

Disturbances within the Earth such as those created by an earthquake are measured and recorded by an instrument known as a *seismograph*. The simple

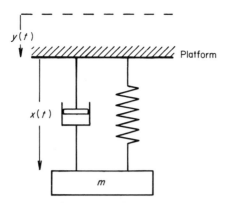

Figure 7.8   A model for a seismograph.

seismograph may be thought of as a mass suspended by a spring and dashpot from a platform fixed to the Earth. Oscillations are transmitted from the platform to the mass. A pointer attached to the mass records its displacement on a scale attached to the platform. Note that the *relative* displacement of the mass is observed (see Figure 7.8).

Let the displacement of the platform be y and let x be the displacement of the mass relative to the platform. The plunger in the dashpot is moving with speed $\dot{y}$ and the pot is descending with speed $(\dot{x} + \dot{y})$. The descent of the mass is therefore resisted by a force $c\dot{x}$. If a is the natural length of the spring, the equation of motion becomes

$$mg - c\dot{x} - k(x - a) = m(\ddot{x} + \ddot{y})$$

or

$$m\ddot{x} + c\dot{x} + kx = mg + ka - m\ddot{y}.$$

Putting $x = x' + mg/k + a$, where x' is the displacement of the mass from its equilibrium position, we find that

$$m\ddot{x}' + c\dot{x}' + kx' = -m\ddot{y}. \tag{17}$$

Essentially x' is the recorded displacement.

Let us suppose that the platform oscillates with simple harmonic motion of amplitude A and angular frequency ω. If we write $y = A \cos(\omega t + \beta)$, Equation (17) reads

$$m\ddot{x}' + c\dot{x}' + kx' = mA\omega^2 \cos(\omega t + \beta)$$

an equation which has already been considered in Section 7.3. The forced oscillation induced in the mass is

$$x' = K_0 \cos(\omega t + \beta - \phi)$$

where

$$K_0 = A\left(\frac{\omega^2}{\Omega^2}\right)\left[\left(1 - \frac{\omega^2}{\Omega^2}\right)^2 + \frac{4b^2\omega^2}{\Omega^2}\right]^{-1/2}$$

$$\cos\phi = K_0(\Omega^2 - \omega^2)/(A\omega^2), \quad \sin\phi = 2b\Omega K_0/(\omega A)$$

The known characteristics of the spring and dashpot enable the observer, in principle, to obtain the frequency and amplitude A of the oscillations of the platform in terms of the recorded value of K_0 and ϕ.

As $\omega/\Omega \to \infty$, $K_0 \to A$, that is K_0 approaches the amplitude of the oscillation of the platform. Consequently for high frequency oscillations the mass remains fixed in space since the phase difference between the oscillations of the platform and the mass approaches π. Let us see how we can achieve the best approximation for modest values of Ω/ω. We can write the amplitude

$$K_0 = A\left[\left(1 - \frac{\Omega^2}{\omega^2}\right)^2 + \frac{4b^2\Omega^2}{\omega^2}\right]^{-1/2} = A\left(1 + 2(2b^2 - 1)\frac{\Omega^2}{\omega^2} + \frac{\Omega^4}{\omega^4}\right)^{-1/2} \quad (18)$$

which suggests the error between K_0 and A will be small for modest Ω/ω if we choose $b^2 = \frac{1}{2}$ or $b \approx 0.7$. For this value of b, the middle term on the right-hand side of (18) disappears which implies that $K_0 - A$ is of the order of the fourth power in Ω/ω. Thus

$$0 < A - K_0 = A\left[1 - \left(1 + \frac{\Omega^4}{\omega^4}\right)^{-1/2}\right]$$

and K_0 is within, say, 5% of the actual amplitude if

$$1 - \left(1 + \frac{\Omega^4}{\omega^4}\right)^{-1/2} < 1/20$$

or

$$\Omega < \frac{(39)^{1/4}}{(19)^{1/2}}\omega = 0.57\omega.$$

In designing the seismograph we wish firstly to make ω/Ω large enough to ensure that the approximation above is appropriate. We thus make $\Omega(=(k/m)^{1/2})$ as small as possible by supporting a large block of concrete (m large) on soft springs (k small). We then adjust the damping so that $b(= c/(2mk)^{1/2}) \approx 0.7$.

If we had designed an instrument so that ω/Ω is small, by using a small mass on a stiff spring, the amplitude of the forced oscillation is approximately $A\omega^2/\Omega^2$. Now Ω^2 is known and $A\omega^2$ is the amplitude of $\ddot{y}$, the acceleration of the platform. Such an instrument is called an accelerometer (see Section 2.4).

7.6 MULTIPLE SYSTEMS AND NORMAL MODES

So far in this chapter we have only investigated the responses of systems with one degree of freedom: one coordinate has been sufficient to represent every configuration. Many models of real devices have multiple degrees of freedom and the system may oscillate in all or some of the representative coordinates. We shall now examine some simple multiple systems for which the linear theory of oscillations seems appropriate and investigate the composition of their oscillatory responses. We shall proceed mainly by example.

Consider two particles A and B each of mass m connected by a spring of stiffness k and natural length a, with each particle linked by two further springs having the same characteristics as the first one to two fixed supports C and D distance $3b$ apart as shown in Figure 7.9(a). The springs are arranged in a smooth tube so that only longitudinal movements of the particles can take place. At time t let the displacements of the particles A and B be x and y from their equilibrium positions, and let T_1, T_2 and T_3 be the forces in the springs as shown in Figure 7.9(b). From Hooke's law

$$T_1 = k(b_1 + x - a), \quad T_2 = k(b_2 + y - x - a), \quad T_3 = k(b_3 - y - a)$$

where b_1, b_2 and b_3 are the equilibrium lengths of the springs ($b_1 + b_2 + b_3 = 3b$). In equilibrium $x = y = 0$ and the tensions T_1, T_2 and T_3 must be equal, so that $b_1 = b_2 = b_3 = b$ as we might expect. Using Figure 7.9(b), we can infer that the equations of motion for A and B are

$$T_2 - T_1 = m\ddot{x}, \quad T_3 - T_2 = m\ddot{y}$$

so that

$$-2kx + ky = m\ddot{x}, \qquad kx - 2ky = m\ddot{y} \tag{19}$$

which are simultaneous linear differential equations for the displacements x and y.

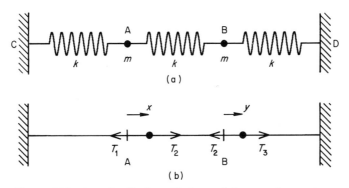

Figure 7.9 Longitudinal oscillations of three-spring system.

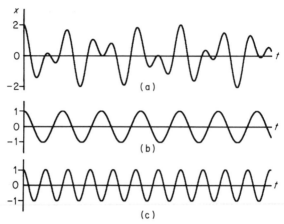

Figure 7.10   (a) shows a solution of (19) for $m = k$ and initial conditions $\dot{x}(0) = 2$, $y(0) = 0$, $\dot{x}(0) = 0$, $\dot{y}(0) = 0$. The solution in (a) is the sum of the solutions in (b) and (c) which have periods 2π and $2\pi/\sqrt{3}$ respectively. They can be isolated by the initial conditions referred to in the text.

Figure 7.10(a) shows the solution for x plotted against time t for the initial data $x(0) = 2$, $y(0) = 0$, $\dot{x}(0) = 0$, $\dot{y}(0) = 0$. The output at first sight appears random but our analysis will show that the solution is the sum of the two sinusoidal functions displayed in Figures 7.10(b) and 7.10(c). Practically, a power spectrum (a technique used particularly in electronics of circuits) of the output would reveal the two frequencies. These components of the output obviously have different frequencies, and our first task is to discover these frequencies. We achieve this by searching for solutions of the form $x = A \cos(\omega t + \varepsilon)$ and $y = B \cos(\omega t + \varepsilon)$ in which both have the same frequency but different amplitudes. Substitute for x and y in (19): the result is the pair of equations

$$(2k - \omega^2 m)A - kB = 0 \tag{20}$$

$$-kA + (2k - \omega^2 m)B = 0 \tag{21}$$

since the cosine cannot be zero for all t. These linear equations have non-trivial solutions (that is, solutions other than $A = B = 0$) if and only if

$$\begin{vmatrix} 2k - \omega^2 m & -k \\ -k & 2k - \omega^2 m \end{vmatrix} = 0 \tag{22}$$

or

$$(2k - \omega^2 m)^2 - k^2 = 0.$$

Thus $2k - \omega^2 m = \pm k$, and the roots are

$$\omega = \omega_1 = (k/m)^{1/2}, \qquad \omega = \omega_2 = (3k/m)^{1/2}.$$

These are known as the *eigenfrequencies* of the system, and ω_1 and ω_2 are the frequencies of the outputs in Figure 7.10(b) and 7.10(c), respectively.

We can now substitute each eigenfrequency back into (20) or (21) and find the non-trivial solutions. If $\omega = \omega_1$, let $A = A_1$ and $B = B_1$ so that (20) implies $B_1 = A_1$. Hence there exist solutions of the form

$$x = A_1 \cos (\omega_1 t + \varepsilon_1), \qquad y = A_1 \cos (\omega_1 t + \varepsilon_1) \tag{23}$$

where we have assigned the phase $\varepsilon = \varepsilon_1$ in this solution. Similarly, if $A = A_2$ and $B = B_2$ for the second frequency, then (20) implies that $B_2 = -A_2$. Thus there also exist solutions

$$x = A_2 \cos (\omega_2 t + \varepsilon_2), \qquad y = -A_2 \cos (\omega_2 t + \varepsilon_2) \tag{24}$$

where we may use a different phase ε_2, say.

Both differential equations in (19) are linear so that the sums of the solutions for x and for y in (23) and (24) are also themselves solutions. Finally, we arrive at the *general* solutions of (19)

$$x = A_1 \cos (\omega_1 t + \varepsilon_1) + A_2 \cos (\omega_2 t + \varepsilon_2) \tag{25}$$

$$y = A_1 \cos (\omega_1 t + \varepsilon_1) - A_2 \cos (\omega_2 t + \varepsilon_2). \tag{26}$$

These solutions contain four unknown constants, A_1, A_2, ε_1 and ε_2, which require four initial values of x, y, $\dot{x}$ and $\dot{y}$. In terms of the model problem this requires the specification of the initial displacement and velocity of both particles.

Each solution in (25) and (26) is a linear combination of two SHMs. In Figures 7.10(b) and 7.10(c) these solutions can be obtained from the two sets of initial data

(i) $x(0) = 1$, $y(0) = 1$, $\dot{x}(0) = 0$, $\dot{y}(0) = 0$ (Figure 7.10(b))
(ii) $x(0) = 1$, $y(0)z - 1$, $\dot{x}(0) = 0$, $\dot{y}(0) = 0$ (Figure 7.10(c)).

In (i) the conditions imply

$$1 = A_1 \cos \varepsilon_1 + A_2 \cos \varepsilon_2, \quad 1 = A_1 \cos \varepsilon_1 - A_2 \cos \varepsilon_2$$

$$0 = -A_1 \left(\frac{k}{m}\right)^{1/2} \sin \varepsilon_1 - A_2 \left(\frac{3k}{m}\right)^{1/2} \sin \varepsilon_2$$

$$0 = -A_1 \left(\frac{k}{m}\right)^{1/2} \sin \varepsilon_1 + A_2 \left(\frac{3k}{m}\right)^{1/2} \sin \varepsilon_2.$$

Since A_1 and A_2 cannot both be zero, try $\sin \varepsilon_1 = \sin \varepsilon_2 = 0$ and choose $\varepsilon_1 = \varepsilon_2 = 0$ (phase angles are always modulo 2π). It follows that $A_1 = 1$ and $A_2 = 0$. Hence this response is given by

$$x = y = \cos \omega_1 t. \tag{27}$$

A similar argument yields the second solution using the initial data in (ii), namely

$$x = -y = \cos \omega_2 t. \tag{28}$$

These particular solution pairs (27) and (28) in which the coordinates vibrate with the same frequency are examples of the *normal modes* of the system. In the first normal mode x and y must oscillate with the single frequency ω_1, which is equivalent to letting $A_2 = 0$ in (25) and (26), and in the second with the frequency ω_2 which can be achieved by letting $A_1 = 0$ in (25) and (26). Often the modes show some geometrical characteristics when activated. In the ω_1 mode, $x = y$ from (27) and the particles both move in phase: in the ω_2 mode $x = -y$ so that the particles oscillate directly out of phase.

7.7 MATRIX METHODS

For systems with more degrees of freedom than the one in the previous section we really require a systematic procedure which shows the structure of the solutions more clearly. Let us reconsider the three spring model and express the equations of motion in matrix form. Introduce the vector and matrices

$$\mathbf{x} = \begin{bmatrix} x \\ y \end{bmatrix}, \quad \mathbf{M} = \begin{bmatrix} m & 0 \\ 0 & m \end{bmatrix}, \quad \mathbf{K} = \begin{bmatrix} 2k & -k \\ -k & 2k \end{bmatrix}.$$

It follows that the derivatives of $\mathbf{x}$ are given by

$$\dot{\mathbf{x}} = \begin{bmatrix} \dot{x} \\ \dot{y} \end{bmatrix}, \quad \ddot{\mathbf{x}} = \begin{bmatrix} \ddot{x} \\ \ddot{y} \end{bmatrix}.$$

The two differential equations in (19) can be combined into the single matrix equation

$$\mathbf{M}\ddot{\mathbf{x}} + \mathbf{K}\mathbf{x} = 0. \tag{29}$$

To solve (29) we now let $\mathbf{x} = \mathbf{p} \cos(\omega t + \varepsilon)$ where $\mathbf{p}$ is a constant column vector. Substitute $\mathbf{x}$ into (29) with the result that

$$-\mathbf{M}\mathbf{p}\omega^2 \cos(\omega t + \varepsilon) + \mathbf{K}\mathbf{p} \cos(\omega t + \varepsilon) = \mathbf{0}$$

or

$$(\mathbf{K} - \omega^2\mathbf{M})\mathbf{p} \cos(\omega t + \varepsilon) = \mathbf{0}.$$

Since $\cos(\omega t + \varepsilon)$ cannot vanish except at isolated values of t, we conclude that

$$(\mathbf{K} - \omega^2\mathbf{M})\mathbf{p} = \mathbf{0}. \tag{30}$$

This is a matrix representation of a homogeneous set of linear equations for the components of $\mathbf{p}$. It will have non-trivial solutions if and only if the determinant of $\mathbf{K} - \omega^2\mathbf{M}$ vanishes, that is

$$\det (\mathbf{K} - \omega^2\mathbf{M}) = 0 \tag{31}$$

which is the same equation as (22). As before we solve this equation for the eigenfrequencies ω_1 and ω_2 (we always choose the positive roots). The *eigenvectors* $\mathbf{p}_1$ and $\mathbf{p}_2$ are obtained by solving (30) for each eigenfrequency ω_1 and ω_2. Each eigenvector contains one arbitrary constant, and we can also associate different phases with each frequency. Finally the linearity of Equation (30) means that the general solution can be written

$$\mathbf{x} = \mathbf{p}_1 \cos (\omega_1 t + \varepsilon_1) + \mathbf{p}_2 \cos (\omega_2 t + \varepsilon_2). \tag{32}$$

As before the two constants in $\mathbf{p}_1$ and $\mathbf{p}_2$, and the constants ε_1 and ε_2 will be determined by the initial data.

The equation set up for the specific example of the three springs is in a form which suggests immediate generalisation. The procedure can be applied to any system for which the equation of motion can be expressed as or approximated by (29), but we need not be restricted to systems with two degrees of freedom. In (29) $\mathbf{x}$ could be a column vector with $\mathbf{M}$ and $\mathbf{K}$ $n \times n$ matrices ($\mathbf{M}$ need not be diagonal). It should be noted that the matrix equation has an affinity with the one-dimensional equation for SHM. The n eigenfrequencies are then given by the positive solutions of the determinant Equation (31). If any solution for ω^2 turns out to be negative then the system is not stable and not all modes are oscillatory. There can arise a complication at this stage; there may occur repeated eigenfrequencies. We shall not go into this diversion in general although a repeated eigenfrequency can be found in Example 9. Suppose that the eigenfrequencies are the distinct and real positive numbers $\omega_1, \omega_2, \ldots, \omega_n$. We can then find the corresponding eigenvectors by solving (30) for each eigenvector. The general solution becomes

$$\mathbf{x} = \sum_{k=1}^{n} \mathbf{p}_k \cos (\omega_k t + \varepsilon_k) \tag{33}$$

whilst the n normal modes can be represented by the particular vibrations

$$\mathbf{x}_k = \mathbf{p}_k \cos (\omega_k t + \varepsilon_k), \qquad k = 1, 2, \ldots, n.$$

Example 8   Find the eigenfrequencies, eigenvectors and general solution for longitudinal vibrations of the four-spring system shown in Figure 7.11. Each particle has the same mass m and each spring has stiffness k and natural length a.

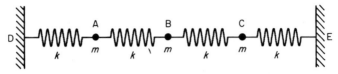

Figure 7.11   Longitudinal oscillations of a four-spring system.

This is the four-spring generalisation of the problem discussed at the beginning of Section 7.6. Let the tensions in the springs be T_1, T_2, T_3 and T_4 reading from the left, and let the displacements of A, B and C at time t be x_1, x_2 and x_3, the positive sense being from left to right. In equilibrium DA = AB = BC = CE = b, say. The equations of motion for A, B and C become

$$T_2 - T_1 = m\ddot{x}_1, \qquad T_3 - T_2 = m\ddot{x}_2, \qquad T_4 - T_3 = m\ddot{x}_3$$

where the tensions, assuming Hooke's law, are given by

$$T_1 = k(b + x_1 - a), \qquad\qquad T_2 = k(b + x_2 - x_1 - a)$$
$$T_3 = k(b + x_3 - x_2 - a), \qquad T_4 = k(b - x_3 - a).$$

Thus the equations of motion are given by

$$m\ddot{x}_1 = -2kx_1 + kx_2$$
$$m\ddot{x}_2 = kx_1 - 2kx_2 + kx_3 \qquad \text{or} \quad \mathbf{M\ddot{x} + Kx = 0}$$
$$m\ddot{x}_3 = kx_2 - 2kx_3$$

where

$$\mathbf{x} = \begin{bmatrix} x_1 \\ x_2 \\ x_3 \end{bmatrix}, \qquad \mathbf{M} = \begin{bmatrix} m & 0 & 0 \\ 0 & m & 0 \\ 0 & 0 & m \end{bmatrix}, \qquad \mathbf{K} = \begin{bmatrix} 2k & -k & 0 \\ -k & 2k & -k \\ 0 & -k & 2k \end{bmatrix}.$$

The frequency determinant equation (31) can now be constructed:

$$\begin{vmatrix} 2k - m\omega^2 & -k & 0 \\ -k & 2k - \omega^2 m & -k \\ 0 & -k & 2k - \omega^2 m \end{vmatrix} = 0.$$

Expansion by the top row implies that

$$(2k - \omega^2 m)[(2k - \omega^2 m)^2 - k^2] - k^2(2k - \omega^2 m) = 0$$

or

$$(2k - \omega^2 m)[(2k - \omega^2 m)^2 - 2k^2] = 0.$$

Hence the three eigenfrequencies can be labelled

$$\omega_1 = (m/2k)^{1/2}, \qquad \omega_2 = [(2 + \sqrt{2})k/m]^{1/2}, \qquad \omega_3 = [(2 - \sqrt{2})k/m]^{1/2}.$$

Let the corresponding eigenvectors be

$$\mathbf{p}_i = \begin{bmatrix} A_i \\ B_i \\ C_i \end{bmatrix} \qquad (i = 1, 2, 3)$$

Then $\mathbf{p}_1$ must satisfy

$$[\mathbf{K} - \omega_1^2 \mathbf{M}]\mathbf{p}_1 = \mathbf{0} \quad \text{or} \quad \begin{bmatrix} 0 & -k & 0 \\ -k & 0 & -k \\ 0 & -k & 0 \end{bmatrix} \begin{bmatrix} A_1 \\ B_1 \\ C_1 \end{bmatrix} = \begin{bmatrix} 0 \\ 0 \\ 0 \end{bmatrix},$$

which is equivalent to the three linear equations

$$-kB_1 = 0, \qquad -kA_1 - kC_1 = 0, \qquad -kB_1 = 0.$$

Thus $B_1 = 0$ and $A_1 = C_1 = -\alpha_1$, say. Hence

$$\mathbf{p}_1 = \begin{bmatrix} \alpha_1 \\ 0 \\ -\alpha_1 \end{bmatrix}.$$

Repeat the method and confirm that

$$\mathbf{p}_2 = \begin{bmatrix} \alpha_2 \\ -\sqrt{2}\alpha_2 \\ \alpha_2 \end{bmatrix}, \qquad \mathbf{p}_3 = \begin{bmatrix} \alpha_3 \\ \sqrt{2}\alpha_3 \\ \alpha_3 \end{bmatrix}.$$

When the individual phase angles are included, the general solution can be expressed in the form

$$\mathbf{x} = \begin{bmatrix} 1 \\ 0 \\ -1 \end{bmatrix} \alpha_1 \cos(\omega_1 t + \varepsilon_1) + \begin{bmatrix} 1 \\ -\sqrt{2} \\ 1 \end{bmatrix} \alpha_2 \cos(\omega_2 t + \varepsilon_2) + \begin{bmatrix} 1 \\ \sqrt{2} \\ 1 \end{bmatrix} \alpha_3 \cos(\omega_3 t + \varepsilon_3).$$

$$(34)$$

The normal modes can be discovered by assigning zeros to all pairs of $\alpha_1, \alpha_2, \alpha_3$. Thus in the first mode $\alpha_2 = \alpha_3 = 0$, $\alpha_1 \neq 0$. Hence, from (34) $x_1 = -x_3$, $x_2 = 0$: in this mode centre particle remains stationary and the particles on either side oscillate with the same amplitude but directly out of phase. The reader should investigate the behaviour of the system in the other modes.

Example 9 *For a certain system with three degrees of freedom the equation of motion is*

$$\mathbf{M\ddot{x}} + \mathbf{Kx} = \mathbf{0}$$

where $\mathbf{M} = \mathbf{I}_3$ *(the 3 × 3 identity) and*

$$\mathbf{K} = \begin{bmatrix} 1 & 0 & 0 \\ 0 & 2 & 1 \\ 0 & 1 & 2 \end{bmatrix}.$$

Find the eigenfrequencies and eigenvectors of the system.

The frequency determinant equation $|\mathbf{K} - \omega^2 \mathbf{M}| = 0$ implies that

$$\begin{vmatrix} 1 - \omega^2 & 0 & 0 \\ 0 & 2 - \omega^2 & 1 \\ 0 & 1 & 2 - \omega^2 \end{vmatrix} = 0.$$

Hence

$$(1 - \omega^2)[(2 - \omega^2)^2 - 1] = 0 \quad \text{or} \quad (1 - \omega^2)^2(3 - \omega^2) = 0.$$

Thus the system has a *repeated* frequency $\omega_1 = 1$ and a second frequency $\omega_2 = \sqrt{3}$. Let the associated eigenvectors be

$$\mathbf{p}_i = \begin{bmatrix} A_i \\ B_i \\ C_i \end{bmatrix}, \quad i = 1, 2.$$

For ω_1, the elements of $\mathbf{p}_1$ satisfy just one equation

$$B_1 + C_1 = 0.$$

Hence we can let $A_1 = \alpha_1$ and $B_1 = -C_1 = \alpha_2$. For ω_2,

$$-2A_2 = 0, \quad -B_2 + C_2 = 0, \cdot B_2 - C_2 = 0$$

from which we infer

$$\mathbf{p}_2 = \begin{bmatrix} 0 \\ \alpha_3 \\ \alpha_3 \end{bmatrix}.$$

Finally the general solution can be written

$$\mathbf{x} = \begin{bmatrix} \alpha_1 \\ \alpha_2 \\ -\alpha_2 \end{bmatrix} \cos (t + \varepsilon_1) + \begin{bmatrix} 0 \\ \alpha_3 \\ \alpha_3 \end{bmatrix} \cos (\sqrt{3}t + \varepsilon_3)$$

in terms of the two frequencies: it still contains six constants to be assigned by the initial data.

The component differential equations for the system show that the equation for x_1 can be uncoupled from the other two, which means that x_2 and x_3 are in equilibrium in one normal mode.

Example 10   The double pendulum shown in Figure 7.12 consists of a string of length a suspended from a fixed point O. A particle of mass m is attached to the free end A together with a further string of length b carrying a second particle of mass M at B. The pendulum oscillates in a fixed vertical plane through O. Find its frequency determinant for oscillations of small amplitude.

The problems involving particles and springs in this and the previous section are exceptional in that the equations of motion are automatically *linear*. In most applications this is not the case: the equations are usually nonlinear but they may be *linearised* in which case this approximation may be justified provided that the amplitudes of the oscillations remain small. The double pendulum is such a case.

For oscillations in a vertical plane the pendulum will have two degrees of freedom (at least as long as the strings remain taut) and every configuration can be specified by the angles θ and ϕ in Figure 7.12. The pendulum is in equilibrium if $\theta = \phi = 0$. By

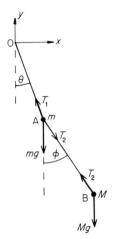

Figure 7.12 The double pendulum.

small amplitude oscillations we mean ones in which θ and ϕ remain small: some measure of just what 'small' means for the simple pendulum was discussed in Section 4.6.

Introduce axes Oxy as shown and let T_1, T_2 be the tensions in the strings OA and AB. Let the coordinates of A and B be (x_A, y_A) and (x_B, y_B) respectively. The horizontal and vertical components of the equations of motion for each particle become

$$T_2 \sin \phi - T_1 \sin \theta = m\ddot{x}_A, \qquad mg + T_2 \cos \phi - T_1 \cos \theta = m\ddot{y}_A \qquad (35)$$

and

$$-T_2 \sin \phi = M\ddot{x}_B, \qquad Mg - T_2 \cos \theta = M\ddot{y}_B. \qquad (36)$$

We now linearise these equations by introducing approximations for x_A, y_A, etc. Now

$$x_A = a \sin \theta \approx a\theta, \qquad y_A = a \cos \theta \approx a$$
$$x_B = a \sin \theta + b \sin \phi \approx a\theta + b\phi, \qquad y_B = a \cos \theta + b \cos \phi \approx a + b$$

using the trigonometric approximations $\sin \theta \approx \theta$, $\cos \theta \approx 1$, etc. Thus Equations (35) and (36) can be approximated by

$$T_2 \phi - T_1 \theta = ma\ddot{\theta}, \qquad mg + T_2 - T_1 = 0$$
$$-T_2 \phi = M(a\ddot{\theta} + b\ddot{\phi}), \qquad Mg - T_2 = 0.$$

Elimination of T_1 and T_2 results in two linear differential equations for θ and ϕ:

$$ma\ddot{\theta} = -(M + m)g\theta + Mg\phi$$
$$M(a\ddot{\theta} + b\ddot{\phi}) = -Mg\phi.$$

In the matrix form (29) these equations can be expressed as

$$\mathbf{M\ddot{\theta} + K\theta = 0}$$

where

$$\boldsymbol{\theta} = \begin{bmatrix} \theta \\ \phi \end{bmatrix} \qquad \mathbf{M} = \begin{bmatrix} ma & 0 \\ Ma & Mb \end{bmatrix} \qquad \mathbf{K} = \begin{bmatrix} (M + m)g & -Mg \\ 0 & Mg \end{bmatrix}.$$

Finally, it follows that the eigenfrequencies of the double pendulum are given by the positive solutions of

$$\begin{vmatrix} (M + m)g - ma\omega^2 & -Mg \\ -Ma\omega^2 & Mg - Mb\omega^2 \end{vmatrix} = 0.$$

Example 10 illustrates the general approach to the linear theory of oscillations as applied to a nonlinear problem. The method may be summarised in the following scheme:

(i) introduce coordinates which define every configuration of the system (sometimes known as *generalised coordinates*) in such a way that they vanish in the equilibrium position being investigated,

(ii) write down the equations of motion for the system,

(iii) linearise the equations in terms of the generalised coordinates and their derivatives, expressing them as second order differential equations,

(iv) construct the corresponding matrix equation, find the constant matrices **M** and **K**, and the frequency determinant equation,

(v) solve the determinant equation for the eigenfrequencies, find the eigenvectors and write down the general solution.

One final remark concerns the potential energy for oscillatory systems. For the pendulum of Example 10, the potential energy can be expressed as

$$\mathscr{V}(\theta, \phi) = mga(1 - \cos \theta) + Mg(a + b - a \cos \theta - b \cos \phi).$$

For small θ and ϕ, the cosine approximation taken to the second term implies that

$$\mathscr{V}(\theta, \phi) \approx \tfrac{1}{2}ag(m + M)\theta^2 + \tfrac{1}{2}Mgb\phi^2$$

which has a *minimum* at $\theta = 0$, $\phi = 0$. In fact, to this approximation, $\mathscr{V}(\theta, \phi)$ is known, in algebra, as a positive definite quadratic form since it is quadratic in θ and ϕ and positive if $(\theta, \phi) \neq (0, 0)$. For a normal mode problem the potential energy must be positive definite in a neighbourhood of the equilibrium state.

Exercises

1. A body of mass 10 kg is supported by a spring of stiffness 200 N/m. The mass is pulled down a distance of 0.5 m from its equilibrium position and released from rest. Find the period, frequency and angular frequency of the oscillations of the body.

2. A particle is held by two springs each of modulus mg and natural length a. The free ends of the springs are attached to a horizontal beam at two points distance $2a\sqrt{3}$ apart. Write down the potential energy of the system when the particle is at a depth x below the mid-point of the spring attachments. Find the equilibrium position of the particle when it is below the beam and the period of small vertical oscillations of the particle about its equilibrium position.

3. A body of mass m is suspended by two springs with stiffness k_1 and k_2. Show that if the springs are in parallel the period of vertical oscillations of the body is $2\pi[m/(k_1 + k_2)]^{1/2}$, and if the springs are in series the period is

$$2\pi[m(k_1 + k_2)/k_1 k_2]^{1/2}.$$

Which arrangement gives the longer period?

4. A bob of mass 5 g suspended by a spring is found to oscillate vertically with period 2 s. Find the stiffness of the spring.

5. Find the frequency of oscillations of the mass in the system shown in Figure 7.13. The pulley has negligible mass and the spring has stiffness k.

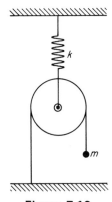

Figure 7.13

6. A particle of mass m moves in the plane $z = 0$ under the attractive force $2m\pi^2 r$ towards the origin, r being the distance from the origin. In addition there is a force of magnitude $m\pi v$ in the direction of $\mathbf{v} \times \mathbf{k}$ where $\mathbf{v}$ is the velocity of the particle and $\mathbf{k}$ is a unit vector perpendicular to $z = 0$. Show that

$$\pi\dot{y} - 2\pi^2 x = \ddot{x} \qquad -\pi\dot{x} - 2\pi^2 y = \ddot{y}.$$

Solve the equations for x and y and show that the motion in the x and y directions is the sum of two simple harmonic oscillations of periods 1 and 2.

7. The displacement x of a spring-mounted mass under the action of Coulomb dry friction satisfies

$$m\ddot{x} + kx = \begin{cases} -F & \dot{x} > 0 \\ F & \dot{x} < 0 \end{cases}$$

where m, k and F are positive constants. The mass is in equilibrium if $\dot{x} = \ddot{x} = 0$ and $|x| < F/k$. If $x = x_0$, $\dot{x} = 0$ at $t = 0$ where $x_0 > 3F/k$, show that at time $t = 2\pi(m/k)^{1/2}$ the displacement of the mass is $x_0 - 4F/k$. (See Section 9.7 for an elaboration of this model.)

8. A particle of mass m lying on a smooth horizontal table is attached by two springs of moduli λ_1, λ_2 and natural lengths a_1 and a_2 to two points on the table distance $a_1 + a_2$ apart. In its equilibrium position the mass is subject to a horizontal force $F_0 \cos \omega t$ along the springs. At what angular frequency does the system resonate?

9. Show that the solution of

$$\ddot{x} + \Omega^2 x = F_0 \cos [(\Omega + \varepsilon)t]$$

with $x = \dot{x} = 0$ when $t = 0$ can be written

$$x = -\frac{2F_0}{\Omega^2 - (\Omega + \varepsilon)^2} \sin \tfrac{1}{2}\varepsilon t \sin (\Omega + \tfrac{1}{2}\varepsilon)t.$$

If $0 < |\varepsilon| \ll \Omega$ this solution reveals the phenomenon of *beats*. The frequency of the second sine is much higher than that of the first and the motion may be thought of as an oscillation of angular frequency $\Omega + \tfrac{1}{2}\varepsilon$ with a slowly varying amplitude of

$$\frac{2F_0|\sin \tfrac{1}{2}\varepsilon t|}{|\Omega^2 - (\Omega + \varepsilon)^2|}.$$

The period of the beats is $4\pi/\varepsilon$. Beats occur when the natural frequency and forcing frequency are close together.

10. Solve the equation

$$\ddot{x} + \Omega^2 x = F_0 \cos \Omega t$$

with $x = \dot{x} = 0$ when $t = 0$. This is the equation of a forced harmonic oscillator in the critical resonating case. Show that the oscillation with period $2\pi/\Omega$ builds up with amplitude increasing linearly with time.

11. The equation of motion of a simple pendulum is given by

$$\ddot{\theta} + \Omega^2 \sin \theta = 0 \qquad \Omega^2 = g/\ell$$

where ℓ is the length of the pendulum and θ is its angular displacement from the downward vertical. By retaining the first *two* terms in the expansion of $\sin \theta$, show that θ satisfies, for small θ, the *non-linear* equation

$$\ddot{\theta} + \Omega^2(\theta - \tfrac{1}{6}\theta^3) = 0.$$

We should expect this equation to give a better approximation to the motion. Substitute $\theta = \theta_0 \sin \omega t$ into the differential equation, and use the identity $4 \sin^3 x = 3 \sin x - \sin 3x$. Rejecting the consequent third harmonic, show that the equation is approximately satisfied if

$$\omega^2 = \Omega^2(1 - \tfrac{1}{8}\theta_0^2).$$

The angular frequency to the second order depends on the amplitude of the oscillations. The method also suggests that the next term in a series expansion for θ will be of the form $\theta_1 \sin 3\omega t$.

12. The spring system shown in Figure 7.14 hits the plane with speed v_0. Before impact the spring is unstrained, has natural length a and modulus λ. If the impact

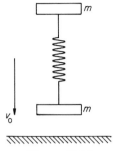

Figure 7.14

between the lower plate and the plane is inelastic show that this plate will subsequently remain in contact with the plane if

$$v_0 \le g(3am/\lambda)^{1/2}.$$

13. A machine of mass 1000 kg contains rotating parts which produce a force on the base of the machine of amplitude F and frequency equal to the frequency of the machine. Its normal running speed is 5000 r.p.m. What stiffness should a spring-mounting have if the transmitted force is to be less than 20% of the applied force?

14. A clock pendulum should have a period of 1 s, but the clock is found to lose 36 s per day. Assuming that the pendulum can be treated as simple, find by how much its length should be altered.

15. An elastic string of modulus λ and natural length a is attached to a fixed point on a smooth horizontal table. The other end of the string is attached to a particle of mass m. The particle is placed on the table with the string of length $\frac{3}{2}a$ and released. Find the period of the oscillations which follow.

16. A spring of stiffness 908 N/m is attached to a fixed support and carries at its free end a block of mass 9 kg. When suspended in equilibrium the block is subject to a downward period force $F_0 \sin \omega t$ where $F_0 = 69$ N. The spring yields if its extension exceeds 1 m. For what frequencies can the periodic force be safely applied?

17. Two masses are linked by a spring as shown in Figure 7.15. The spring which has modulus λ and natural length a is compressed to $\frac{3}{4}$ of its natural length and released from rest. Determine completely the subsequent displacements of both masses. (This system has *two* degrees of freedom—the displacements, with respect to a fixed point, of both masses. There are therefore two equations of motion to be solved simultaneously.)

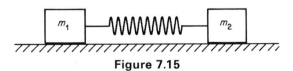

Figure 7.15

18. A block of mass m stands on a smooth horizontal plane and is attached by a horizontal spring of stiffness k to a fixed point of the plane. When in equilibrium the block receives a horizontal blow of magnitude I in the direction of the spring. Find the amplitude of the resulting oscillations of the block.

19. A circular cylinder of radius a and height h, and composed of material of uniform density ρ_1, floats in a liquid of density ρ_2 $(>\rho_1)$. Find the period of vertical oscillations of the cylinder. (The upthrust on the cylinder equals the weight of liquid displaced.)

20. Classify the system represented by

$$m\ddot{x} + c\dot{x} + kx = 0$$

as strongly or weakly damped in the following cases:

(i) $m = 12$ kg $c = 10$ kg/s $k = 2$ N/m
(ii) $m = 10$ kg $c = 8$ kg/s $k = 2$ N/m
(iii) $m = 5$ kg $c = 4$ kg/s $k = 3$ N/m
(iv) $m = 2$ kg $c = 5$ kg/s $k = 3$ N/m

21. Write down the equations of motion for the spring–dashpot systems shown in Figure 7.16.

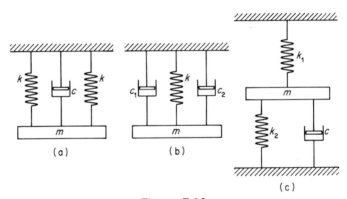

(a)   (b)   (c)

Figure 7.16

22. The displacement of a linear damped oscillator is given, for weak damping, by

$$x = Ke^{-b\Omega t} \sin(\beta t + \varepsilon)$$

(see Section 7.3). Show that x has a maximum at times given by $\beta t = 2n\pi + d$ where n is an integer and d gives the position of the first maximum. Show also that the ratio of successive maxima is $\exp(2b\Omega\pi/\beta)$. The quantity $2b\Omega\pi/\beta$ is called the *logarithmic decrement* and is a measure of the rate of decay of the oscillation. Note that it is independent of n and K.

Such an oscillator is found to have successive maxima of 2.1 cm and 1.3 cm. If the damping coefficient is 10 g/s, find the stiffness of the spring given that the oscillator supports a mass of 5 g.

23. The displacement x of a linear damped oscillator satisfies

$$m\ddot{x} + c\dot{x} + kx = 0.$$

Multiplying this equation by $\dot{x}$ and integrating from $t = 0$ to $t = \tau$ show that the energy dissipated by the system is

$$c \int_0^\tau \dot{x}^2 \, dt$$

in time τ. In mechanical systems this energy is transformed into heat by the friction of the dashpot.

24. A body of mass 10 g is suspended by a spring of stiffness 0.25 N/m and subject to damping which is 1% of critical. After approximately how many oscillations will the amplitude of the system be halved?

25. The frictional resistance acting on the block in Figure 7.17 is assumed to be proportional to its speed. The block has mass 5 kg and is set oscillating. It is found that successive maximum displacements of the block from the wall diminish in the ratio 0.85 and that the period of oscillations is 0.8 s. Find the resistance to motion and the stiffness of the spring.

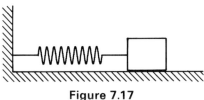

Figure 7.17

26. Show that for damping which is less than 10% of critical, the undamped natural frequency, the damped frequency and the resonant frequency for forced oscillations agree to within 1%.

27. A linear damped system with $m = 10$ kg, $k = 200$ N/m and $c = 25$ kg/s is subject to a periodic force $F_0 \cos(\omega t + \beta)$ where $F_0 = 1.83$ N. For what value of ω are the sustained oscillations of maximum amplitude and what is their phase shift?

28. A jet of water issues vertically upwards from a nozzle with speed v and strikes a ball of mass M. If the vertical velocity of the water relative to the ball is destroyed on striking the ball, show that the equilibrium height of the ball is $(k^2v^2 - g^2)/2gk^2$ where kM is the mass of water issued per unit time.

Show that a small vertical motion of the ball is a critically damped free oscillation. (Neglect the square of the velocity of the ball.)

29. A particle of mass m moves along the straight line OX. It is attracted to O by a force $17mk^2x/2$ and its motion is retarded by a force $3mk\dot{x}$, x being the distance from O. A force $mA \cos \omega t$ is applied to the particle along OX. Show that the greatest amplitude of the forced oscillations occurs for $\omega = 2k$ and is $2A/15k^2$.

30. The mounting of an electric motor is equivalent to a single spring of stiffness 1.1×10^5 N/m and a dashpot with a damping coefficient 20% of the value required for critical damping. The motor is subject to an alternating force which has the same frequency as that of its rotation. The mass of the motor is 454 kg. Show that, if the speed of rotation is less than 210 r.p.m. the amplitude of the force transmitted to the foundation is greater than the amplitude of the exciting force. Find the range of values of the speed of rotation for which the transmitted force is less than 50% of the exciting force.

31. Find the displacement x of the damped oscillator

$$m\ddot{x} + c\dot{x} + kx = 0, \qquad \beta^2 = 4km - c^2 > 0$$

given that $x = 0$, $\dot{x} = v$ at $t = 0$. After the first and subsequent cycles the kinetic energy of the system is increased instantaneously by T_0. Show that periodic motion takes place if

$$T_0 = \tfrac{1}{2}mv^2[1 - \exp(2c\pi/m\beta)].$$

This system represents a possible model for a clock mechanism. Each cycle the system is impulsively excited by giving it extra speed which maintains the periodic oscillations.

32. Obtain the Fourier expansion of the function

$$F(t) = \begin{cases} 0 & 0 \le t < \dfrac{\pi}{2\omega} \\[2mm] F_0 & \dfrac{\pi}{2\omega} < t < \dfrac{3\pi}{2\omega} \\[2mm] 0 & \dfrac{3\pi}{2\omega} < t \le \dfrac{2\pi}{\omega} \end{cases}$$

with $F(t) = F[t + (2\pi/\omega)]$. Find the sustained response of the linear system

$$m\ddot{x} + c\dot{x} + kx = F(t)$$

to this input.

33. Consider a microphone as a movable vertical plate of mass m separated by an air gap from a fixed vertical plate. The effect of the mounting of the movable plate may be modelled by two forces acting on the centre of mass:

(i) a force $5mk^2 \times$ (deflection) restoring the plate to its equilibrium position,

(ii) a force $2mk \times$ (velocity) resisting the motion.

Output from the microphone is undistorted provided the separation between the plates is greater than ra ($0 < r < 1$), where a is the equilibrium separation. Suppose that the sound waves exert a horizontal force $P \cos \omega t$ on the centre of mass and show that the output due to these forced oscillations is undistorted for all ω provided

$$2k \ge \{P/[ma(1 - r)]\}^{1/2}.$$

Find the work done per cycle by the sound waves.

34. In the three-spring system shown in Figure 7.9(a), let the springs CA, AB, BD have stiffnesses k_1, k_2, k_3, natural lengths a_1, a_2, a_3, and let the masses of A and B be m_1, m_2. Determine the equilibrium positions of A and B. Find the quadratic equation for ω^2 which determines the eigenfrequencies of the system for longitudinal oscillations of A and B.

It is observed that the system has a mode in which the length of the spring AB remains unchanged. Show that $m_1/m_2 = k_1/k_3$.

35. A molecule consisting of three atoms, two being of mass m and one of mass nm, is modelled by the dynamical system shown in Figure 7.18. The springs are equal, each with stiffness k, and the longitudinal vibrations are assumed to take place on a smooth horizontal table. If x_1, x_2, x_3 are the displacements from equilibrium, show that one eigenfrequency is zero and find the other two.

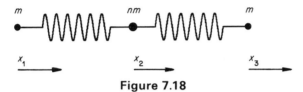

Figure 7.18

36. Two particles of masses m_1 and m_2 are suspended by two equal springs as shown in Figure 7.19. Each spring has stiffness k and natural length a. Find the equilibrium position of each particle. Construct the equations of motion for vertical oscillations of each particle, and show that the eigenfrequencies are the positive roots of

$$m_1 m_2 \omega^4 - k(m_1 + 2m_2)\omega^2 + k^2 = 0.$$

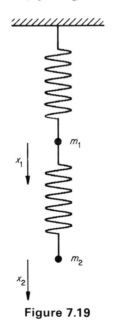

Figure 7.19

37. The system solved in Example 8 is extended to $n + 1$ equal springs and n equal masses. Let $x_1, x_2, \ldots, x_n$ be the displacements of the particles in the same sense. Write down the equations of motion for all the particles in matrix form, and show that the frequencies are the positive roots of the $n \times n$ *tridiagonal* determinant equation

$$\Delta_n = \begin{vmatrix} 2k - m\omega^2 & -k & 0 & \cdots & 0 & 0 \\ -k & 2k - m\omega^2 & -k & \cdots & 0 & 0 \\ \cdots & \cdots & \cdots & \cdots & & \cdots \\ 0 & 0 & 0 & \cdots & 2k - m\omega^2 & -k \\ 0 & 0 & 0 & \cdots & -k & 2k - m\omega^2 \end{vmatrix} = 0.$$

Show by induction that $\Delta_n = -k \sin (n + 1)\theta/\sin \theta$ where $\cos \theta = (m\omega^2 - 2k)/(2k)$ Hence find a formula for the eigenfrequencies.

Show that there exists a mode in which all the particles oscillate in phase with the same amplitude.

38. In a certain oscillator with two degrees of freedom the equations of motion are

$$\ddot{x} = -10x + 18y, \quad \ddot{y} = -3x + 5y.$$

Express the equations in matrix form, and find the eigenfrequencies and eigenvectors of the system. Show that, in the normal modes, $x = 2y$ and $x = 3y$.

Initially $x(0) = \alpha$, $\dot{x}(0) = y(0) = \dot{y}(0) = 0$. Find x and y subsequently.

39. A light string of length $3a$ is suspended from one end and has particles of masses $6m$, $2m$ and m attached to it at distances a, $2a$, $3a$ from this end. Find the eigenfrequencies of this triple pendulum for oscillations of small amplitude in a vertical plane through its point of suspension.

40. A pendulum has a bob of mass m and is suspended from a fixed beam by a linear spring of stiffness k and unstretched length a. The pendulum oscillates in a fixed vertical plane with the spring straight throughout the motion. Let θ be the inclination of the spring to the downward vertical and let z be the extension of the spring from its vertical equilibrium state. For oscillations of small amplitude, show that the equations for θ and z uncouple. Solve the equations, and confirm that synchronous oscillations in the two variables are not possible.

41. In the three-spring system shown in Figure 7.9 the right-hand support D is made to oscillate harmonically as $C \cos \Omega t$, thus creating forced oscillations in the system. Write down the equations of motion for the two particles A and B, and show that they can be expressed in the matrix form

$$\mathbf{M\ddot{x}} + \mathbf{Kx} = \mathbf{p} \cos \Omega t, \quad \mathbf{p} = \begin{bmatrix} kC \\ 0 \end{bmatrix}.$$

Confirm that the forced part of the solution is

$$\mathbf{x} = [\mathbf{K} - \Omega^2 \mathbf{M}]^{-1} \mathbf{p} \cos \Omega t$$

except for two values of Ω. Explain carefully what happens to the system at or near these exceptional frequencies.

42. In the three-spring system shown in Figure 7.9, the particles are allowed to vibrate transversely only, that is perpendicular to the spring. Set up the equations of motion for x and y, where x and y are now the transverse displacements of A and B. Find the eigenfrequencies of the system for oscillations of small amplitude, assuming that $b > a$. How does the system oscillate in its normal modes?

43. In the three-spring system shown in Figure 7.9, let A have mass m and B have mass $2m$. Suppose that the springs CA, AB, BD have stiffnesses k, $2k$ and $3k$ respectively. Find the eigenfrequencies of the system. Confirm that in one normal mode the centre of mass of the particles remains fixed.

8
Orbits

8.1 CENTRAL FORCES

In the solar system the planets and asteroids have a combined mass which is less than 0.2% of the mass of the Sun, with the result that the perturbative effects of the planets on the Sun are very small. Hence, unless extreme accuracy is required, many orbits can be analysed on the reasonable assumption that the Sun is fixed in space. As defined in Section 3.6, the gravitational force exerted by the Sun on a planet is proportional to the product of the masses of the Sun and the planet and inversely proportional to the square of the distance between the two. Since the dimensions of any planet are small compared with its distance from the Sun, we can reasonably suppose that the planet occupies a point in space. The gravitational force is directed towards the Sun which we have agreed is fixed. Such a force directed towards a fixed point is called a *central force*. The path described by the planet is called its *orbit*. Central forces occur in other contexts, for example in the case of a bob attached by an elastic string to a fixed point, the tension in the string always being directed towards that point.

We shall consider the general case first, in which a particle is subject to a central force alone, the force being a function of the radial distance only. It is clear that if, in such circumstances, the particle is fired with velocity $\mathbf{v}_A$ from a point A with position vector $\mathbf{r}_A$ then the motion will take place in the plane formed by the vectors $\mathbf{r}_A$ and $\mathbf{v}_A$. This follows since no linear momentum is created in the direction perpendicular to this plane. In other words, any orbit under the action of a central force must be a plane orbit.

Let the central *attractive* force be $F(r)$ and suppose that the particle occupies the point with position vector $\mathbf{r}$ at time t. From the remarks in the previous paragraph we can choose $\mathbf{r}$ to be a plane vector by adopting a suitable coordinate frame. In vector form, the force is given by $-F(r)\mathbf{r}/r$ and the equation of motion becomes

$$-F(r)\frac{\mathbf{r}}{r} = m\ddot{\mathbf{r}}$$

where m is the mass of the particle. Usually the most convenient coordinate system to choose is the polar one (r, θ) where θ is the angle between the position

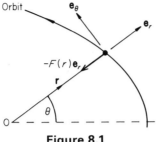

Figure 8.1

vector and some fixed direction (Figure 8.1). The velocity and acceleration in polar coordinates were derived in Section 1.10 and we repeat them here

$$\dot{\mathbf{r}} = \dot{r}\mathbf{e}_r + r\dot{\theta}\mathbf{e}_\theta, \qquad \ddot{\mathbf{r}} = (\ddot{r} - r\dot{\theta}^2)\mathbf{e}_r + \frac{1}{r}\frac{d}{dt}(r^2\dot{\theta})\mathbf{e}_\theta.$$

The equation of motion becomes

$$-F(r)\mathbf{e}_r = m(\ddot{r} - r\dot{\theta}^2)\mathbf{e}_r + \frac{m}{r}\frac{d}{dt}(r^2\dot{\theta})\mathbf{e}_\theta$$

or, in terms of its components,

$$m(\ddot{r} - r\dot{\theta}^2) = -F(r) \qquad (1)$$

$$\frac{d}{dt}(r^2\dot{\theta}) = 0.$$

The second of these equations implies that

$$r^2\dot{\theta} = \text{constant} = h \quad \text{say.} \qquad (2)$$

Note that this condition could have been obtained directly since it simply represents conservation of angular momentum, the central force having no moment about the origin. The elimination of $\dot{\theta}$ between (1) and (2) yields the following equation relating r and t:

$$m\ddot{r} - \frac{mh^2}{r^3} = -F(r). \qquad (3)$$

With the identity $\ddot{r} = \dot{r}\,d\dot{r}/dr$, the equation becomes

$$m\dot{r}\frac{d\dot{r}}{dr} - \frac{mh^2}{r^3} = -F(r)$$

which can be integrated once to give

$$\tfrac{1}{2}m\dot{r}^2 + \frac{mh^2}{2r^2} = -\int F(r)\,dr + \text{constant.} \qquad (4)$$

The reader should note that this equation expresses conservation of energy since the indefinite integral $\int F(r)\,dr$ is the potential of the force $F(r)$ and the two terms on the left-hand side of the equation represent the kinetic energy of the particle.

The actual orbit in the plane of motion is determined by a relation between the polar coordinates r and θ. Return to Equation (3) and eliminate the time in the derivative of r by using the change of derivative

$$\frac{d}{dt} = \frac{d\theta}{dt}\frac{d}{d\theta} = \frac{h}{r^2}\frac{d}{d\theta}.$$

Thus

$$m\frac{h}{r^2}\frac{d}{d\theta}\left(\frac{h}{r^2}\frac{dr}{d\theta}\right) - \frac{mh^2}{r^3} = -F(r) \tag{5}$$

where we now view r as a function of θ rather than t.

Example 1   A particle of mass m lies on a smooth horizontal table and is attached by a linear spring to a fixed point on the table. The particle is fired with speed V perpendicular to the unstrained spring along the table. If the spring has modulus of elasticity $3mV^2/4a$, where a is the natural length of the spring, show that subsequently the particle always lies within a circle of radius 2a.

The only horizontal force acting on the particle is that due to the tension in the spring which is directed towards a fixed point on the table. This represents a central force

$$F(r) = 3mV^2(r - a)/4a^2$$

in the notation above.

Here r is the polar distance from the point of attachment O and θ is measured from the initial radius.

The initial moment of momentum is maV so that $h = aV$. Thus the energy Equation (4) in this case becomes

$$\tfrac{1}{2}m\dot{r}^2 + \frac{ma^2V^2}{2r^2} = -\int \frac{3mV^2}{4a^2}(r - a)\,dr + C$$

$$= -\frac{3mV^2}{4a^2}(\tfrac{1}{2}r^2 - ar) + C \tag{6}$$

where the constant C is determined by the initial condition $\dot{r} = 0$ when $r = a$. Therefore

$$C = \frac{1}{2a^2} + \frac{3}{8a^2} - \frac{3}{4a^2} = \frac{1}{8a^2}.$$

When the particle reaches its greatest distance from O, r will have a minimum value for which the necessary condition $\dot{r} = 0$ must hold. From Equation (6) this must occur where

$$3r^4 - 6ar^3 - a^2r^2 + 4a^4 = 0.$$

It can be verified that this quartic equation can be factorised into

$$(r - a)(r - 2a)(3r^2 + 3ar + 2a^2) = 0$$

and that its only real roots are $r = a$ and $r = 2a$. Equation (6) can be recast as

$$\dot{r}^2 = -\frac{V^2}{4a^2r^2}(r - a)(r - 2a)(3r^2 + 3ar + 2a^2).$$

For the actual motion $\dot{r}^2 \geq 0$ and this will be the case if and only if $a \leq r \leq 2a$ since the term in the last bracket is always positive. The particle therefore oscillates within this annulus on the table.

8.2 GRAVITATIONAL CENTRAL FORCE

It is assumed that the satellite body does not significantly affect the position of the central body owing to the large difference in mass as, for example, in the cases of a planet moving round the Sun and an artificial satellite circling the Earth. To fix ideas we will consider a satellite of mass m moving round the Sun. By Newton's law of gravitation the satellite will experience a central attractive force

$$F(r) = \frac{\gamma m_s m}{r^2}$$

where γ is the gravitational constant, m_s the mass of the Sun and r the distance of the satellite from the Sun. The satellite's path must satisfy Equation (5) which now reduces to

$$\frac{d}{d\theta}\left(\frac{1}{r^2}\frac{dr}{d\theta}\right) - \frac{1}{r} = -\frac{\gamma m_s}{h^2}.$$

We can transform this into a second-order differential equation with constant coefficients by putting $u = 1/r$. The result is

$$\frac{d^2u}{d\theta^2} + u = \frac{\gamma m_s}{h^2}.$$

The corresponding homogeneous equation

$$\frac{d^2u}{d\theta^2} + u = 0$$

has a solution $A \cos (\theta - \theta_0)$ where A and θ_0 are constants whilst it is easy to verify that the constant $\gamma m_s / h^2$ is a particular integral.

The full solution reads

$$u = \frac{1}{r} = A \cos (\theta - \theta_0) + \frac{\gamma m_s}{h^2}. \tag{7}$$

One particular orbit corresponds to the condition $A = 0$, in which case $r = h^2 / \gamma m_s$, a constant. This orbit is a circle with its centre coinciding with the centre of force. If a is the radius of the orbit and the orbit is described with speed V, then $h = aV$ and $a = a^2 V^2 / \gamma m_s$ implying $a = \gamma m_s / V^2$. Since $r^2 \dot{\theta} = h$ for any orbit and $r = a$, $\dot{\theta}$ must be constant which implies that the orbit can only be described with constant speed V. If a and V are known quantities from observation and if the satellite takes time T to complete one orbit of the Sun in a sidereal frame,

$$\gamma m_s = aV^2 = 4\pi^2 a^3 / T^2 \tag{8}$$

since $VT = 2\pi a$. For the Earth, whose orbit is almost circular, $a = 1.5 \times 10^{11}$ m and $T = 3.15 \times 10^7$ s, so that $\gamma m_s = 1.3 \times 10^{20}$ m^3/s^2.

The quantity γm_s is called the *gravitational mass* of the Sun (see Section 3.6) and it is relatively easy to calculate it from a knowledge of the orbital radius and period of a planet. Once the gravitational constant has been found the *mass* of the Sun can be found. Every body has associated with it a gravitational mass. For example, the gravitational mass of the Earth can be computed from the behaviour of the Moon, or, with greater accuracy, from the orbit of an artificial satellite. The mass of a planet with no satellites is more difficult to obtain and can usually only be discovered by analysing its effects on the orbits of neighbouring planets.

Let us express the orbit given by (7) in the standard form

$$\frac{\ell}{r} = e \cos (\theta - \theta_0) + 1 \tag{9}$$

where $\ell = h^2 / \gamma m_s$ and $e = Ah^2 / \gamma m_s$ (without loss of generality we can assume that $A > 0$). Equation (9) is the polar equation of a *conic section*. Three cases need to be distinguished: $0 \le e < 1$, $e = 1$ and $e > 1$.

Elliptic orbit, $e < 1$. Since $|\cos (\theta - \theta_0)| \le 1$, it follows that the right-hand side of (9) must always be positive so that to each value of θ in $0 \le \theta < 2\pi$ there exists a corresponding positive value for r. Further ℓ/r is a periodic function of θ with period 2π so that the orbit is a closed curve called an *ellipse*.

The line $\theta = \theta_0$ is called the *major axis* and it cuts the ellipse at the points with polar coordinates $[\ell/(1 + e), \theta_0]$ and $[\ell/(1 - e), \theta_0 + \pi]$. The length of the semi-major axis, denoted by a, is given by

$$a = \tfrac{1}{2}\left(\frac{\ell}{1+e} + \frac{\ell}{1-e}\right) = \frac{\ell}{1-e^2}.$$

The centre of the ellipse is located at the mid-point of the major axis; it has polar coordinates $[e\ell/(1-e^2),\ \theta_0 + \pi]$. The *minor axis* is perpendicular to the major axis and passes through the centre of the ellipse. The coordinates of the points common to the minor axis and the ellipse are $[(b^2 + a^2e^2)^{1/2},\ \theta_0 \pm (\pi - \alpha)]$ where $\cos\alpha = ae/(b^2 + a^2e^2)^{1/2}$ and b is the length of the semi-minor axis. Since this point lies on the ellipse

$$\frac{\ell}{(b^2 + a^2e^2)^{1/2}} = -\frac{ae^2}{(b^2 + a^2e^2)^{1/2}} + 1$$

or

$$\ell + ae^2 = (b^2 + a^2e^2)^{1/2}$$

and, since $\ell = a(1 - e^2)$,

$$a = (b^2 + a^2e^2)^{1/2} \quad \text{and} \quad b = a(1 - e^2)^{1/2}.$$

The number e is called the *eccentricity* of the ellipse; if $e = 0$ the ellipse becomes a circle. The main features of the ellipse are shown in Figure 8.2. The point O to which the force is directed is called a *focus* of the ellipse and ℓ is the length of the *semi-latus rectum*. Note that the angle θ_0 merely indicates the orientation of the ellipse to the fixed axis.

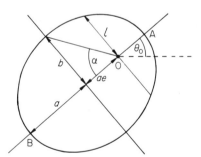

Figure 8.2 The elliptic orbit.

The closest approach A of a planet or satellite to the Sun at O is called the *perihelion* and the furthest point B is called the *aphelion*. The corresponding points for elliptic orbits of the Earth (and planets in general) are called the *perigee* and *apogee*. In general, a point where the velocity is perpendicular to its position vector is called an *apse*.

Hyperbolic orbit, e > 1. The right-hand side of Equation (9) will vanish when $\cos(\theta - \theta_0) = -1/e$. Let β be such that $\cos\beta = 1/e$ with $0 < \beta < \frac{1}{2}\pi$. As $\theta \to \theta_0 \pm (\pi - \beta)$, $r \to \infty$: the graph of the orbit is shown in Figure 8.3. The curve, which is one branch of a hyperbola, is symmetric about the line $\theta = \theta_0$ and cuts this line where $r = \ell/(1 + e)$.

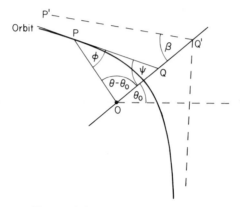

Figure 8.3 The hyperbolic orbit.

Consider a point P on the curve and let the tangent to the curve there be the line PQ. Let $O\hat{P}Q = \phi$ and $O\hat{Q}P = \psi$. The cotangent of the angle ϕ must be the ratio of the radial and transverse components of velocity at P:

$$\cot\phi = \frac{\dot{r}}{r\dot{\theta}} = \frac{1}{r}\frac{dr}{d\theta} = \frac{re}{\ell}\sin(\theta - \theta_0)$$

using Equation (9). By the sine rule in triangle OPQ,

$$\frac{OQ}{\sin\phi} = \frac{r}{\sin\psi} = \frac{r}{\sin(\theta - \theta_0 + \phi)}.$$

Therefore

$$OQ = \frac{r\sin\phi}{\sin(\theta - \theta_0)\cos\phi + \cos(\theta - \theta_0)\sin\phi}$$

$$= \frac{r}{\sin(\theta - \theta_0)\cot\phi + \cos(\theta - \theta_0)}$$

$$= \frac{r\ell}{re\sin^2(\theta - \theta_0) + \ell\cos(\theta - \theta_0)}.$$

With P on the section of the hyperbola shown in Figure 8.3, $\theta - \theta_0 \to \pi - \beta$ as $r \to \infty$, with the result that

$$OQ \to \frac{\ell}{e \sin^2 \beta} = \frac{\ell e}{e^2 - 1} = OQ' \quad \text{say}$$

in Figure 8.3. The tangent QP approaches the line Q'P' which is called an *asymptote* of the hyperbola. There is a second asymptote which is the image of the first in the line OQ.

Parabolic orbit, e $= 1$. The equation of the orbit becomes

$$\frac{\ell}{r} = \cos (\theta - \theta_0) + 1. \tag{10}$$

The parabola is the locus of the points which are equidistant from the focus O and a straight line D called the *directrix* (Figure 8.4). This is a special case of the more general relationship for the three curves considered in that they are the locus of points for which the ratio (distance of point from focus/distance of point from directrix) $= e$. The distance between the focus and the directrix is ℓ and it is easy to verify geometrically from the figure that the equation of a parabola is given by (10).

Of the three orbits the ellipse is the only closed orbit. In the case of the parabola and hyperbola the orbiting body escapes from the influence of the Sun. In the solar system the planets all move in approximately elliptic orbits. However the orbits are not strictly closed paths because of the perturbative

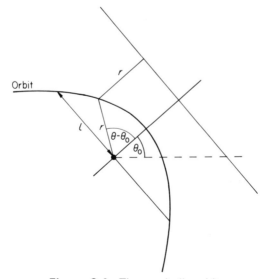

Figure 8.4 The parabolic orbit.

effects of other planets. Certain comets appear to be moving in parabolic or hyperbolic orbits although this is often difficult to ascertain from a small section of the orbit plotted in the neighbourhood of the Sun. Furthermore the orbit of any comet can be changed substantially by the influence of planets which lie close to its orbit.

The total energy $\mathscr{E}$ of an orbiting particle is given by

$$\mathscr{E} = \tfrac{1}{2}mv^2 - \frac{\gamma m_s m}{r} \tag{11}$$

where v is the speed of the particle, and the second term is the gravitational potential energy. Whether the orbit is an ellipse, hyperbola or parabola, it must have an apse at $\theta = \theta_0$ with an apsidal distance of $\ell/(1 + e)$. Let V be the speed at this apse. Since the energy is conserved,

$$\mathscr{E} = \tfrac{1}{2}mV^2 - \gamma m_s m(1 + e)/\ell \tag{12}$$

the value of the energy at the apse. The angular momentum per unit mass

$$h = \ell V/(1 + e) = (\ell \gamma m_s)^{1/2}$$

by definition. Substituting for V from this equation into (12), we find that

$$\mathscr{E} = \gamma m_s m(e^2 - 1)/2\ell. \tag{13}$$

We conclude that the orbit is an ellipse, hyperbola or parabola according as the energy is negative, positive or zero. Put another way, a body will escape from the Sun's influence if $\mathscr{E} \geq 0$, that is, if $V^2 \geq 2\gamma m_s(1 + e)/\ell$.

The energy equation provides a simple formula relating speed and distance. From (11) and (13),

$$\tfrac{1}{2}v^2 - \frac{\gamma m_s}{r} = \frac{\gamma m_s(e^2 - 1)}{2\ell}$$

or

$$\frac{v^2}{\gamma m_s} = \frac{2}{r} + \frac{e^2 - 1}{\ell}.$$

If the orbit is elliptic, the semi-major axis $a = \ell/(1 - e^2)$ and

$$\frac{v^2}{\gamma m_s} = \frac{2}{r} - \frac{1}{a}. \tag{14}$$

It follows from this relationship that the maximum and minimum speeds of a body circling the Sun occur at the perihelion and aphelion respectively.

Example 2   What is the minimum period of an Earth satellite if it is assumed that the Earth has no atmosphere? The gravitational mass of the Earth is about $4.0 \times 10^{14}\ m^3/s^2$ and the Earth's radius is about 6400 km.

The minimum period occurs when the satellite is moving in a circular orbit slightly larger than the Earth's radius. The required period T is therefore given by Equation (8):

$$T = \frac{2\pi a^{3/2}}{(\gamma m_e)^{1/2}}$$

where m_e is the mass of the Earth. Thus

$$T = \frac{2\pi(6.4 \times 10^6)^{3/2}}{2 \times 10^7} \text{ s}$$

$$\approx 5100 \text{ s} = 85 \text{ min}.$$

The presence of the atmosphere makes the actual minimum orbit have a slightly larger radius which, in any case, is no longer a circle but becomes distorted into a spiral by the continuous drag of the atmosphere on the satellite.

Example 3   A two-stage rocket is fired with speed V in a direction at an angle α to the vertical from a launching site on the Earth. When at the apogee of its initial orbit the second stage is ignited to put the payload into orbit. Obtain the additional speed required at the apogee to make this orbit circular.

We shall ignore the rotational effect of the Earth and atmospheric resistance. The fuel burning times are assumed to be of short duration so that the rocket can effectively change speed instantaneously. Let μ_e be the gravitational mass of the Earth and c its radius.

Equation (14) immediately gives the semi-major axis a:

$$\frac{1}{a} = \frac{2}{c} - \frac{V^2}{\mu_e}. \tag{15}$$

The angular momentum per unit mass,

$$h = r^2\dot{\theta} = Vc \sin \alpha = (\mu_e \ell)^{1/2} \tag{16}$$

where ℓ is the semi-latus rectum. This, in turn, is related to the semi-major axis through

$$\ell = a(1 - e^2). \tag{17}$$

If we take the equation of the initial orbit to be

$$\frac{\ell}{r} = e \cos (\theta - \theta_0) + 1$$

we must have

$$\frac{\ell}{c} = e \cos \theta_0 + 1 \tag{18}$$

since this orbit must pass through the launching site. The four Equations (15), (16), (17) and (18) enable us to determine the four parameters a, ℓ, e and θ_0 for the orbit in terms of μ_e, V, c and α.

At the apogee $r = a(1 + e)$ and by Equation (14) again the apogee speed v_a is given by

$$v_a^2 = \mu_e\left(\frac{2}{a(1 + e)} - \frac{1}{a}\right) = \frac{\mu_e(1 - e)}{a(1 + e)}.$$

To put the satellite into a circular orbit we need to increase its speed to v_c, say, where

$$v_c^2 = \mu_e/a(1 + e)$$

the radius of the circular orbit being $a(1 + e)$ (see Figure 8.5). Consequently the speed at the apogee must be increased by

$$v_c - v_a = [\mu_e/a(1 + e)]^{1/2}[1 - (1 - e)^{1/2}].$$

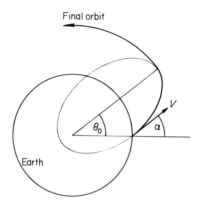

Figure 8.5   A two-stage rocket path for a circular orbit placement of a satellite.

For the Earth, $\mu_e = 4.0 \times 10^{14}\ \text{m}^3/\text{s}^2$ and $c = 6400$ km. A few quick calculations yield the additional boost the rocket needs if it is fired initially with speed 6 km/s at an angle $\alpha = 30°$. From the equations listed above,

$$a = 4.5 \times 10^6\ \text{m} \qquad \ell = 9.2 \times 10^5\ \text{m}$$

$$e = 0.89 \qquad v_a = 2.3 \times 10^3\ \text{m/s}$$

$$v_c = 6.8 \times 10^3\ \text{m/s}.$$

Thus the satellite requires an additional speed of 4.5 km/s at its apogee.

8.3   ORBITAL PERIOD: KEPLER'S LAWS

Kepler's first law of planetary motion asserts that the planets move in elliptic orbits with the Sun at one focus of the ellipse. We have already seen that gravitational central orbits must be *conics* (ellipse, hyperbola or parabola); for the planets the eccentricities of their orbits are considerably less than unity (see the table on p. 204)

The conservation of angular momentum embodied in the equation $r^2\dot{\theta} = h$ for central-force orbits can be used to find the time taken by a particle to cover a part of its orbit. Suppose the position vector of the particle sweeps through the angle between $\theta = \theta_1$ and $\theta = \theta_2$ in a time t_1. The time-integral of $r^2\dot{\theta} = h$ over this angle gives

$$\int_{\theta=\theta_1}^{\theta=\theta_2} r^2 \, d\theta = h \int_0^{t_1} dt = ht_1.$$

By the usual techniques of elementary integration in polar coordinates, the left-hand side of this equation represents twice the area enclosed by the orbit and the extreme position vectors (see Figure 8.6). The implication of this result

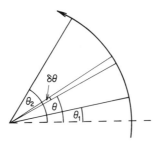

Figure 8.6

is that the position vector sweeps out equal areas in equal times. This is *Kepler's second law* of planetary motion. It should be noted that it holds for any central orbit and is not restricted to gravitational orbits. For a particle moving in an elliptic orbit, the period T of the orbit will be

$$T = \frac{2}{h} \, (\text{area of the ellipse})$$

$$= \frac{2\pi ab}{h}$$

where a and b are the semi-axes of the ellipse. The semi-latus rectum $\ell = h^2/\gamma m_s = a(1 - e^2)$ and $b^2 = a^2(1 - e^2)$ so that

$$T^2 = 4\pi^2 a^3/\gamma m_s$$

which is equivalent to the statement of *Kepler's third law*: the ratio of the square of the orbital period to the cube of the semi-major axis is the same constant for all planets.

Kepler's laws pre-date Newton's laws of motion and gravitation and were obtained empirically on the basis of data collected for planets adjacent to the Earth.

Some data concerning the planets are given in the following table.

Name	Mass (mass of Earth = 1 unit = 5.974 × 10²⁴ kg)	Period of revolution about Sun (period of Earth = 1 unit = 365.256 days)	Semi-major axis of orbit (Distance of Earth = 1 unit = 149.5 × 10⁶ km)	Eccentricity of orbit	Mean diameter (Earth diameter = 1 unit = 12 735 km)	Sidereal period of rotation† (days)
Sun	331 950.0	—	—	—	109.187	25.37
Moon	0.012	—	—	0.017	0.273	27.32
Mercury	0.05	0.241	0.387	0.206	0.366	58.64
Venus	0.81	0.615	0.723	0.007	0.960	−243.01
Earth	1	1	1	0.017	1	0.997
Mars	0.11	1.88	1.524	0.093	0.531	1.026
Jupiter	317.9	11.862	5.203	0.048	10.969	0.414
Saturn	95.2	29.458	9.539	0.051	9.036	0.437
Uranus	14.5	84.015	19.182	0.047	3.715	−0.65
Neptune	17.2	164.788	30.058	0.007	3.538	0.768
Pluto	0.0025	247.697	39.518	0.253	0.236	−6.387

N.B. Gravitational constant $\gamma = 6.67 \times 10^{-11}$ m³/kg s².
　† Minus sign indicates retrograde rotation.

Example 4 By using the table above find the acceleration due to gravity at the surface of Jupiter. By how much is this reduced at the equation of Jupiter by its rotation?

The acceleration due to gravity at the surface of the Earth,

$$g_e = \frac{\gamma m_e}{r_e^2} = \frac{4\gamma m_e}{d_e^2}$$

where m_e is the mass of the Earth and d_e is its diameter.
　Similarly the acceleration due to gravity at the surface of Jupiter

$$g_j = \frac{4\gamma m_j}{d_j^2} = \frac{m_j d_e^2}{m_e d_j^2} g_e$$

$$= \frac{318.4}{(10.969)^2} g_e$$

$$= 2.645 g_e$$

$$= 25.94 \text{ m/s}^2.$$

Due to the rotation the acceleration at the equator of Jupiter is reduced by an amount

$$r_j \Omega_j^2 = \left(\frac{2\pi}{0.411 \times (86\,400 - 240)} \right)^2 \times \frac{10.969 \times 12\,735 \times 10^3}{2}$$

$$= 2.20 \text{ m/s}^2 \text{ or } 8.5\%.$$

8.4 ORBITAL TRANSFER

A straightforward application of the results of the previous two sections is provided by the problem of finding an optimal path for a rocket to be transferred from one circular planetary orbit to another circular planetary orbit. Figure 8.7 shows the orbits of Earth and Mars as coplanar circles with centres at the Sun (this is a first approximation since we shall ignore the small eccentricities of the orbits and their mutual inclination). We shall suppose that the rocket is fired some distance from the Earth and neglect the gravitational forces on the rocket due to the Earth and Mars. After the initial firing the rocket moves in a free orbit about the Sun.

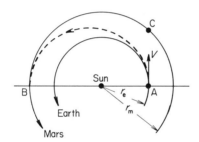

Figure 8.7 An orbital transfer path.

Since the Earth is moving around the Sun, any rocket fired from the Earth will itself have this velocity in addition to its projected speed. In order to take maximum advantage of this fact we shall fire the rocket in a direction tangential to the Earth's orbit as shown in Figure 8.7. Let the rocket start from the point A of the Earth's orbit with absolute speed V. The point A must be an apse of the orbit whose equation can be written

$$\frac{\ell}{r} = e \cos \theta + 1.$$

When $\theta = 0$, $r = r_e$ so that

$$\ell = r_e(e + 1). \tag{19}$$

The angular momentum per unit mass

$$h = (\ell \mu_s)^{1/2} = V r_e$$

where μ_s is the gravitational mass of the Sun. Thus $V = (\ell \mu_s)/r_e$. We shall consider the optimum orbit which makes V as small as possible which, in turn, means that the initial kinetic energy required to fire the rocket is a minimum. It is easy to see that we must make ℓ as small as possible consistent with the rocket reaching the orbit of Mars, which by (19) implies that e must be minimised.

Suppose the rocket reaches the orbit of Mars where $\theta = \alpha$ so that

$$\frac{\ell}{r_m} = e \cos \alpha + 1. \tag{20}$$

Eliminating ℓ between (19) and (20), we find that the eccentricity

$$e = (r_m - r_e)/(r_e - r_m \cos \alpha)$$

which is clearly a minimum when $\cos \alpha = -1$ or $\alpha = 180°$. This means that the aphelion of the orbit (known as the *Hohmann transfer orbit*) must lie on the orbit of Mars as shown in Figure 8.7. It is easy to verify that the minimum firing speed is then

$$V_{min} = [2 r_m \mu_s / r_e (r_e + r_m)]^{1/2}.$$

By Kepler's third law, the time τ taken by the rocket is equal to half the period of the orbit:

$$\tau = \pi [\tfrac{1}{2}(r_e + r_m)]^{3/2}/\mu^{1/2}.$$

In order to compute the firing velocity and flight time we require the following data:

$$\mu_s = 1.3 \times 10^{20} \text{ m}^3/\text{s}^2 \qquad r_e = 1.5 \times 10^{11} \text{ m} \qquad r_m = 1.5 r_e.$$

A straightforward calculation gives $V_{min} = 3.4 \times 10^4$ m/s and $\tau = 250$ days. A more realistic value for the firing speed is really the difference between V_{min} and the orbital speed of the Earth which is 3.0×10^4 m/s. The relative speed of firing is therefore about 3.7×10^3 m/s.

Of course the transfer orbit must meet Mars at B which means that the two planets should be in the correct conjunction (A and C in Figure 8.7) initially.

8.5   MUTUAL ORBITS

When the masses of two bodies are comparable, each influences the other and both bodies describe orbits. In the two-body system the centre of mass moves as a fictitious particle with the total external force acting on it. Let G

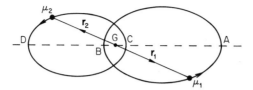

Figure 8.8 Mutual orbits.

be the mass-centre of the system with $\mathbf{r}_1$ and $\mathbf{r}_2$ the position vectors relative to G of the two bodies idealized as particles (Figure 8.8). If μ_1 and μ_2 are the gravitational masses of the two bodies, then, by definition,

$$\mu_1\mathbf{r}_1 + \mu_2\mathbf{r}_2 = 0 \qquad (21)$$

the two particles being disposed on either side of the mass-centre in a straight line through the mass-centre. Under the influence of no other forces, the equations of motion for the system are

$$\ddot{\mathbf{r}}_1 = \mu_2(\mathbf{r}_2 - \mathbf{r}_1)/|\mathbf{r}_2 - \mathbf{r}_1|^3, \qquad \ddot{\mathbf{r}}_2 = \mu_1(\mathbf{r}_1 - \mathbf{r}_2)/|\mathbf{r}_1 - \mathbf{r}_2|^3. \qquad (22)$$

Eliminating in turn $\mathbf{r}_2$ and $\mathbf{r}_1$ between (21) and (22), we obtain the separate equations for $\mathbf{r}_1$ and $\mathbf{r}_2$,

$$(\mu_1 + \mu_2)^2\ddot{\mathbf{r}}_1 = -\frac{\mu_2^3\mathbf{r}_1}{r_1^3}, \qquad (\mu_1 + \mu_2)^2\ddot{\mathbf{r}}_2 = -\frac{\mu_1^3\mathbf{r}_2}{r_2^3}.$$

These equations when compared with the central orbit given by $\ddot{\mathbf{r}} = -\mu\mathbf{r}/r^3$ indicate that both particles describe conics with respect to the centre of mass G. In, say, the first of these equations the orbit can be thought of as that due to a central force of magnitude $\mu_2^3/(\mu_1 + \mu_2)^2 r_1^2$ per unit mass.

Both orbits lie in the same fixed plane and, if they are ellipses as shown in Figure 8.8 with semi-major axes a_1 and a_2 respectively, then AB $= 2a_1$ and CD $= 2a_2$ in the figure. Since the magnitudes of the position vectors are in the same proportion, the two ellipses are similar and have the same eccentricity; thus $a_1/a_2 = \mu_2/\mu_1$.

The periods of both particles must be the same since they always remain in line with the mass-centre. By analogy with the central force problem, the period T is given by

$$T^2 = \frac{4\pi^2 a_1^3(\mu_1 + \mu_2)^2}{\mu_2^3} = \frac{4\pi^2 a_2^3(\mu_1 + \mu_2)^2}{\mu_1^3}.$$

The orbit of μ_1 relative to μ_2 can be determined easily by letting $\mathbf{r} = \mathbf{r}_1 - \mathbf{r}_2$ and taking the difference between Equations (22) so that

$$\ddot{\mathbf{r}} = -(\mu_1 + \mu_2)\mathbf{r}/r^3$$

This equation can again be interpreted as an equivalent central force problem, the force being $(\mu_1 + \mu_2)/r^2$ this time. The orbit of μ_2 relative to μ_1 satisfies the same equation; this can be verified by writing $-\mathbf{r}$ for $\mathbf{r}$. Both relative orbits are conics. In the case of mutual elliptic orbits the semi-major axis of the relative orbit is, in Figure 8.8,

$$\tfrac{1}{2}(AD + BC) = a_1 + a_2.$$

Kepler's third law (Section 9.3) now becomes

$$T^2 = 4\pi^2 a^3/(\mu_1 + \mu_2)$$

which means that for the planets the ratio T^2/a^3 is not quite constant. If μ_1 is the mass of the Sun and μ_2 that of a planet there is a slight variation of T^2/a^3 between the planets.

Mutual orbits usually take place under the influence of a distant force centre as, for example, in the case of the Earth-Moon system in the Sun's gravitational field. Let $\bar{\mathbf{r}}$ be the position vector of the mass-centre of the Earth and Moon in a sidereal frame, and let $\mathbf{r}_e$ and $\mathbf{r}_m$ be the position vectors of the Earth and the Moon relative to this mass-centre. Let the gravitational force on the Earth and Moon due to the Sun be $m_e \mathbf{F}_e$ and $m_m \mathbf{F}_m$ respectively where m_e and m_m are their masses. The separate equations of motion for the two bodies are:

$$\ddot{\mathbf{r}}_e + \ddot{\bar{\mathbf{r}}} = \mu_m \frac{(\mathbf{r}_m - \mathbf{r}_e)}{|\mathbf{r}_m - \mathbf{r}_e|^3} + \mathbf{F}_e$$

$$\ddot{\mathbf{r}}_m + \ddot{\bar{\mathbf{r}}} = \mu_e \frac{(\mathbf{r}_e - \mathbf{r}_m)}{|\mathbf{r}_m - \mathbf{r}_e|^3} + \mathbf{F}_m.$$

The position vector of the Moon relative to the Earth, $\mathbf{r} = \mathbf{r}_m - \mathbf{r}_e$, satisfies the equation obtained by taking the difference between the equations above:

$$\ddot{\mathbf{r}} = -(\mu_e + \mu_m)\frac{\mathbf{r}}{r^3} + \mathbf{F}_m - \mathbf{F}_e.$$

Since the Sun is remote from both the Earth and the Moon, the difference $\mathbf{F}_m - \mathbf{F}_e$ is small and the relative orbit satisfies, to a first approximation, the relative equation for the isolated two-body problem.

8.6 A SIMPLE MODEL FOR DRAG ON A SATELLITE

We consider the case of a satellite orbit whose apogee is at a much greater height than its perigee. This implies that the eccentricity e differs appreciably from zero. If the perigee is within the Earth's upper atmosphere, the satellite then experiences a drag on a small part of its orbit which we can replace in our mathematical model by a small impulse at the perigee which causes its velocity there to be reduced on each orbit by a factor ε (see Figure 8.9).

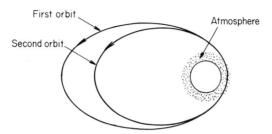

Figure 8.9   Eccentric satellite orbit passing through the Earth's atmosphere (its thickness is exaggerated in the figure).

If μ is the gravitational mass of the Earth, ℓ_0 is the semi-latus rectum of the first orbit and h_0 is the angular momentum of the first orbit, we have, from definitions in Section 8.2, that

$$a(1 - e_0^2) = \ell_0 = h_0^2/\mu$$

where e_0 is the eccentricity of the first orbit.

Therefore, since the distance to the perigee $r_p = a(1 - e_0)$, then

$$h_0^2 = \mu r_p(1 + e_0)$$

which rearranged gives

$$e_0 = \frac{h_0^2}{\mu r_p} - 1. \tag{23}$$

Now $0 < e_0 < 1$ for an ellipse and therefore

$$2 > \frac{h_0^2}{\mu r_p} > 1. \tag{24}$$

If subscript n applies to the nth orbit we also have

$$e_n = \frac{h_n^2}{\mu r_p} - 1. \tag{25}$$

Now, as indicated above, the model assumes that the satellite returns to the same perigee on each orbit but that its velocity there is reduced due to drag. That is, we assume $r_p = \text{constant}$, and $v_n = v_0(1 - n\varepsilon)$ using the binomial expansion and the assumption that $n\varepsilon$ is a small quantity whose higher powers may be neglected. Thus we have $h_n = r_p v_0(1 - n\varepsilon)$ and from Equation (25),

$$e_n = \frac{r_p v_0^2(1 - 2n\varepsilon)}{\mu} - 1 = e_0 - \frac{2n r_p v_0^2}{\mu}\varepsilon$$

indicating that e_n is becoming smaller and that the orbit is tending to a circle.

The period for each successive orbit decreases however, as we shall now show, and the speed at the apogee is increased. We have

$$\text{period for } n\text{th orbit} = \frac{\text{Area}}{\text{Areal velocity}} = \frac{\pi a_n b_n}{h_n/2} = \frac{2\pi a_n^2 (1 - e_n^2)^{1/2}}{h_n}$$

$$= \frac{2\pi r_p^2}{h_n (1 - e_n^2)^{3/2}} \tag{26}$$

where a_n and b_n are the semi-major and semi-minor axes respectively and where relationships derived in Section 9.2 have been used.

Now,

$$h_n(1 - e_n^2)^{3/2} = h_0(1 - n\varepsilon)\left[1 - \left(\frac{h_0^2(1 - 2n\varepsilon)}{\mu r_p} - 1\right)^2\right]^{3/2}$$

from Equation (25), and by rearrangement as a series in ε, becomes

$$h_n(1 - e_n^2)^{3/2} = \frac{h_0^4}{(\mu r_p)^{3/2}}\left[2 - \frac{h_0^2}{\mu r_p} + n\varepsilon\left(\frac{7h_0^2}{\mu r_p} - 8\right)\right]$$

to the first order. Therefore provided $h_0^2/\mu r_p > 8/7$, the implication of which we will examine shortly, the denominator of the right-hand side of (26) increases with each orbit and the period therefore decreases. From (23), $e_0 < 1/7$ if $h_0^2/\mu r_p < 8/7$. Such an eccentricity makes the orbit virtually a circle for with $e_0 = 1/7$, and the ratio of the minor to major axes $[= (1 - e^2)^{1/2}]$ is approximately 0.99. By the time this orbit has been attained the model will no longer be of value since drag would occur over a considerable portion of the orbit.

Since the angular momentum is constant throughout each orbit,

$$(\text{speed at apogee}) \times a_n(1 + e_n) = v_n \times a_n(1 - e_n).$$

Therefore

$$\text{speed at apogee} = v_n\left(\frac{1 - e_n}{1 + e_n}\right)$$

$$= v_0(1 - n\varepsilon)\left(\frac{2 - h_n^2/\mu r_p}{h_n^2/\mu r_p}\right)$$

$$= \frac{\mu}{v_0 r_p}\left[2 - \frac{h_0^2}{r_p \mu} + n\varepsilon\left(2 + \frac{h_0^2}{r_p \mu}\right)\right].$$

Thus the speed of the satellite at the apogee increases with each successive orbit.

The two results we have just found are, in a sense, paradoxical, for we have found a dynamical system in which the introduction of drag caused a speeding up of the processes. Closer examination, however, indicates that both the path length and the total energy are decreasing with each successive orbit.

Exercises

[Any additional data required should be taken from the table on p. 204]

1. Show that, if variations of gravity are taken into account, the time in which a particle falls to the Earth's surface from a height h is

$$\left(\frac{2h}{g}\right)^{1/2}\left(1 + \frac{5h}{6a}\right)$$

approximately, h being much smaller than the Earth's radius a.

2. A particle moves under an attractive central force k/r^{α} per unit mass. Show that the orbit of the particle satisfies the differential equation

$$\frac{d^2u}{d\theta^2} + u = \frac{k}{h^2}u^{\alpha-2}$$

where h is the angular momentum per unit mass of the particle. Verify that this equation has a solution $r = a$, where $a = (h^2/k)^{1/(3-\alpha)}$. Substitute $u = a^{-1} + x$ into the differential equation and show that approximately

$$\frac{d^2x}{d\theta^2} + (3 - \alpha)x = 0$$

for small x.
   What do you infer about the stability of the circular orbit?

3. At the end of the launching process a satellite is travelling with velocity 29 600 km/h parallel to the Earth's surface at a height of 225 km. Find the subsequent greatest height achieved, the eccentricity and period of the orbit.

4. A particle is describing an ellipse of eccentricity e and latus rectum 2ℓ under a gravitational central force. When at the nearer apse, it is given a small radial velocity v. Show that the apse line is turned through an angle $\ell v/eh$ where h is the angular momentum.

5. The orbit of the Earth about the Sun is assumed to be a circle of radius R. The coplanar orbit of a comet about the Sun is parabolic and the distance between them when they are closest is $kR(k < 1)$. Prove that the comet is within the Earth's orbit for

$$(\sqrt{2/3\pi})(1 + 2k)(1 - k)^{1/2}$$

years.

6. A satellite is describing an ellipse of eccentricity e about the Earth as focus. At its greatest height above the surface, when its speed is v, it collides with a meteorite. As a result of the collision the satellite acquires an extra velocity v/n at an angle β to v and in such a sense as to increase its distance from the Earth. If n is very large show that the eccentricity is reduced by $(2/n)(1 - e)\cos\beta$ approximately.

7. Halley's comet moves in a very eccentric orbit with $e = 0.97$ in a period of about 76 years. Find its semi-major axis and the distance of its perihelion from the Sun.

8. A particle moves in a parabolic orbit under the influence of a gravitational central force. Show that its speed at a distance r from the centre is $\sqrt{2}$ times its speed in a circular orbit of radius r.

9. A satellite moves in an elliptic orbit of major axis $2a$ and eccentricity e about the Earth. When at its apogee a small retrorocket on the satellite is ignited which reduces its speed from v to $(1 - \varepsilon)v$ where $\varepsilon \ll 1$. Ignoring any changes of mass in the rocket, show that the major axis of the orbit is reduced by

$$4\varepsilon a(1 - e)/(1 + e).$$

10. A communications satellite is placed in a circular orbit in the Earth's equatorial plane in such a way that the satellite appears stationary relative to the rotating Earth. Find the height of the satellite. What is the minimum number of communications satellites required for every point on the equator to be in view of at least one satellite?

11. A ballistic missile is fired with initial speed V from a point on the Earth's surface at an angle β to the vertical. Show that the angle α between the radius to the launching site and the apogee of the missile's orbit is given by

$$\tan\alpha = \kappa\sin\beta\cos\beta/(1 - \kappa\sin^2\beta),$$

where $\kappa = RV^2/\mu_e$, R is the Earth's radius and μ_e the gravitational mass of the Earth, provided $\kappa < 1$. Show that the range of the missile is a maximum when

$$\cos 2\beta = \kappa/(\kappa - 2).$$

12. A rocket is fired from the vicinity of the Earth in the direction in which the Earth is moving in its orbit round the Sun. Find the minimum speed with which the rocket must be fired if it is to reach the orbit of Venus. Calculate the time of flight. Assume that the Earth and Venus move in coplanar circular orbits centred at the Sun and that the rocket is influenced only by the gravitational attraction of the Sun.

13. Two particles with gravitational masses μ_1 and μ_2 move in elliptic orbits under the influence of their mutual gravitational forces. Show that their orbits intersect if their eccentricity is greater than $|\mu_1 - \mu_2|/(\mu_1 + \mu_2)$.

14. A satellite is circling the Earth at a height b above its surface. A rocket is fired from a launching site in the plane of the satellite's orbit with speed V at an inclination β to the vertical. If the rocket and satellite rendezvous at the apogee of the rocket's path show that

$$R^3 V^2 \sin^2\beta = (R + b)[R(R + b)V^2 - 2b\mu_e]$$

where R is the radius of the Earth and μ_e the Earth's gravitational mass, provided

$$\frac{2(R + b)\mu_e}{R(2R + b)} > V^2 > \frac{2b\mu_e}{R(R + b)}.$$

15. A satellite is moving in a circular orbit with speed V at a height b above the Earth's surface. Retrorockets are fired which reduce the satellite's speed to kV $(0 < k < 1)$. Show that the satellite will collide with the Earth if

$$k^2 \leq 2R/(2R + b),$$

where R is the radius of the Earth. Show also that the impact velocity V_c is then given by

$$V_c^2 = \mu_e(2b + Rk^2)/(R + b)R,$$

where μ_e is the Earth's gravitational mass.

16. What is the minimum orbiting time for a Moon satellite?

17. A particle has an equation of motion

$$\ddot{\mathbf{r}} + n^2\mathbf{r} = 0$$

where $\mathbf{r} = x\mathbf{i} + y\mathbf{j}$ (the motion is simple harmonic in two-dimensions). Show that the orbit of the particle is an ellipse with its centre as the centre of force.

18. The Earth passes through a cloud of meteorites with relative speed V. By assuming a model in which the Earth is stationary and the cloud of particles approach from infinity, show that the Earth's gravitational field causes a cylinder of particles of radius

$$R(V^2 + 2gR)^{1/2}/V$$

to be deposited on the Earth's surface where g is the acceleration due to gravity at the Earth's surface and R is the radius of the Earth. Evaluate this radius when the relative velocity is 3050 m/s.

19. A satellite is describing a circular orbit about the Earth with speed V. The satellite is separated into two pieces by explosive bolts which give them an equal but opposite velocity v relative to and tangential to the orbit of the original satellite. Find the two new orbits.

20. Two particles A and B have masses m and $2m$. Initially A is at rest and B is given a speed V perpendicular to AB. If AB $= d$, find the eccentricity and period of the mutual orbits for this two-body problem.

21. Show that part of the orbit of Pluto lies inside the orbit of Neptune.

22. Find the acceleration due to gravity at the surface of Uranus and its percentage reduction at the equation of Uranus due to the rotation of the planet.

23. Three bodies with gravitational masses μ_1, μ_2, μ_3 move under their mutual gravitational forces only, and lie at the vertices of an equilateral triangle of side-length a. Relative to their combined mass-centre G their position vectors are $\mathbf{r}_1, \mathbf{r}_2, \mathbf{r}_3$. Show

that this triangular configuration can persist if it turns in its own plane at an angular rate Ω about G where $a^3\Omega^2 = \mu_1 + \mu_2 + \mu_3$. (It is thought that some asteroids on the orbit of Jupiter are in such a configuration with the Sun).

24. A satellite of the Sun is in a circular orbit of period T. Show that, if the satellite were suddenly stopped in its orbit, it would fall into the Sun in a time $\sqrt{2}\,T/8$ (really the centre of the Sun since its radius has been neglected in this formula). How long would it take for the Earth to fall into the Sun in this unlikely event?

9

Non-linear Dynamics

9.1 INTRODUCTION

Most applications in dynamics are really non-linear. Linear systems are the exception although they often receive prominence in text books since they are usually easier to solve analytically. Of the applications we have looked at so far, the following list indicates some of those which exhibit linearity:

   (i) projectiles, simple harmonic motion and some resisted motion in Chapter 4;

  (ii) the Hookean springs in Section 5.5;

 (iii) the rocket equation in Chapter 6;

 (iv) the linear oscillation theory of Chapter 7 largely by definition;

  (v) the equation for gravitational orbits of Section 8.2 (but not their time-wise evolution).

Some of these applications can be thought of as exactly linear since no approximation is involved as in the case of the planetary orbits. Others do require some approximation as in the derivation of simple harmonic motion from the pendulum equation. In this case the approximation can only be considered valid if the amplitude of the oscillations remains 'small'.

Linear problems have the significant property that different outputs combine by addition. Thus if x_1 and x_2 are two independent solutions of a linear dynamical system then $Ax_1 + Bx_2$ is also a solution for any constants A and B. This means that any solution can be expressed in terms of its 'component' solutions as we saw in Chapter 7. In non-linear systems no such order generally prevails: apart from both being solutions of the differential equation of a non-linear dynamical system, solutions starting from different initial data may have no common features.

Simple analytic techniques become scarce for even quite elementary non-linear dynamical applications so that we have to resort to various approximations, qualitative methods and straight numerical solutions. In this chapter we intend to describe a few of the more readily available approaches under these headings.

9.2   THE SIMPLE PENDULUM AGAIN

As an initial working example let us consider again the equation of the simple pendulum given by Equation (22) of Section 4.6, namely

$$mga \sin \theta + ma^2 \ddot{\theta} = 0 \tag{1}$$

where a is the length of the pendulum and θ its inclination to the downward vertical (see Figure 4.9). As we saw in the same section the equation can be integrated once with respect to θ to produce

$$\tfrac{1}{2}ma^2 \dot{\theta}^2 - mga \cos \theta = C, \quad \text{a constant.} \tag{2}$$

Equation (2) displays the principle of conservation of energy: the sum of the kinetic and potential equations is constant. It is not possible to integrate (2) again in terms of elementary functions. However, we can obtain much qualitative information about the simple pendulum from Equations (1) and (2).

In Section 5.6 we defined equilibrium of a particle as the state in which the forces acting on the particle balanced and the particle was at rest. By Newton's second law of motion this is equivalent to the acceleration and velocity both ·being zero for a single particle. For the simple pendulum, equilibrium is given by the two conditions $\dot{\theta} = 0$ and $\ddot{\theta} = 0$. Suppose that we now introduce a $(\theta, \dot{\theta})$ plane and plot on it the family of curves defined by (2) for a selection of values of the constant C: in this case C is the particular energy level of the curve plotted. From (1) equilibrium occurs where $\sin \theta = 0$ or $\theta = n\pi(n = 0, \pm 1, \ldots)$. These are *solutions* of (1) which appear as points on the θ-axis in the $(\theta, \dot{\theta})$ plane (see Figure 9.1). In (2) they correspond to $C = \pm mga$. However, the periodicity of the problem means that, for example, in the equilibrium states $\theta = 0, \pm 2\pi, \pm 4\pi, \ldots$, the pendulum is physically in the same position with its bob below the point of suspension.

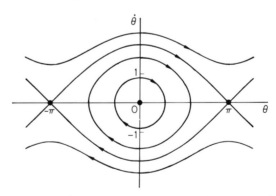

Figure 9.1   Computed phase diagram for the pendulum equation (1). The individual paths are given by (2) but it is easier to compute the paths by numerical integration of (1). In the graph shown $g/a = 1$.

With $C = -mga$ Equation (2) only has the real solutions corresponding to $\cos \theta = 1$, $\dot{\theta} = 0$, since $|\cos \theta| \le 1$ for all θ. On the other hand for $C = mga$ the paths in the $(\theta, \dot{\theta})$ plane become

$$\tfrac{1}{2}a\dot{\theta}^2 - g \cos \theta = g \quad \text{or} \quad a\dot{\theta}^2 = 2g(1 + \cos \theta).$$

Between $\theta = -\pi$ and $\theta = \pi$, $\dot{\theta} = \pm[2g(1 + \cos \theta)/a]^{1/2}$ with the result that the corresponding paths (known as *separatrices*) join the equilibrium points at $\theta = \pm\pi$ as shown in Figure 9.1 both above and below the θ axis. For $|C| < mga$, $\dot{\theta}$ must vanish at pairs of values where $\cos \theta = -C/mgla$. These paths appear as closed trajectories surrounding the origin. Finally for $C > mga$, $\dot{\theta}$ is never zero with the result that the paths remain either entirely above or below the θ axis as shown in Figure 9.1. From the periodicity of $\dot{\theta}$ for any fixed C we infer that copies of the diagram in the interval $|\theta| \le \pi$ are reproduced in each of the intervals $(2n - 1)\pi \le \theta \le (2n + 1)\pi$ $(n = \pm 1, \pm 2, \ldots)$.

By examining the qualitative nature of Figure 9.1 we can interpret possible motions of the pendulum. Equilibrium at $\theta = 0$ corresponds to the bob at rest below its point of suspension whilst $\theta = \pi$ (or $\theta = -\pi$) is a position of (unstable) equilibrium above. Closed paths indicate oscillations of the pendulum about equilibrium at $\theta = 0$. The paths joining the equilibrium points at $\theta = -\pi$ and $\theta = \pi$ correspond to the swing of the pendulum in which it is disturbed from its upturned equilibrium and just (theoretically) returns to that state. The paths totally above or below the θ axis correspond to whirling motions of the pendulum in one direction or the other.

Figure 9.1 displays what is known as a *phase diagram* of the simple pendulum. The individual curves are known as *phase paths, trajectories*, or *orbits*. The sense in which the phase paths are described (by the arrows) in Figure 9.1 follows since, for $\dot{\theta} > 0$, θ must increase with time whilst for $\dot{\theta} < 0$, θ decreases with time. Hence for $\dot{\theta} > 0$ the phase paths must be followed from left to right with the contrary direction for $\dot{\theta} > 0$.

9.3   THE PHASE PLANE

Let us now move away from the particular example of the previous section and consider the phase plane in more general cases. The one-dimensional motion of particle is usually represented by a second order differential equation of the form

$$\ddot{x} = f(x, \dot{x}, t) \tag{3}$$

where we can interpret x as displacement (although it could be an angle, length on a curve, area, etc in any actual problem), $\dot{x}$ as velocity and $\ddot{x}$ as acceleration. Equation (3) can be thought of as representing Newton's equation of motion in which $f(x, \dot{x}, t)$ is viewed as force per unit mass. If f is not a function

t explicitly, that is, $\partial f/\partial t = 0$, then the system described by (3) is said to be *autonomous*: if t is present then the system is said to be *nonautonomous* or *forced*. The phase plane is particularly useful for autonomous applications and for the remainder of this section we shall assume that

$$\ddot{x} = f(x, \dot{x}). \tag{4}$$

The system is said to be in *equilibrium* when its velocity and acceleration both vanish, that is, $\dot{x} = \ddot{x} = 0$. Hence *equilibrium* (or *critical*) *points* are solutions of the equation $f(x, 0) = 0$. Now set up the *phase plane* with axes x and $y = \dot{x}$. It follows that equilibrium points must lie on the x axis. The introduction of y means that we can replace (4) by the pair of *first-order* equations

$$\dot{x} = y, \qquad \dot{y} = f(x, y). \tag{5}$$

The *phase paths* of the system are given by the family of solutions of the first-order equation

$$\frac{dy}{dx} = \frac{\dot{y}}{\dot{x}} = \frac{f(x, y)}{y}. \tag{6}$$

Effectively time has now been removed.

We can now list some general observations about phase paths and diagrams:

(i) for $y > 0$ the paths must be described from left to right (since x must be increasing with time) and for $y < 0$ from right to left,

(ii) except possibly at equilibrium points, phase paths cut the x axis at right angles,

(iii) the direction of motion on the phase paths is known but the rate at which phase paths are covered cannot be inferred from the phase diagram,

(iv) except possibly at equilibrium points there will be only one phase path passing through a given point in the phase plane,

(v) phase paths for autonomous systems are not sensitive to the initial time: in other words the time scale can be translated by a constant shift without affecting either the equations or the phase diagram.

Example 1   *Find the equilibrium points and phase paths of the dynamical system governed by the equation*

$$\ddot{x} = x - x^3.$$

This is equivalent to the first-order system

$$\dot{x} = y, \qquad \dot{y} = x - x^3.$$

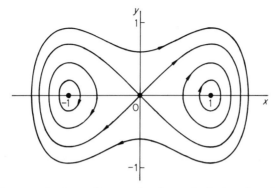

Figure 9.2 Computed phase diagram for Example 1 showing three equilibrium points.

The equilibrium points are defined by $x - x^3 = x(1 - x)(1 + x) = 0$ and hence they must occur at $(-1, 0)$, $(0, 0)$, $(1, 0)$ as shown by the dots in Figure 9.2. The phase paths are given by the solutions of

$$\frac{dy}{dx} = \frac{\dot{y}}{\dot{x}} = \frac{x - x^3}{y} \qquad y \neq 0.$$

This is a first-order separable differential equation with general solution

$$\int y\,dy = \int (x - x^3)\,dx + C \quad \text{or} \quad \tfrac{1}{2}y^2 = \tfrac{1}{2}x^2 - \tfrac{1}{4}x^4 + C.$$

Hence

$$y = \pm(x^2 - \tfrac{1}{2}x^4 + 2C)^{1/2} \tag{7}$$

which implies that the paths are symmetrically disposed about the x axis. A computed phase diagram is shown in Figure 9.2 but before embarking on the computation we really require some clues as to general features which we might expect. In the next two sections we shall look particularly at the classification of equilibrium points. It should be observed in this example that the equilibrium points at $x = \pm 1$ are surrounded by nests of oscillations.

9.4 CONSERVATIVE SYSTEMS

Systems in which f is also independent of $\dot{x}$ and depends on x only are said to be *conservative*, and they are particularly significant in dynamics. They are characterised by

$$\ddot{x} = f(x) \quad \text{or} \quad \dot{x} = y,\ \dot{y} = f(x). \tag{8}$$

The equilibrium points occur where $f(x) = 0$ and the phase paths are given by solutions of the separable equation

$$\frac{dy}{dx} = \frac{f(x)}{y}.$$

Hence, after integration, the paths can be represented by the family

$$\int y \, dy = \int f(x) \, dx + C \quad \text{or} \quad \tfrac{1}{2}y^2 = \int f(x) \, dx + C. \tag{9}$$

We can interpret (9) as conservation of energy for a particle of unit mass with x as displacement and $\dot{x}$ velocity (see Section 5.7). Hence

$$\mathcal{T} = \tfrac{1}{2}y^2, \qquad \mathcal{V} = -\int f(x) \, dx$$

can be thought of as the kinetic and potential energies of the particle (this is why these particular systems are called conservative).

Since $d\mathcal{V}/dx = f(x)$, equilibrium points must occur where the potential energy is stationary. We can associate three types of stationary value with a function of a real variable: the minimum, the maximum and the point of inflection. Each has its characteristic local phase diagram. We can solve (9) in the form

$$y = \pm(2C - 2\mathcal{V}(x))^{1/2} \tag{10}$$

and the phase paths can be constructed as follows. Suppose that two planes are chosen, $2\mathcal{V}(x)$ against x and y against x with the first immediately above the other as shown in Figure 9.3. The upper diagrams show typical stationary values for $2\mathcal{V}(x)$. In Figure 9.3(a) choose a value of $2C$ greater than the stationary minimum value of $2\mathcal{V}(x)$. The phase path, given by (10), will cut the

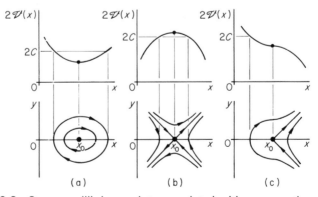

Figure 9.3 Some equilibrium points associated with conservative systems.

x-axis where x satisfies $\mathscr{V}(x) = C$ and y can only be real for x lying between the two solutions of this equation. The result is a closed path surrounding the equilibrium point at $x = x_0$. Other values of C (greater than $\mathscr{V}(x)$) will also generate further closed orbits as shown. Locally this part of the phase diagram is known as a *centre*. Figure 9.3(b) shows a maximum of $2\mathscr{V}(x)$. If we select values for C and compute the square root of the difference between $2C$ and $2\mathscr{V}(x)$, then we expect the phase diagram below the maximum to be characteristic locally of the energy maximum. Such phase diagrams are known as *saddle points*. Any path which either enters or leaves the equilibrium point is known as a *separatrix*. Following the remarks in Section 5.6, a centre indicates *stable* equilibrium whilst a saddle point must be *unstable*. The phase diagram associated with a point of inflection is shown in Figure 9.3(c).

The phase diagram for the simple pendulum shown in Figure 9.1 has centres at $\theta = 0, \pm 2\pi, \pm 4\pi, \ldots$ and saddle points at $\theta = \pm \pi, \pm 3\pi, \ldots$.

The potential energy arising in Example 1 should now be investigated.

Example 2   Find the equation of the phase paths and sketch the phase diagram for the simple harmonic oscillator with equation of motion $\ddot{x} + \omega^2 x = 0$ (Equation (19), Section 4.5).

Let $\dot{x} = y$. The differential equation for the phase paths is

$$y \frac{dy}{dx} = -\omega^2 x$$

which is first order separable with solution

$$\int y \, dy = -\omega^2 \int x \, dx + C \quad \text{or} \quad \omega^2 x^2 + y^2 = 2C.$$

The system is conservative with potential energy $\mathscr{V}(x) = \frac{1}{2}\omega^2 x^2$, which has a minimum at $x = 0$. The simple harmonic oscillator has one equilibrium point at $x = 0$, and the phase paths are ellipses centred on the origin as shown in Figure 9.4.

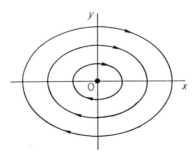

Figure 9.4   Centre for the simple harmonic oscillator with $\omega = 1.4$.

Example 3   *Sketch the phase paths for $\ddot{x} - \omega^2 x = 0$, $(\omega > 0)$.*

In this case the phase paths are given by the family

$$y^2 - \omega^2 x = C$$

with one equilibrium point at $x = 0$. The trajectories are branches of a hyperbola which has asymptotes or separatrices $y = \pm \omega x$. Some typical paths are shown in Figure 9.5.

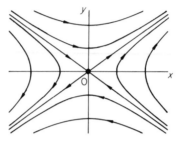

Figure 9.5   The saddle point of Example 3 with $\omega = 0.7$.

9.5   DAMPED SYSTEMS

The introduction of friction or damping into a dynamical system has a marked effect on the phase diagram of an otherwise conservative system. For example, the equation of motion of a simple pendulum with linear damping could be represented by the equation of motion (we have now taken the mass of the bob, $m = 1$, without loss of generality)

$$\ddot{x} + c\dot{x} + k \sin x = 0 \tag{11}$$

where $c > 0$, $k > 0$ are constants (we shall now use the standard notation x, $y = \dot{x}$ for the phase plane rather than the angle θ). The equilibrium points of (11) still occur where $\sin x = 0$ at $x = 0$, $\pm \pi$, $\pm 2\pi$, ... However, the differential equation for the phase paths, namely

$$\frac{dy}{dx} = \frac{-cy - k \sin x}{y} \tag{12}$$

has now no simple solutions. We have to resort to numerical methods of solution: it is usually easier to program the integration of the first-order equations

$$\dot{x} = y, \quad \dot{y} = -cy - k \sin x \tag{13}$$

from an initial point rather than (12) (see the Appendix for details of numerical routines). The resulting phase diagram for this damped system is shown in Figure 9.6, and it should be compared with Figure 9.1 for the undamped case. Almost all paths now terminate in spirals (if $c^2 < 4k$) which approach the equilibrium points at $x = 2n\pi$ $(n = 0, \pm 1, \ldots)$. The exceptions are the separatrices which terminate at the saddle points at $x = (2n + 1)\pi$ $(n = 0, \pm 1, \ldots)$. Note that the unstable saddle points persist but that the centres locally have undergone a substantial structural change into spirals.

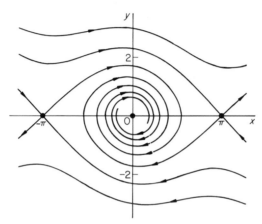

Figure 9.6 Phase diagrams for the damped pendulum given by (11) with $c = 0.15$, $k = 1$.

We can investigate the phase diagram in the neighbourhood of the origin (and similarly, any other stable equilibrium points) by *linearising* (11) for small $|x|$. Since $\sin x \approx x$, Equation (11) becomes, for small $|x|$,

$$\ddot{x} + c\dot{x} + kx = 0 \tag{14}$$

and (13)

$$\dot{x} = y, \qquad \dot{y} = -cy - kx. \tag{15}$$

We could integrate dy/dx given by (15) but the answer is not illuminating: it is more helpful to solve (14) instead for x and $\dot{x}$ in terms of t. Also the solutions are to hand in Section 7.3, namely

$$x = A \exp(-p_1 t) + B \exp(-p_2 t) \tag{16}$$

$$y = \dot{x} = -p_1 A \exp(-p_1 t) - p_2 B \exp(-p_2 t) \tag{17}$$

where

$$p_1 = \tfrac{1}{2}[c - (c^2 - 4k)^{1/2}], \qquad p_2 = \tfrac{1}{2}[c + (c^2 - 4k)^{1/2}]$$

and A and B are determined by the initial conditions.

If $c^2 < 4k$ then weak damping occurs (see Section 7.3 again) and the roots p_1, p_2 are complex with positive real part. The solutions show damped oscillatory behaviour which leads to the *spiral* equilibrium point (sometimes known as a *focus*) shown typically in Figure 9.7(a).

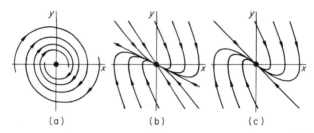

(a)   (b)   (c)

Figure 9.7   Linear models for asymptotically stable equilibrium points: (a) the spiral, (b) the node, (c) the critical node.

If $c^2 > 4k$, then strong damping occurs since p_1 and p_2 are both real and positive. The phase paths defined parametrically by (16) and (17) now show no oscillatory behaviour as displayed in Figure 9.7(b). Such an equilibrium point is known as a *node*. Four paths are straight lines corresponding to initial data for which $A = 0$ or $B = 0$. Thus if $A = 0$, $B \neq 0$ then two paths into the origin lie on $y = -p_2 x$, and if $A \neq 0$, $B = 0$ then two paths lie on $y = -p_1 x$. In the critical case for which $c^2 = 4k$ these two lines merge and the phase diagram is that of the *critical node* shown in Figure 9.7(c).

Both the spiral and the node are stable but they have the additional property of what is known as *asymptotic stability* since $x \to 0$ as $t \to \infty$. For the damped pendulum the suspended equilibrium state of the bob is a spiral (Figure 9.6) for weak damping which means that physically the to-and-fro motion will be progressively damped out.

It is possible to have unstable spirals and nodes when *negative damping* occurs. This happens when $c < 0$ with k still positive. The phase paths are now directed outwards from the equilibrium point (see Figure 9.8).

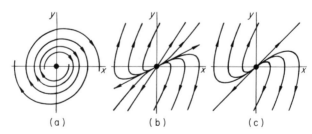

(a)   (b)   (c)

Figure 9.8   Unstable equilibrium points: (a) the unstable spiral, (b) the unstable node, (c) the unstable critical node.

Returning to the original Equation (11) we expect the system to behave as a spiral or node close to $x = 0$ as predicted by the linearisation and confirmed by the computed phase diagram in Figure 9.6. Near to the unstable equilibrium point at $x = \pi$, let $x = \pi + x'$. Then $\sin x = \sin(\pi + x') = -\sin x' \approx -x'$ for $|x'|$ small. Thus close to this equilibrium point x' satisfies

$$\ddot{x}' + c\dot{x}' - kx' = 0.$$

The characteristic equation is

$$p^2 + cp - k = 0$$

with roots

$$p_1 = \tfrac{1}{2}[-c + (c^2 + 4k)^{1/2}], \qquad p_2 = \tfrac{1}{2}[-c - (c^2 + 4k)^{1/2}].$$

For $c > 0$, both roots are real but of opposite sign which implies that this equilibrium point still remains a saddle as in the undamped case.

This approach using linearisation of the equation close to an equilibrium point enables us to identify the type of equilibrium point within the categories of centres, saddle points, spirals or nodes. Thus, if

$$\dot{x} = y, \qquad \dot{y} = f(x, y)$$

and the system has an equilibrium point at $x = x_0$, $y = 0$, that is $f(x_0, 0) = 0$, and $f(x, y) \approx a(x - x_0) + by$ (the leading term in its Taylor expansion), then the linearised equations become

$$\dot{x}' = y, \qquad \dot{y} = ax' + by$$

where $x' = x - x_0$. The classification of the equilibrium point depends on the three parameters a, b and $\Delta = b^2 + 4a$. The various types are set out in the parameter diagram Figure 9.9. For example, if $a < 0$, $b < 0$ and $\Delta > 0$ then the equilibrium point is a stable node, and so on.

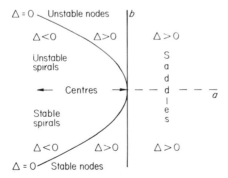

Figure 9.9 Classification and stability diagram for linear equilibrium points.

Example 4   A particle of unit mass with displacement x is subject to a damping force $k\dot{x}$ $(k > 0)$ and a restoring force $-x + x^2$. Find and classify the equilibrium points of the system.

The equation of motion is given by

$$\ddot{x} + k\dot{x} - x + x^2 = 0 \tag{18}$$

or, as first-order system,

$$\dot{x} = y, \qquad \dot{y} = -ky + x - x^2. \tag{19}$$

In equilibrium, $y = 0$ and $x(1 - x) = 0$. Hence there are two equilibrium points at $x = 0$ and $x = 1$. We now linearise (19) in the neighbourhood of each of these points.

Near the origin, $a = 1$, $b = -k$ and $\Delta = k^2 + 4$. Reading from Figure 9.9 we see that the origin is a saddle point.

Near $x = 1$, let $x = x' + 1$ in (19) and retain only first degree terms. Thus

$$\dot{x}' = y, \qquad \dot{y} = -ky + x' + 1 - (x' + 1)^2 \approx -x' - ky.$$

In this case, $a = -1$, $b = -k$ and $\Delta = k^2 - 4$. Hence this point is a stable node or spiral according as $k > 2$ or $k < 2$.

The full phase diagram, computed numerically, is shown in Figure 9.10 for $k = 0.1$. The shaded region shows the set of initial values for x and y from which solutions ultimately reach the stable spiral at $x = 1$. This is known as the *domain of asymptotic stability* of $x = 1$. Stable nodes and spirals are also known as *attractors*: unstable ones are *repellors*.

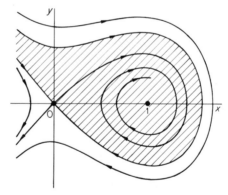

Figure 9.10   Phase diagram for Example 4 with damping coefficient $k = 0.1$. The shaded region shows the domain of asymptotic stability of the spiral at $x = 1$.

Note that linearisation might fail to predict correctly in the case of centres as implied by the following counter-example. Compare, near the origin, the computed phase diagram of

$$\ddot{x} + \dot{x}(x^2 + \dot{x}^2) + x = 0$$

(a spiral) with its linear approximation which is a centre.

9.6 THE TOPOLOGY OF PHASE DIAGRAMS

Phase diagrams can give details of the behaviour of a system but they are particularly useful for the qualitative picture they suggest. For example, we may think of the centre as the collection of phase diagrams which are locally and topologically the same as the phase diagram of the simple harmonic oscillator (Example 2, Section 9.4). By 'topological' equivalence of phase diagrams we mean that in some neighbourhood of the equilibrium one phase diagram can be continuously deformed into or mapped onto the other. If a centre appears in part of a phase diagram then we can infer that locally the system will have features in common with the simple harmonic oscillator: that is, the equilibrium state will be surrounded by a nest of closed paths indicating a periodic motion. In the same way we can regard the saddle point, spiral and node for the linear equation in the previous section as exemplars of their types.

As we have already remarked, closed orbits in the phase plane generally indicate periodic motion, and they are a particularly interesting feature in non-linear dynamics. Consider the following model problem with x satisfying

$$\ddot{x} - \varepsilon(1 - x^2 - \dot{x}^2)\dot{x} + x = 0 \qquad (20)$$

where $\varepsilon > 0$ is a constant. Mechanically the system experiences a linear restoring force but the nonlinear damping is negative inside the circle $x^2 + y^2 = 1$ but positive outside it. These two competing effects create an isolated closed orbit in the phase diagram. The corresponding first-order system is

$$\dot{x} = y, \qquad \dot{y} = -x + \varepsilon(1 - x^2 - y^2)y. \qquad (21)$$

Clearly the system has just one equilibrium point at $x = 0$, which according to the linear theory is an unstable spiral if $\varepsilon < 2$. We cannot solve (20) or (21) in general, but we can easily verify that $x = \cos t$, $y = -\sin t$ satisfies (21). The phase path generated by this particular solution is the circle $x^2 + y^2 = 1$. Computed phase paths (Figure 9.11) indicate that this closed path is *isolated* with no neighbouring closed orbits, but with phase paths spiralling into it from both inside and outside. Such an isolated closed orbit is known as a *limit cycle*. Apart from the unstable equilibrium at the origin, all other solutions will approach this oscillation.

This equation is unusual in that the limit cycle solution can be obtained explicitly: in most applications we have to rely on numerical methods backed up by qualitative analysis. Limit cycles are genuinely nonlinear phenomena which means that they cannot be generated by linear systems. Fortunately initial value problems are fairly easy to handle numerically and phase paths can be readily displaced on VDU screens using graphical software (see the Appendix).

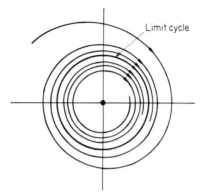

Figure 9.11   Phase diagram for (21) with $\varepsilon = 0.1$ showing the limit cycle.

9.7   COULOMB DRY FRICTION

In this section we shall examine a mechanical device which generates a limit cycle (see Exercise 7, Chapter 7 for an earlier reference to this subject). Dry (or Coulomb) friction occurs when two dry surfaces are in contact and in relative motion. Without lubrication the surfaces tend to stick together for a time and then suddenly slip. The following device illustrates the main characteristics of dry friction. A continuous belt is driven by rollers at a constant speed u as shown in Figure 9.12. A block of mass m sits on the belt, and is restrained by a spring of stiffness k. If F is the frictional force between the block and spring, then the equation of motion of the block is

$$-kx - F = m\ddot{x} \tag{22}$$

where x is the displacement of the block beyond the natural length of the spring. The usual Coulomb assumption is that the force depends on the *slip velocity*, $\dot{x} - u$. We might expect the force to be an *odd function* of $\dot{x} - u$ since it will reverse if the slip velocity reverses, and also that $F(0) = 0$. Experimental evidence indicates that for $\dot{x} - u > 0$ and small, $F'(\dot{x} - u)$ is large and positive

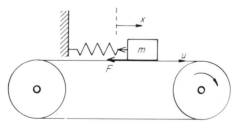

Figure 9.12   Block restrained by a spring and driven by a constant-speed belt with Coulomb dry friction.

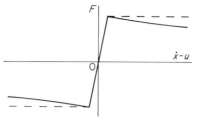

Figure 9.13 Frictional force plotted against slip velocity showing actual graph (——) and approximation (---).

until suddenly it changes sign as $(\dot{x} - u)$ increases, and $F(\dot{x} - u)$ slowly decreases and approaches a positive constant value as shown in Figure 9.13, at least within the operating limits on the device.

Let $y = \dot{x}$. The block has one equilibrium point as $x = -F(-u)/m$. Consider the perturbation $x = x' - F(-u)/m$. Then

$$\dot{y} = -\frac{kx}{m} - \frac{F(y-u)}{m} \approx -\frac{kx'}{m} - \frac{F'(-u)y'}{m}$$

so that in the notation of Section 9.5

$$a = -\frac{k}{m}, \quad b = -\frac{F'(-u)}{m}, \quad \Delta = \frac{F'(u)^2}{m^2} - \frac{4k}{m}.$$

The classification of this equilibrium point depends on the signs of b and Δ. In practice $F'(-u)$ is usually in the domain where $F'(-u) < 0$: hence the equilibrium point is an unstable spiral or node depending on the sign Δ.

Figure 9.14 shows a phase diagram for a piecewise linear friction force in which the curves in Figure 9.13 are replaced by pairs of straight lines shown

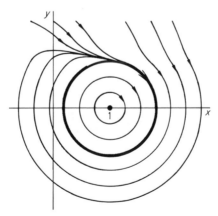

Figure 9.14 Phase diagram for the Coulomb dry friction approximation with $k = 1$, $v = 0.2$, $\lambda = 5$.

by the broken lines. The actual formula for $F(\dot{x} - u)$ is defined below

$$F(\dot{x} - u) = \lambda(\dot{x} - u) \qquad \begin{array}{ll} \lambda v & \dot{x} - u > v \\ & |\dot{x} - u| \le v \\ -\lambda v & \dot{x} - u < v \end{array}$$

with $k = 1$, $v = 0.2$, $\lambda = 5$ in the figure. In this idealised form the system has a *centre* at $x = \lambda v/k$ which is bounded by a critical closed path: all paths which start outside this periodic solution approach it asymptotically. For the more realistic model in which F decreases for $\dot{x} - u > v$ the centre becomes an unstable spiral. The bunch of paths which approach it very close to the inclined line from the left are in states in which the block is almost adhering to the belt.

This model was originally put forward by Lord Rayleigh (1842–1919) for the bowing of a violin string: the block is a section of the string, the spring corresponds to the transverse restoring effect of the string and the belt models the bow.

9.8   CHAOTIC OSCILLATIONS

So far in this and earlier chapters all the systems investigated have shown outputs which have some regular behaviour in the long term: either the system approaches equilibrium or a regular oscillation, or the limiting state is a linear combination of oscillations. Some of these states may be unstable, but at least theoretically, they show equilibrium or periodic characteristics. In the past two decades with the widespread use of computers it has been realised that theoretical models of quite simple mechanical devices can display highly irregular or seemingly random outputs.

Consider the forced spring–dashpot system shown in Figure 9.15. A block of mass m is attached to a vibrating support by a spring and dashpot. If $F(x)$

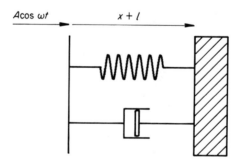

Figure 9.15   Block restrained by a nonlinear spring and dashpot which are attached to a vibrating support.

is the restoring force in the spring and x is the extension of the spring, then the equation of motion becomes

$$-F(x) - c\dot{x} = m \frac{d^2}{dt^2} (x + A \cos \omega t) = m(\ddot{x} - A\omega^2 \cos \omega t).$$

Thus x satisfies

$$m\ddot{x} + c\dot{x} + F(x) = mA\omega^2 \cos \omega t. \tag{23}$$

In the application to the seismograph in Section 7.5, the spring was assumed to be one satisfying Hooke's law (Section 5.5) which led to a linear differential equation. For non-Hookean springs the relation between restoring force and the extension of the spring becomes nonlinear. As explained in Section 5.5 springs for which the restoring effect grows at a slower rate than the extension are called *soft*, and those which grow at a higher rate are known as *hard* springs (Figure 9.16). We shall briefly investigate here the case of hard springing in which the restoring force has the cubic behaviour given by $F(x) = \beta x^3$, where β is a positive constant. We can reduce the number of parameters by rescaling displacement x and the time t. Let $t = s/\omega$ and $x = \omega z(m/\beta)^{1/2}$. Then Equation (23) becomes

$$\ddot{z} + \alpha\dot{z} + z^3 = \Gamma \cos s \tag{24}$$

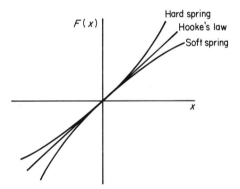

Figure 9.16   Force plotted against extension for hard and soft springs.

where $\alpha = C/(m\omega)$ and $\Gamma = (A/\omega)(\beta/m)^{1/2}$ ('dots' now refer to differentiation with respect to the dimensionless variable s). In this non-dimensional form the equation and its solutions are characterised by just two parameters α and Γ, which we can suppose are non-negative. As we shall see this model has extremely complex behaviour. Equation (24) is a special case of *Duffing's equation* and its solutions have been extensively mapped by Ueda.

Consider first the limiting case in which damping and forcing are absent, that is, $\alpha = \Gamma = 0$. The equation reduces to the autonomous system

$$\ddot{z} + z^3 = 0. \tag{25}$$

If we let $y = \dot{z}$, then the equation for the phase paths becomes

$$\frac{dy}{dz} = \frac{\dot{y}}{\dot{z}} = -\frac{z^3}{y}$$

which is of first-order separable type. Thus

$$\int y \, dy = -\int z^3 \, dz + C \quad \text{or} \quad \tfrac{1}{2}y^2 + \tfrac{1}{4}z^4 = C.$$

The system has one equilibrium point at the origin, and the phase paths are closed curves about this point. This classifies the origin as a centre as shown in Figure 9.17(a). As a matter of interest, linearisation fails near the origin since the equation would reduce to $\ddot{z} = 0$, which implies that $y = \text{constant}$.

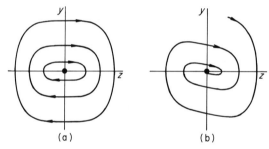

(a)   (b)

Figure 9.17   (a) the undamped, unforced centre of (25) at the origin, (b) the damped, unforced spiral with $\alpha = 0.2$.

The introduction of damping through $\alpha > 0$ but still with $\Gamma = 0$ perturbs the centre into a stable spiral (Figure 9.17(b)). We shall investigate how the output behaves for a fixed value of α as we increase the forcing amplitude from zero. Whilst a complete description can only be achieved by a comprehensive computer scan we can show one feature by estimating the amplitude of the *first harmonic* of any period solutions. The existence of stable oscillations of the system is of particular interest.

Let

$$z = a \cos s + b \sin s \tag{26}$$

where a and b are constants. Then, we can infer from substitution in (24) that

$$\ddot{z} + \alpha \dot{z} + z^3 - \Gamma \cos s$$
$$= -a \cos s - b \sin s - \alpha a \sin s + \alpha b \cos s + (a \cos s + b \sin s)^3 - \Gamma \cos s.$$
$$(27)$$

Now

$$(a \cos s + b \sin s)^3 = a^3 \cos^3 s + 3a^2 b \cos^2 s \sin s + 3ab^2 \cos s \sin^2 s$$
$$+ b^3 \sin^3 s$$
$$= (a^3 - 3ab^2) \cos^3 s + (b^3 - 3a^2 b) \sin^3 s$$
$$+ 3ab^2 \cos s + 3a^2 b \sin s$$
$$= \tfrac{3}{4} a(a^2 + b^2) \cos s + \tfrac{3}{4} b(a^2 + b^2) \sin s$$
$$+ \tfrac{1}{4} a(a^2 - 3b^2) \cos 3s + \tfrac{1}{4} b(3a^2 - b^2) \sin 3s \qquad (28)$$

using the identities

$$\cos^3 s = \tfrac{3}{4} \cos s + \tfrac{1}{4} \cos 3s, \quad \sin^3 s = \tfrac{3}{4} \sin s - \tfrac{1}{4} \sin 3s.$$

Thus (27) and (28) imply

$$\ddot{z} + \alpha \dot{z} + z^3 - \Gamma \cos s = [-a + \alpha b + \tfrac{3}{4} a(a^2 + b^2) - \Gamma] \cos s$$
$$+ [-b - \alpha a + \tfrac{3}{4} b(a^2 + b^2)] \sin s + \text{higher harmonics.}$$

Hence we can say that (24) is satisfied as far as the first harmonics are concerned if the coefficients of $\cos s$ and $\sin s$ are zero. Thus a and b must satisfy

$$-a + \alpha b + \tfrac{3}{4} a(a^2 + b^2) = \Gamma \qquad (29)$$
$$-b - \alpha a + \tfrac{3}{4} b(a^2 + b^2) = 0. \qquad (30)$$

Square and add these equations and let $r = (a^2 + b^2)^{1/2}$; the result is that r^2 satisfies the cubic equation

$$r^2(1 + \alpha^2 + \tfrac{9}{16} r^4 - \tfrac{3}{2} r^2) = \Gamma^2. \qquad (31)$$

If we introduce a phase ε into (26) by putting $a = r \sin \varepsilon$, $b = r \cos \varepsilon$, then $z = r \sin (s + \varepsilon)$. It follows that r is the amplitude of the first harmonic.

We shall put $\alpha = 0.2$ throughout although not all possibilities by any means are covered by this choice. For fixed α, Equation (31) relates the forcing amplitude Γ to the output amplitude r. Figure 9.18 shows the graph of r against Γ. For the particular value of α chosen there appear to be *three* output amplitudes for $0.229 \ldots < \Gamma < 0.464 \ldots$.

These first harmonic estimates encourage us to search for computed periodic solutions in the neighbourhood of their predicted locations. This can be done by starting solutions at (a, b) (which can be calculated from (29), (30) and (31)) at $s = 0$, and letting the computer program run until it settles into a periodic

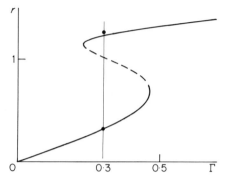

Figure 9.18 Output amplitude r plotted against forcing amplitude Γ for Ueda's equation with $\alpha = 0.2$ (see also Figure 9.19).

solution. Only two of the three oscillations in the critical interval of Γ seem to be stable and these correspond to the solid lines in Figure 9.18. These stable harmonics coexist at the same amplitude: a pair is shown in Figure 9.19 for $\Gamma = 0.3$. The dots on the closed orbits indicate the phase of the oscillations and their periods: the solutions pass through these points when $s = 0$ and multiples of 2π.

As we have already remarked the structure of solutions of (24) is surprisingly complex, and can only really be investigated numerically. Figure 9.20 shows just some more of the possible outputs for different values of Γ. These include periodic solutions, *subharmonics* (that is, periodic solutions with periods $2\pi n$ ($n = 2, 3, \ldots$) and , in Figure 9.21, a solution which appears to be bounded but not periodic even after any transience has died down. An interval of time output

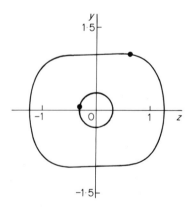

Figure 9.19 Two computed stable 2π-periodic solutions for equation (24) with $\alpha = 0.2$ and $\Gamma = 0.3$. There amplitudes are shown by the dots on the line through $\Gamma = 0.3$ in the previous figure, where they may be compared with the first harmonic.

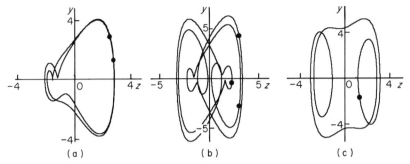

Figure 9.20   Three periodic solutions of (24) all with damping coefficient $\alpha = 0.2$: (a) shows a period 2 solution for $\Gamma = 5.5$; (b) is period 3 for $\Gamma = 10$; (c) is period 1 for $\Gamma = 13.5$.

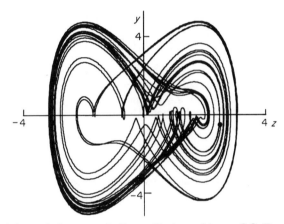

Figure 9.21   A bounded non-periodic oscillation with $\alpha = 0.2$, $\Gamma = 7$. The starting point is at $(2.4748, -0.5304)$ at time $s = 0$ (indicated by the dot), abd 20 cycles are shown.

of the orbit in Figure 9.21 is shown in Figure 9.22(a). A comprehensive catalogue of orbits and their parameter domains has been compiled by Ueda (see Further Reading).

The output shown in Figure 9.21 is described as *chaotic*, but we have to exercise care in the use of this term. We need to distinguish chaotic behaviour from that in other systems which might appear to show similar characteristics. It is possible to design even linear systems which have bounded and wandering orbits. For example, the damped linear oscillator

$$m\ddot{x} + c\dot{x} + kx = A_1 \cos \omega_1 t + B_1 \cos \omega_2 t \qquad (32)$$

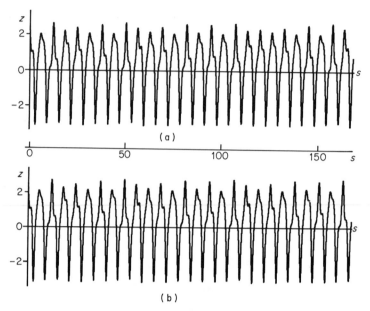

Figure 9.22 The upper oscillation (a) shows the (z, s) output with initial data $z = 2.474\,80$, $y = -0.530\,40$ at $s = 0$. The lower diagram shows the output if $z = 2.47\,481$ but y and s are unchanged. The solutions appear to be the same until $s \approx 120$ when they diverge.

driven by two harmonic forcing terms with frequencies ω_1 and ω_2 such that ω_1/ω_2 is not a rational number will exhibit what appears to be a similar phenomenon.

What characterises the difference between the systems? In the solution of (32) just two frequencies are present, namely, ω_1 and ω_2. A *frequency spectrum* of this solution would reveal them. The solution in 9.22(a), on the other hand, turns out to have a continuous frequency spectrum implying that all frequencies in some interval are present. We shall say more about this in the next section.

One more feature which chaotic oscillations possess is sensitivity to initial data: a slight change in the initial conditions leads after a few cycles to a completely different output. A typical example is shown in Figure 9.22. In any real problem initial data can only be approximate anyway so that after a short time, the link between an output and its initial data is effectively broken. This is the so-called *butterfly effect* in dynamical systems; particularly in the case of the weather the flapping of a butterfly's wing can affect the weather in a month's time. It has the result of making predictions about the future behaviour of certain mechanical systems impossible. It is thought that the satellite *Hyperion* of Saturn rotates or tumbles in chaotic manner. Even though Newtonian mechanics is theoretically *deterministic*, errors (which always occur in any data) make the answers to certain applications in dynamics effectively random.

9.9 POINCARÉ MAPS

The tangled orbit shown in Figure 9.21 does not immediately convey any underlying structure. One help in this regard is the Poincaré or first return map of solutions in which there exists a known forcing frequency of ω. We still use the phase plane but only mark the solutions at times $t = 0$ and integer multiples of $2\pi/\omega$. This defines the Poincaré map of the solution. Thus, for example, in Figure 9.19 we delete the phase path and just retain the dots on both paths. Since $\omega = 1$ in Equation (24), the returns arrive at the dot after each period: in other words a periodic solution in the fundamental period has a Poincaré map which is a single point on the plane. This is known as the *fixed point* of the oscillation.

Example 5   Find the Poincaré map of period 2π for the solution of

$$64\ddot{x} + 16\dot{x} + 65x = 64 \cos t$$

which starts from $(0, 0)$ at time $t = 0$. Calculate also the coordinates of the fixed point of the periodic solution.

The roots of the characteristic equation are $p_1 = -\frac{1}{8} + i$ and $p_2 = -\frac{1}{8} - i$. To find the forced periodic response let

$$x = A \cos t + B \sin t$$

so that

$$64\ddot{x} + 16\dot{x} + 65x - 64 \cos t = (A + 16B - 64) \cos t + (B - A) \sin t = 0$$

if A and B satisfy

$$A + 16B - 64 = 0, \quad B - 16A = 0.$$

Hence $A = 0.249$, $B = 3.984$. The full solution is therefore

$$x = e^{-t/8}(C \cos t + D \sin t) + A \cos t + B \sin t$$

The initial conditions imply that $C = -0.249$, $D = -4.015$. The first returns of this solution are given by

$$x_n = A(1 - \exp(-\tfrac{1}{4}n\pi)), \quad y_n = B(1 - \exp(-\tfrac{1}{4}n\pi)), \quad n = 1, 2, \dots.$$

Figure 9.23 shows the sequence of first returns from the origin. As $n \to \infty$ the sequence of dots approaches the fixed point at (A, B).

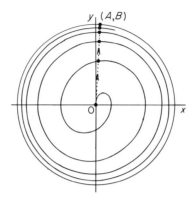

Figure 9.23   Solution and sequence of first returns for Example 5.

Other solutions have characteristic Poincaré maps. For example, a sub-harmonic of order 2 given by $x = \sin t + \sin \frac{1}{2} t$ will have a Poincaré map of two points (Figure 9.24(a)) in which the solution alternates between them. A subharmonic of order 3 will have three points and so on. The two frequency example (32) has the Poincaré map shown in Figure 9.24(b), all the returns lying on a closed curve. Poincaré maps effectively delete unnecessary informa-tion on the phase plane for periodically forced systems.

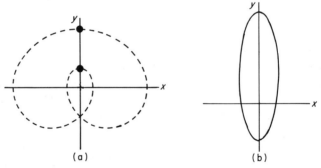

Figure 9.24   Poincaré maps for (a) a period 2 oscillation, (b) the forced solution of the two-frequency system $\ddot{x} + \dot{x} + x = \cos t + \cos \pi t$.

The chaotic orbit displayed in Figure 9.21 has the Poincaré map shown in Figure 9.25 after initial transience has died down. The diagram shows 5000 returns which are tightly scattered in a set in the phase plane. Given any point in this set there will be future returns which will be arbitrarily close to that point. Sets of this type are known as *strange attractors*.

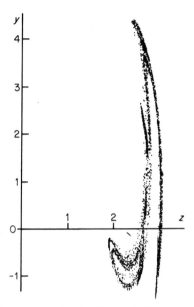

Figure 9.25 Poincaré map of orbit in Figure 9.21 (equation (24) with $\alpha = 0.2$, $\Gamma = 7$) after 5000 returns.

9.10   THE BOUNCING BALL: PERIOD DOUBLING

Another way in which chaotic responses can develop is through the phenomenon known as *period doubling*. We shall illustrate this route to chaos by an example drawn from impulsive motion.

Let a ball be fired vertically with speed u_1 from a table at $z = 0$. In the absence of air resistance, the ball will return to the table with the same speed u_1 after an elapsed time $t_1 = 2u_1/g$, having reached a maximum height $u_1^2/2g$ (see Section 2.3 for these results). Suppose that the table is designed so that it can move vertically, and is controlled in such a way that the next impact still takes place at $z = 0$, but with the table moving upwards with velocity $v(t)$ (Figure 9.26). Let the coefficient of restitution between the ball and the table be e. The inertia of the table is large so that its velocity $v(t)$ is unaffected by the ball.

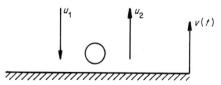

Figure 9.26   First impact of the ball bouncing on a vibrating table.

Newton's law of restitution (Section 3.10) implies that, at time $t = t_1$,

$$v(t_1) - u_2 = -e(v(t_1) - (-u_1))$$

where u_2 is the upward speed of the ball after the impact. Hence

$$u_2 = eu_1 + (1 + e)v(t_1).$$

Continue bouncing the ball so that the impact always takes place at $z = 0$. After the second bounce at time $t_2 = t_1 + (2u_2/g)$ the ball rebounds with velocity

$$u_3 = eu_2 + (1 + e)v(t_2).$$

Thus after n bounces at time

$$t_n = t_{n-1} + (2u_n/g) \tag{33}$$

the rebound velocity is

$$u_{n+1} = eu_n + (1 + e)v(t_n). \tag{34}$$

The question is: does the bouncing ball develop a regular pattern of vertical oscillations? This may appear as rather a contrived illustration but practically it is reminiscent of bouncing a ball on a tennis racket in which the ball always hits the racket at the same height but the racket is made to have a prescribed velocity at impact.

We can observe various aspects of the oscillation including the sequence $\{u_n^2/2g\}$ $(n = 1, 2, \ldots)$ of maximum heights of the ball. Essentially there are two sequences $\{u_n\}$ and $\{t_n\}$ which are related through (33) and (34). However we can simplify (33) and (34) a little by introducing dimensionless variables

$$U_n = u_n/v_0, \qquad T_n = gt_n/2v_0$$

where v_0 is a representative speed of $v(t)$. Thus (33) and (34) can be replaced by

$$U_{n+1} = eU_n + (1 + e)v(2v_0 T_n/g)/v_0 \tag{35}$$

and

$$T_n = T_{n-1} + U_n. \tag{36}$$

These equations are examples of simultaneous *difference equations* for the sequences $\{U_n\}$ and $\{T_n\}$.

Example 6   *Let the impact velocity of the table be a constant v_0. Find U_n.*

In this case $v = v_0$. Hence (35) becomes

$$U_n - eU_{n-1} = (1 + e) \tag{37}$$

For $e \neq 1$, this *first-order* difference equation has the constant solution $(1 + e)/(1 - e)$, as may be verified. Let

$$U_n = \frac{1 + e}{1 - e} + W_n.$$

Then W_n satisfies

$$W_n - eW_{n-1} = 0.$$

By forward iteration

$$W_2 = eW_1, \qquad W_3 = eW_2 = e^2 W_1 \quad \text{etc.}$$

Hence

$$W_n = e^{n-1} W_1$$

and the general solution of (37) is

$$U_n = e^{n-1} W_1 + \frac{1 + e}{1 - e} = e^{n-1} U_1 + \frac{1 + e}{1 - e} (1 - e^{n-1})$$

where U_1 is the initial (dimensionless) speed of the ball. Since $0 \leq e < 1$, $e^n \to 0$ as $n \to \infty$. Hence the sequence of rebound speeds approaches the constant $(1 + e)/(1 - e)$ irrespective of the initial speed as $n \to \infty$. Between successive impacts the ball's displacement is given by the usual formula (Section 2.3) relating displacement and time under constant acceleration. Thus, between the nth and $(n + 1)$th impacts the displacement is

$$X = U_n T - T^2, \qquad T_n \leq T \leq T_{n+1}$$

where $X = gx/(2v_0^2)$. Figure 9.27 shows a typical displacement–time graph for the system.

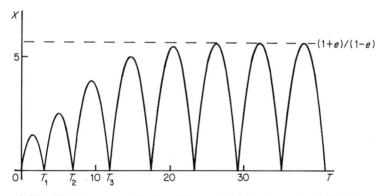

Figure 9.27 Displacement (X)-time (T) graph for the bouncing ball with constant table velocity or impact and $e = 0.7$. (Example 6).

The previous example shows a regular progression to a steady state. Suppose that we now let the table move harmonically with the response velocity

$$v(t) = \beta(1 + \sin \omega t). \tag{38}$$

Thus Equations (35) and (36) become

$$U_{n+1} = eU_n + b(1 + \sin T_n) \tag{39}$$

$$T_n = T_{n-1} + U_n \tag{40}$$

where $v_0 = g/(2\omega)$ and $b = 2\omega(1 + e)\beta/g$.

The dynamical system defined by (39) and (40) depends on the two parameters e and b. Eliminate U_n between the two difference equations so that $\{T_n\}$ satisfies

$$T_{n+1} - (1 + e)T_n + eT_{n-1} = b(1 + \sin T_n) \tag{41}$$

where $T_0 = 0$ and $T_1 = U_1$. Suppose that we now search for regular behaviour in (41). Are oscillations of the ball of period 2π possible? For a 2π-periodic solution, the impact times must satisfy $T_n = 2n\pi + \alpha$, which is a solution of (41) if the constant α is given by

$$b(1 + \sin \alpha) = 2(1 - e)\pi. \tag{42}$$

Solutions will exist if $0 < (1 - e)\pi \le b$, in which case we could choose α to be given by the fundamental value

$$\alpha = \alpha_1 = \sin^{-1}\{[2(1 - e)\pi - b]/b\}. \tag{43}$$

Periodic behaviour is possible if $\alpha = \pi - \alpha_1$. Although we have shown that a periodic oscillation of the bouncing ball is possible in the forcing frequency it does not follow that this oscillation will automatically be visible in an experiment or in a computer simulation. We still need to know whether the oscillation is *stable* or *unstable*. We shall not pursue this topic here but content ourselves with a computer run showing what happens.

In the numerical case presented here $e = 0.1$ (heavily damped oscillation), and b was steadily increased from its minimum value at $b = (1 - e)\pi = 2.827$ .... A stable oscillation of period 2π exists for impact time sequence $T_n = 2(n + 1)\pi - \alpha_1$, as b is increased until $b = 3.240$ approximately where the oscillation *bifurcates* into an oscillation of period 4π. A system is said to go through a *bifurcation* with a parameter change if the qualitative behaviour of the system has suffered a sudden change. This could be an equilibrium point appearing or disappearing, or, as in this problem, the period suddenly doubling. The 4π oscillation in turn develops until $b = 3.432$ where the period doubles again to 8π. This process continues with ever-decreasing intervals in b. The sequence tends to a limit, and lengthy computer outputs at $b = 3.5$ show no regular cycles in either the sequence of impact times or the maximum height reached by the ball between each impact. Thus there exists a sequence of bifurcation values for b at which *period doubling* takes place. Period doubling develops from amplitude modulation of the original output so that ultimately the ball performs irregular bounded non-periodic oscillations and *chaos* is visible in the bouncing ball.

The computations were compiled through an interaction of (39) and (40) starting with $b = 3.2$ and incrementing b after a run of 4000 iterations for each increment of its values. Initially U remains at 2π with the bifurcations becoming visible in the output at the values of b mentioned in the previous paragraph. Figure 9.28 shows the period doubling in a b–U diagram. At $b = 3.5$ approximately the period doubling is complete, and as b increases beyond this value chaotic dynamics ensues, that is the ball continues to bounce but in a non-repeatable manner.

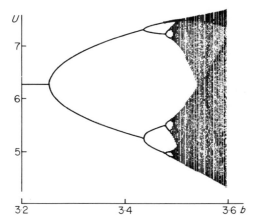

Figure 9.28 Period doubling route to chaos for (39) and (40) with $e = 0.1$. Where possible steady oscillations were sought for $3.2 < b < 3.6$ at increments of 0.002 in b.

Figure 9.29 shows the displacement–time graphs for the bouncing ball in three states (after any transience has died down in each case) all with $e = 0.1$.

   (i) with $b = 3.20$ (Figure 9.29a) showing 2π-periodic oscillations,
   (ii) with $b = 3.40$ (Figure 9.29b) showing period 2 oscillations,
   (iii) with $b = 3.58$ (Figure 9.29c) showing chaotic output.

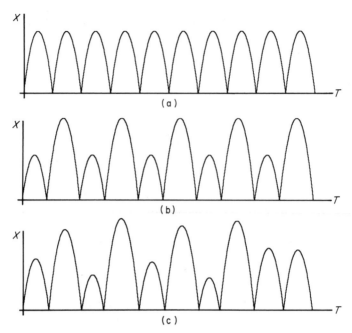

Figure 9.29 Displacement–time graphs for the bouncing ball with $e = 0.1$ for (a) $b = 3.20$ (period 1), (b) $b = 3.40$ (period 2), (c) $b = 3.58$ (chaos).

A careful study of the numerical output shows that the period doubling takes place at a sequence $\{b_n\}$ of values of b where the first few terms are given in the table below:

n	b_n	Period
1	3.240	2
2	3.432	4
3	3.480	8
4	3.491	16

In order to proceed beyond period 16 more decimal places are rapidly required. The sequence may appear at first sight to be just another sequence but it was discovered by Feigenbaum in the mid-seventies that all period doubling sequences have the *universal* property that

$$\frac{b_n - b_{n+1}}{b_{n+1} - b_n} \to \delta$$

where δ is always the same constant $\delta = 4.6692 \ldots$, which is known as *Feigenbaum's constant*. The first few terms usually give a good estimate for this constant. From the table

$$\frac{b_3 - b_2}{b_4 - b_3} = \frac{0.048}{0.011} \approx 4.4.$$

The dynamics of the bouncing ball is extremely complex, and we have really only touched upon some of its properties. There can also coexist for some parameter values, larger amplitude oscillations which have periods 4, 6, $\ldots$, which can be triggered by selecting the appropriate initial conditions. Each of these can bifurcate into period doubling sequences. A close scrutiny of the heavily shaded region of Figure 9.28 between $b = 3.5$ and $b = 3.6$ indicates that there do exist within what seems to be chaos, small intervals of values of b where periodic motion reappears briefly. For example, close to $b = 3.52$ a period 9 oscillation can be found by computation.

Exercises

1. Show that the phase diagram in Figure 9.1 applies for all parameter values in the pendulum Equation (1) with a suitable change of scale.

2. Redraw the phase diagram of Figure 9.1 on a piece of paper. Cut the paper through $\theta = \pi$ and $\theta = -\pi$ parallel to the $\dot{\theta}$ axis. Join the two cut edges to form a circular cylinder. Explain why this form of the phase diagram is equally valid for the pendulum.

3. Find the equilibrium points of the system governed by the equation

$$\ddot{x} + \cos x = 0$$

Sketch or compute its phase diagram.

4. Find the equilibrium points and the equations of the phase paths for the systems governed by the following equations. Sketch (or plot using a computer program) phase diagrams in each case.

(i) $\ddot{x} = x - x^3$

(ii) $\ddot{x} = x\dot{x}$

(iii) $\ddot{x} = -x^2$

(iv) $\ddot{x} = -x\,e^{-x}$

5. A bead can slide on a smooth horizontal wire. The bead is attached by a spring of stiffness k and unstretched length a to a fixed point P distance b above the wire (see Figure 9.30). If x is the perpendicular distance of the bead from the point on the wire immediately below P, show that the potential energy of the bead is

$$\mathcal{V}(x) = \tfrac{1}{2}k[(x^2 + b^2)^{1/2} - a]^2.$$

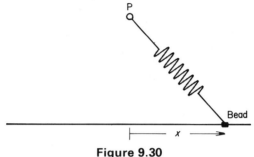

Figure 9.30

Sketch the energy $\mathscr{V}(x)$ against x distinguishing the cases $b > a$ and $b < a$. In both cases sketch also the corresponding phase diagram using the method of Section 9.4.

6. Classify the equilibrium points of the following linear equations using Figure 9.9:

(i) $\ddot{x} - 2\dot{x} - x = 0$

(ii) $\ddot{x} + 2\dot{x} - x = 0$

(iii) $\ddot{x} - 2\dot{x} + x = 0$

(iv) $\ddot{x} + 5x = 0$

(v) $\ddot{x} + 2\dot{x} + 3x = 0.$

7. Suppose that friction is introduced into Exercise 5 with the assumption that motion of the bead (of mass m) is resisted by a force $c \times$ (speed). Write down the equation of motion of the bead in terms of x. Classify the equilibrium points according to their linear approximations in the two cases $b > a$ and $b < a$. Sketch or compute typical phase diagrams in the two cases, and indicate the domains of asymptotic stability of stable equilibrium points.

8. A block of mass m on a plane is attached by a spring of stiffness k to a fixed point P on the plane. The frictional force between the block and the plane is $c \times$ (speed)^2, where c is a constant. If the block moves on a fixed line through P, write down its equation of motion in terms of x, the extension of the spring. The linear approximation predicts a centre at $x = 0$. Explain physically why equilibrium must be asymptotically stable.

9. Find the equilibrium points of the following autonomous systems and classify them according to their linear approximations:

(i) $\ddot{x} - x + x^3 = 0$

(ii) $\ddot{x} + x - x^3 = 0$

(iii) $\ddot{x} + \dot{x} - x + x^3 = 0$

(iv) $\ddot{x} + \dot{x} + \sin x = 0$

(v) $\ddot{x} + (1 - x^2)\dot{x} + x - x^2 = 0.$

10. A circular cylindrical shell with internal radius a rotates about its horizontal axis at a constant rate ω. A small shoe slides inside the cylinder perpendicular to the axis (Figure 9.31). The frictional force F between the shoe and the cylinder is assumed to be proportional to the slip velocity, that is $F = F_0 a(\dot{\theta} - \omega)$ where F_0 is a constant. Find the equation of motion of the shoe. Find also the equilibrium position of the shoe, and classify it according to its linear approximation.

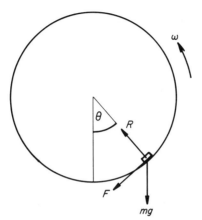

Figure 9.31

11. A smooth parabolic wire with equation $y = ax^2$ $(a > 0)$ rotates at a constant rate ω about the y axis. A bead slides on the wire. Show that its equation of motion is

$$\ddot{x}(1 + 4a^2x^2) + (2ag - \omega^2 + 4a^2\dot{x}^2)x = 0.$$

Locate the equilibrium point of the system and investigate its type by using the linear approximation. Confirm this by computing phase diagrams.

12. A simple pendulum of length a is designed so that it can only oscillate in a vertical plane about an axle perpendicular to the plane. The horizontal axle is made to rotate at a constant rate ω about the vertical through the point of suspension. Show that the inclination θ of the pendulum to the downward vertical satisfies

$$a\ddot{\theta} - (a\omega^2 \cos\theta - g)\sin\theta = 0.$$

Find the equilibrium values of θ and analyse the phase plane of the system.

13. Investigate using a computer program the phase diagrams of the following equations:

(i) $\ddot{x} + \mu(x^2 - 1)\dot{x} + x = 0$ (van der Pol's equation),
(ii) $\ddot{x} - \mu(\dot{x} - \frac{1}{3}\dot{x}^3) + x = 0$ (Rayleigh's equation)

for $\mu > 0$ in both cases.

14. A block of mass m on a smooth horizontal plane oscillates between two equal springs, each of stiffness k and natural length a (Figure 9.32). Both springs are assumed to be inertialess. There is a deadband between the springs of length b. Draw a sketch showing the equilibrium positions of the block and the phase paths in the phase plane. What is the period of an oscillation of amplitude A where $b < A < a + b$? How would friction affect the phase diagram?

Figure 9.32

15. In Example 3, Chapter 7, the linear theory for the transverse oscillation of a particle between two taut elastic strings is investigated. Repeat the same problem using the exact equation and phase plane qualitative methods.

16. A bead slides on a smooth circular wire of radius a, which is forced to rotate about a vertical diameter at a constant rate ω. Show that the equation of motion of the bead is

$$a\ddot{\theta} = a\omega^2 \sin \theta \cos \theta - g \sin \theta$$

where θ is the angle between the bead and the downward vertical. Find the equilibrium points of the system and the equations of the phase paths, and sketch the phase diagrams. What is the bifurcation value of the parameter $a\omega^2/g$? Which equilibrium points are stable?

17. Show that the equation

$$\ddot{x} - (1 - x^2 - \dot{x}^2)\dot{x} + \omega^2 x = \Gamma \cos t$$

has a periodic solution $x = \cos t$ if $\Gamma = \omega^2 - 1$.

18. Show that the equation

$$\ddot{x} - (1 - x^2 - \dot{x}^2)\dot{x} + x = 0$$

has a periodic solution $x = \cos t$. Classify the equilibrium point at the origin. Using a computer investigation to test whether the periodic solution is a limit cycle in the phase plane.

19. A pendulum of length a is suspended from a support which is oscillating horizontally with displacement $\Gamma \cos \omega t$. Show that θ the angle of inclination of the pendulum from the downward vertical satisfies

$$a\ddot{\theta} + g \sin \theta = \Gamma\omega^2 \cos \omega t \cos \theta.$$

Friction is present which introduces a term $c\dot{\theta}$ on the left-hand side. Show that for small θ, θ is given by

$$a\ddot{\theta} + c\dot{\theta} + g\theta = \Gamma\omega^2 \cos \omega t.$$

Find the coordinates in the $(\theta, \dot{\theta})$ plane of the fixed point of the Poincaré map (period $2\pi/\omega$, $t = 0$) of the periodic solution of this equation.

20. (Project) A forced hard spring is governed by Ueda's equation (Equation (24)). Investigate the long term outputs of the system, after transience has declined, for $\alpha = 0.2$ and the following selected values of Γ: (i) $\Gamma = 2$, (ii) $\Gamma = 4$, (iii) $\Gamma = 10$, (iv) $\Gamma = 11.4$, (v) $\Gamma = 14$, (vi) $\Gamma = 24$. In particular search for periodic solutions, including subharmonics and chaotic solutions.

21. (Project) Compute the Poincaré map of Ueda's equation in the previous Exercise at $\alpha = 0.2$, $\Gamma = 11.4$ in the section $s = 0$ (*modulo* 2π). A graphical output over a large number of first returns should reveal a strange attracting set. Compare this with the Poincaré maps at sections $s = \pi/2$, π and $3\pi/2$.

22. Two stars with gravitational masses μ_1 and μ_2 are orbiting each other under their mutual gravitational forces (as in Section 8.5). A satellite of relatively negligible mass is moving along a straight line through the mass-centre G, and perpendicular to the mutual orbits of the binary system. Explain why (apart from questions of stability) the satellite will continue on this line. If z is its displacement from G show that z satisfies

$$\ddot{z} = -\frac{\mu_1 z}{(r_1^2 + z^2)^{3/2}} - \frac{\mu_2 z}{(r_2^2 + z^2)^{3/2}}$$

where

$$\mu_1 \mathbf{r}_1 + \mu_2 \mathbf{r}_2 = 0, \qquad (\mu_1 + \mu_2)^2 \ddot{\mathbf{r}}_1 = -\mu_2 \frac{\mathbf{r}_1}{r_1^3}$$

and $\mathbf{r}_1$, $\mathbf{r}_2$ are the plane position vectors of the stars.

Let the stars have equal masses and assume that their orbits are circular, each of radius a. Verify that z satisfies

$$\ddot{z} = -\frac{2\mu z}{(a^2 + z^2)^{3/2}}.$$

Using the potential energy method, show that the equilibrium point at $z = 0$ is a global centre.

23. (Project) Express the equation in the final part of the previous problem in dimensionless form by writing $z = aZ$ and $t = \tau(a^3/\mu)^{1/2}$. Confirm that

$$Z'' = -2Z/(1 + Z^2)^{3/2}.$$

Compute a phase diagram for the system about the origin.

What is the period of the binary stars in terms of τ? Compute the amplitude of the satellite oscillation which is synchronous with the stars.

24. (Project) An extensive computer investigation of the full equations in Exercise 22 could be undertaken in the case of elliptic orbits but equal masses. The equations will be equivalent to a first-order system in the variables z, $\dot{z}$, r_1, $\dot{r}_1$, say. The main interest would be in the oscillatory behaviour of the satellite, both periodic and non-periodic.

25. (Project) Here is a slightly easier problem than Exercise 24.
Suppose we let a be a function of time in

$$\ddot{z} = -\frac{2\mu z}{(a^2 + z^2)^{3/2}}$$

with period equal to that of the binary system. With this in view consider the forcing term $a = a_0(1 + \varepsilon \cos \omega t)$ where $\omega = 2(\mu/a_0^3)^{1/2}$. This simulates the oscillatory effect of the gravitational force acting on the satellite due to the elliptic orbits. Find the dimensionless equations using a_0 now. Search for periodic orbits which are multiples of the periods of the stars.

26. A block lies on a smooth table and is restrained by a spring of natural length a attached to a fixed point. The block is fired with speed u_0 in a direction which extends the spring from $x = a$. At each return it is periodically struck by a hammer of large mass which is controlled by a servo-mechanism in such a way that it moves with speed ku_n at impact where u_n is the return speed after the nth impact (Figure 9.33). The coefficient of restitution e depends on the relative impact speed v, that is $e = e(v)$. If u_n is the speed of the block immediately after the nth impact show that

$$u_n = \{k + (k + 1)e[(k + 1)u_{n-1}]\}u_{n-1}.$$

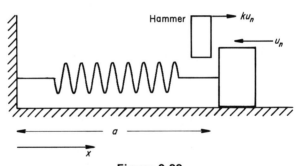

Figure 9.33

Assume that $e(v) = a - bv$ $(b > 0)$, that is the coefficient decreases linearly with v (at least for $v \le a/b$). Show that the sequence of speeds satisfy

$$u_n = \alpha u_{n-1} - \beta u_{n-1}^2 \quad \alpha > 0 \quad \beta > 0 \quad n = 1, 2, \ldots$$

where α and β are constants. Put $u_n = \alpha U_n/\beta$ so that in dimensionless form U_n satisfies the non-linear difference equation

$$U_n = \alpha U_{n-1}(1 - U_{n-1}).$$

Given $U_n > 0$, show that this system has two equilibrium values $U_n = 0$, $U_n = (\alpha - 1)/\alpha$ for $\alpha > 1$. How does the block behave in these cases?

27. (Project) The previous hammer-driven spring may appear to be a contrived example but the resulting non-linear difference equation has some interesting properties. Con-

struct a computer program to investigate the solutions of the quadratic difference equation

$$U_n = \alpha(U_n - U_n^2) \qquad \alpha > 1$$

of the previous example (it is known as the *logistic* equation). In particular search for the sequence of values of α at which period doubling takes place. Start at $\alpha = 3$ and increase α by suitable increments but do not go beyond $\alpha = 4$. The Feigenbaum limit can be approximated by the first few values of the sequence. Beyond $\alpha > \alpha_\infty \approx 3.568$ the spring behaves chaotically having passed through the period doubling sequence. Since the amplitude of the spring is SHM between impacts the amplitude is proportional to the impact speed, and the period doubling will be shown in the experiment. Figure 9.34 shows examples of period 1 ($\alpha = 2.8$) and period 2 ($\alpha = 3.2$) oscillations.

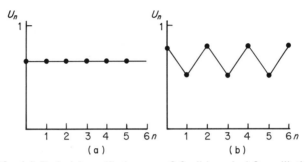

Figure 9.34 (a) Period 1 oscillation, $\alpha = 2.8$; (b) period 2 oscillation, $\alpha = 3.2$.

28. Let $f(U) = U - U^2$. The graph of $V = f(U)$ against U is shown in Figure 9.35. The figure can be used to generate the sequence $\{U_n\}$ of the previous example. Choose a value of α such that $1 < \alpha < 3$, and draw the straight line $V = U/\alpha$ (in the figure $\alpha = 2.8$). Draw the line $U = U_0$ and mark where it cuts the parabola, and then mark the intersection of the lines $V = V_0 = f(U_0)$ and $V = U/\alpha$. This point defines U_1. Now repeat the procedure to find U_2, U_3, etc. For this value of α, U_n approaches the equilibrium

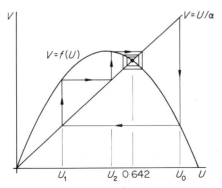

Figure 9.35 Cobweb for $\alpha = 2.8$.

value at P, where $U = 0.642\ldots$. The direction of the sequence indicates that the equilibrium which corresponds to a period 1 solution is stable. This sequence of lines is known as the *cobweb* method. Design a computer program to generate cobwebs for different values of α and the initial state U_0. Explain why the period 1 oscillation will be stable for $|f'(U)| < 1$, but unstable for $|f'(U)| > 1$. What is the critical value of α? Period 2 and period 4 cycles are shown in Figure 9.36.

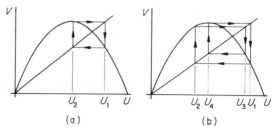

(a)   (b)

Figure 9.36   (a) period 2 oscillation, $\alpha = 3.2$; (b) period 4 oscillation, $\alpha = 3.5$.

29. Find the equation which represents the phase paths of the non-linear equation

$$\ddot{x} + \dot{x}^2 + x = 0.$$

Show that the origin is a centre separated from open paths by the path

$$2y^2 = -2x + 1.$$

30. A simple harmonic oscillator experiences damping which is proportional to the square of its speed. The equation governing the motion is

$$\ddot{x} + c|\dot{x}|\dot{x} + \dot{x} = 0$$

where $c > 0$. Compute solutions of the equation in the phase plane, and show that the origin is a spiral for $c = 0.2$ but a centre for $c = 0$.

31. The displacement of a mechanical oscillator satisfies

$$m\ddot{x} + kx = 0$$

where the restoring coefficient, $k = k_1$ for $x > 0$ and k_2 for $x < 0$ (this could be achieved by a block vibrating between two springs but attached to neither). Find the period of oscillations of this bilinear oscillator.

32. The inclination θ of a simple pendulum satisfies the equation

$$\ddot{\theta} + \sin\theta = 0.$$

If a is the amplitude of its oscillation, show that its period T is given by

$$T = 4\int_0^{\pi/2} \frac{d\phi}{(1 - \beta\sin^2\phi)^{1/2}}, \quad \beta = \sin^2\tfrac{1}{2}a.$$

Expand the integral in powers of β by the binomial theorem and integrate term-by-term. Confirm that

$$T = 2\pi\left[1 + (\tfrac{1}{2})^2\beta + \left(\frac{1 \times 3}{2 \times 4}\right)^2 \beta^2 + \cdots\right].$$

Further, by inserting the power series for $\sin \frac{1}{2}a$ in terms of a deduce

$$T = 2\pi(1 + \tfrac{1}{16}a^2 + O(a^4)).$$

It is possible to find the next term in the series although the algebra involved becomes more complicated. (Project) Compute the period–amplitude graph using this approximation and compare it with Figure 4.10.

33. The equation of the forced pendulum is approximated by the Duffing equation

$$\ddot{x} + c\dot{x} + x - \tfrac{1}{6}x^3 = \Gamma \cos \omega t$$

where c, Γ and ω are constants. Investigate numerically and graphically solutions and Poincaré maps of this equation in particular when the forcing amplitude $\Gamma = 0.580$, the forcing frequency $\omega = 0.555$ and $c = 0.400$. Show that an oscillation of period $2\pi/\omega$ with amplitude approximately 0.91 and a chaotic attractor co-exist. (Try the initial data sets $x(0) = -0.34$, $\dot{x}(0) = 0.50$ and $x(0) = -2.35$, $\dot{x}(0) = -0.138$.)

34. Consider the oscillator with Duffing equation

$$\ddot{x} + c\dot{x} - x + x^3 = \Gamma \cos \omega t$$

which has a negative linear and positive cubic restoring terms for $x > 0$. Investigate numerically the case with $\omega = 1.2$, $c = 0.3$ by progressively increasing Γ from $\Gamma = 0.24$ to $\Gamma = 0.34$ by steps 0.005. Compute the Poincaré map in each case. Show that the oscillator goes through a period doubling sequence resulting eventually in a strange attractor. (Use initial data $x_0(0) \approx 0.5$, $\dot{x}_0(0) \approx 0.5$ and allow transience to die down.)

35. A damped pendulum oscillating with small amplitude satisfies the linear equation

$$\ddot{\theta} + c\dot{\theta} + \theta = 0$$

where θ is the angle between the pendulum and the downward vertical. Assuming that $0 < c < 2$, obtain the general solution of the equation. Choose the particular spirals which satisfy $\theta = 0$ at $t = 0$. When $\theta = 0$ and $\dot{\theta} > 0$, the pendulum receives an impulse which instantaneously increases $\dot{\theta}$ by ε. Show that there exists a periodic solution for the pendulum. Confirm that immediately before the impulse $\dot{\theta} = \varepsilon/(\exp(\pi c/\omega) - 1)$. Sketch the phase path of this periodic solution in the phase plane. (This is a possible autonomous model for a clock mechanism.)

10

Rotating Frames of Reference

10.1 THE EARTH AS A ROTATING FRAME

As we described in Section 3.1, a frame of reference fixed in the Earth is not an inertial frame but a rotating one with period equal to one day. The axis of rotation is the line joining the poles. Taking the Earth to be a sphere, we can imagine the origin to be at the centre of the Earth with $OXYZ$ as the fixed or inertial frame (we assume that the axes are parallel to those of a sidereal frame) and $Oxyz$ as the rotating frame, these axes being specified directions in the Earth as indicated in Figure 10.1. Strictly speaking $OXYZ$ is not an inertial frame since the Earth orbits the Sun and the motion of the origin O along this path induces further accelerations in $OXYZ$. However, as we saw in Section 3.1, this acceleration is smaller than that due to the spin of the Earth.

Before we derive the equations for motion relative to a rotating frame, it is interesting to look at the effect of the Earth's spin on a simple pendulum suspended at the Earth's surface. Suppose the pendulum is suspended from the

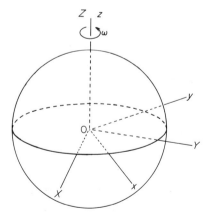

Figure 10.1 Inertial and rotating frames in the earth; OZ is the polar axis.

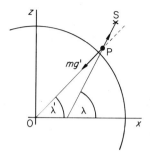

Figure 10.2

point S with its bob at P in the Oxz plane as shown in exaggerated form in Figure 10.2. We should expect correctly that the bob would be thrown outwards due to the spin of the Earth and to appear inclined as in Figure 10.2. Let T be the tension in the string, λ' the angle which the radius to the pendulum makes with Ox and λ the angle which the string makes with Ox. The bob experiences a force mg' which is the gravitational attraction of the Earth, that is

$$mg' = \frac{\gamma m M_{\mathrm{E}}}{a^2}$$

where m is the mass of the bob, M_{E} the mass of the Earth, a the radius of the Earth and γ the gravitational constant. The quantity mg' is not the 'weight' which we measure since we measure weight, in practice, on the rotating Earth.

With the pendulum in equilibrium the forces acting on it must balance the accelerations. The Earth is spinning with angular speed ω, say, about the z-axis and the bob must therefore be subject to an acceleration $\omega^2 a \cos \lambda'$ towards the z-axis. Resolving parallel to the x- and z-axes, we conclude that

$$mg' \cos \lambda' - T \cos \lambda = m\omega^2 a \cos \lambda' \qquad (1)$$

$$T \sin \lambda = mg' \sin \lambda'. \qquad (2)$$

The tension in the string is what we usually understand as the weight of the bob, mg, where g is the acceleration due to gravity—the acceleration which we measure at the Earth's surface. Putting $T = mg$ in Equations (1) and (2) we obtain

$$g' \cos \lambda' - g \cos \lambda = \omega^2 a \cos \lambda' \qquad (3)$$

$$g \sin \lambda = g' \sin \lambda'. \qquad (4)$$

The quantities g and λ can be measured and these equations give g' and λ' provided a (or $a \cos \lambda'$ if the oblateness of the Earth is taken into account) is known.

Eliminating g' between (3) and (4) we have

$$g(\sin \lambda \cos \lambda' - \sin \lambda' \cos \lambda) = \omega^2 a \sin \lambda' \cos \lambda'$$

or

$$\sin (\lambda - \lambda') = \frac{\omega^2 a}{g} \sin \lambda' \cos \lambda'.$$

The ratio $\omega^2 a/g$ is small with the result that $\sin (\lambda - \lambda') \approx \lambda - \lambda'$. Thus

$$\lambda - \lambda' \approx \frac{\omega^2 a}{g} \sin \lambda' \cos \lambda'.$$

The angle λ which measures the inclination of the pendulum is the *astronomical* or *geographic latitude* and is what is generally understood by the term latitude. The angle λ' is the *geocentric latitude* and indicates the direction of the true vertical at a point on the Earth. The maximum deviation between λ and λ' occurs where $\sin \lambda' \cos \lambda' = \frac{1}{2} \sin 2\lambda'$ takes its greatest value, namely at $\lambda' = 45°$. Taking $a = 6370$ km (the mean radius of the Earth) and $g = 981$ cm/s^2, and computing $\omega = 7.29 \times 10^{-5}$ rad/s, we find that $a\omega^2/g = 3.450 \times 10^{-3}$. The maximum deviation at latitude $45°$ is therefore

$$\frac{1}{2} \frac{a\omega^2}{g} = 1.725 \times 10^{-3} \text{ rad}$$

which is about 6 min of arc. However, it must be emphasised that the oblateness of the Earth is a factor which should be taken into account although the estimate of the angular difference between the true vertical and apparent vertical (indicated by a plumb-line) is fairly good. A more accurate treatment requires consideration of the gravitational field associated with an oblate spheroid which is the surface obtained by spinning an ellipse about its minor axis. For the earth the polar radius is about 6357 km and the equatorial radius 6378 km.

10.2 FRAME ROTATING ABOUT A FIXED AXIS

As before let OXY represent the fixed frame and Oxy the rotating frame as shown in Figure 10.3 with the common z-axis and Z-axis both pointing out from the paper. The Z-axis is the fixed axis about which Oxy is rotating. The angle θ is the angular displacement of the rotating frame and will be a function of time. Let $\mathbf{I}$ and $\mathbf{J}$ be fixed unit vectors in the directions OX and OY and let $\mathbf{i}$ and $\mathbf{j}$ be unit vectors in the directions Ox and Oy. We can see from the diagram that, by the triangle law for addition of vectors,

$$\mathbf{i} = \mathbf{I} \cos \theta + \mathbf{J} \sin \theta, \quad \mathbf{j} = -\mathbf{I} \sin \theta + \mathbf{J} \cos \theta. \tag{5}$$

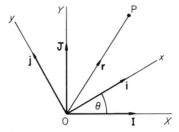

Figure 10.3   Unit vectors in the fixed and rotating frames.

Note that the vectors **i** and **j** will be time-dependent, so that

$$\frac{d\mathbf{i}}{dt} = (-\mathbf{I}\sin\theta + \mathbf{J}\cos\theta)\dot{\theta} = \mathbf{j}\dot{\theta}$$

$$\frac{d\mathbf{j}}{dt} = (-\mathbf{I}\cos\theta - \mathbf{J}\sin\theta)\dot{\theta} = -\mathbf{i}\dot{\theta}.$$

If **r** is the position vector of a point P in the plane, then we can write both

$$\mathbf{r} = X\mathbf{I} + Y\mathbf{J} \quad \text{and} \quad \mathbf{r} = x\mathbf{i} + y\mathbf{j}.$$

The velocity of P as it appears to an observer in the fixed frame of reference will be

$$\frac{d\mathbf{r}}{dt} = \dot{\mathbf{r}} = \dot{X}\mathbf{I} + \dot{Y}\mathbf{J}$$

but to an observer moving with the rotating frame, P has velocity

$$\frac{\delta\mathbf{r}}{\delta t} = \dot{x}\mathbf{i} + \dot{y}\mathbf{j}.$$

We adopt the notation $\delta/\delta t$ for time-derivatives within a rotating frame relative to an observer moving with that frame, for it is certainly not true, in general, that $d\mathbf{r}/dt$ and $\delta\mathbf{r}/\delta t$ represent the same function of time. However we can easily find the relation between the two derivatives. Clearly,

$$\frac{d\mathbf{r}}{dt} = \dot{x}\mathbf{i} + \dot{y}\mathbf{j} + x\frac{d\mathbf{i}}{dt} + y\frac{d\mathbf{j}}{dt}$$

$$= \frac{\delta\mathbf{r}}{\delta t} + x\omega\mathbf{j} - y\omega\mathbf{i}$$

where $\dot{\theta}$ has been replaced by ω—the angular rate of rotation of the rotating frame. By the definition of a vector product (Section 1.4), we can verify that

$$\frac{d\mathbf{r}}{dt} = \frac{\delta\mathbf{t}}{\delta t} + \boldsymbol{\omega} \times \mathbf{r} \qquad (6)$$

where $\boldsymbol{\omega} = \omega\mathbf{k}$ is the *angular velocity* of the rotating frame.

Whilst this formula has been obtained specifically for the position vector, it is true for any vector. Thus, if $\mathbf{a}$ is a function of time then

$$\frac{d\mathbf{a}}{dt} = \frac{\delta\mathbf{a}}{\delta t} + \boldsymbol{\omega} \times \mathbf{a}.$$

The result we have really shown is the identity of the two operators d/dt and $(\delta/\delta t + \boldsymbol{\omega} \times)$ operating on a vector function of time.

If $\mathbf{r}$ is fixed in the rotating frame $\delta\mathbf{r}/\delta t$ must vanish and

$$\frac{d\mathbf{r}}{dt} = \boldsymbol{\omega} \times \mathbf{r}.$$

Example 1   A bead slides so that its displacement along a straight wire is given by $A \cos \Omega t$. The wire rotates with constant angular velocity ω about an axis which is perpendicular to the wire and passes through the mean position O of the bead. Find the actual velocity and acceleration of the bead.

Let Ox and Oy be axes along and perpendicular to the wire and let OX and OY be fixed axes, the two sets of axes coinciding at, say, time $t = 0$ (Figure 10.3 should be consulted again). The bead moves with simple harmonic motion along the wire so that

$$\mathbf{r} = \mathbf{i}A \cos \Omega t.$$

The velocity relative to the rotating frame is given by

$$\frac{\delta\mathbf{r}}{\delta r} = -\mathbf{i}A\Omega \sin \Omega t.$$

Using Equation (6) with $\boldsymbol{\omega} = \omega\mathbf{k}$, the velocity $\mathbf{v}$ relative to the fixed frame is given by

$$\mathbf{v} = \frac{d\mathbf{r}}{dt} = -\mathbf{i}A\Omega \sin \Omega t + \omega\mathbf{k} \times \mathbf{i}A \cos \Omega t$$

$$= -\mathbf{i}A\Omega \sin \Omega t + \mathbf{j}A\omega \cos \Omega t$$

in components parallel to the moving axes. To obtain the velocity components referred to the fixed axes, we use the transformation (Equations (5)) relating the pairs of unit vectors with $\theta = \omega t$. Thus

$$
\begin{aligned}
\mathbf{v} &= -A\Omega \sin \Omega t (\mathbf{I} \cos \omega t + \mathbf{J} \sin \omega t) \\
&\quad + A\omega \cos \Omega t (-\mathbf{I} \sin \omega t + \mathbf{J} \cos \omega t) \\
&= -\mathbf{I}A(\Omega \sin \Omega t \cos \omega t + \omega \cos \Omega t \sin \omega t) \\
&\quad + \mathbf{J}A(\omega \cos \Omega t \cos \omega t - \Omega \sin \Omega t \sin \omega t).
\end{aligned}
$$

The acceleration $\mathbf{f}$ is found by repeating the operator for moving axes on the velocity:

$$
\mathbf{f} = \frac{d\mathbf{v}}{dt} = \frac{\delta \mathbf{v}}{\delta t} + \boldsymbol{\omega} \times \mathbf{v}
$$

or by straightforward differentiation of the velocity in terms of its components in the fixed coordinate system. Adopting the first approach, we have

$$
\begin{aligned}
\mathbf{f} &= -\mathbf{i}A\Omega^2 \cos \Omega t - \mathbf{j}A\omega\Omega \sin \Omega t + \omega \mathbf{k} \times (-\mathbf{i}A\Omega \sin \Omega t + \mathbf{j}A\omega \cos \Omega t) \\
&= -\mathbf{i}A(\Omega^2 + \omega^2) \cos \Omega t - 2\mathbf{j}A\Omega\omega \sin \Omega t
\end{aligned}
$$

in components parallel to the rotating axes. Consider now a particle of mass m placed on the point P and subject to a plane force $\mathbf{F}$. In the fixed or inertial frame, we must have

$$
\mathbf{F} = m\ddot{\mathbf{r}}
$$

$$
= m\left(\frac{\delta}{\delta t} + \boldsymbol{\omega} \times\right)\left(\frac{\delta}{\delta t} + \boldsymbol{\omega} \times\right)\mathbf{r}
$$

$$
= m\left(\frac{\delta}{\delta t} + \boldsymbol{\omega} \times\right)\left(\frac{\delta \mathbf{r}}{\delta r} + \boldsymbol{\omega} \times \mathbf{r}\right)
$$

$$
= m\left(\frac{\delta^2 \mathbf{r}}{\delta t^2} + \boldsymbol{\omega} \times \frac{\delta \mathbf{r}}{\delta t} + \frac{\delta}{\delta t}(\boldsymbol{\omega} \times \mathbf{r}) + \boldsymbol{\omega} \times (\boldsymbol{\omega} \times \mathbf{r})\right)
$$

$$
= m\left(\frac{\delta^2 \mathbf{r}}{\delta t^2} + 2\boldsymbol{\omega} \times \frac{\delta \mathbf{r}}{\delta t} + \frac{\delta \boldsymbol{\omega}}{\delta t} \times \mathbf{r} + \boldsymbol{\omega} \times (\boldsymbol{\omega} \times \mathbf{r})\right).
$$

Since the rate of change of angular velocity

$$
\dot{\boldsymbol{\omega}} = \frac{\delta \boldsymbol{\omega}}{\delta t} + \boldsymbol{\omega} \times \boldsymbol{\omega} = \frac{\delta \boldsymbol{\omega}}{\delta t}
$$

it follows that

$$
\mathbf{F} = m\left(\frac{\delta^2 \mathbf{r}}{\delta t^2} + 2\boldsymbol{\omega} \times \frac{\delta \mathbf{r}}{\delta t} + \dot{\boldsymbol{\omega}} \times \mathbf{r} + \boldsymbol{\omega} \times (\boldsymbol{\omega} \times \mathbf{r})\right). \tag{7}
$$

This is the equation for motion in a rotating frame of reference. Note that the force $\mathbf{F}$ is not proportional to the relative acceleration $\delta^2\mathbf{r}/\delta t^2$ in such a frame. Remember also that $\mathbf{F}$ and $\mathbf{r}$ are plane vectors; however the vector equation for a frame rotating about a fixed axis suggests its generalization to a frame rotating in an unrestricted manner. We shall look at the general case in Section 10.6.

Equation (7) can be rearranged into

$$F - 2m\omega \times \frac{\delta r}{\delta t} - m\dot{\omega} \times r - m\omega \times (\omega \times r) = m \frac{\delta^2 r}{\delta t^2} \tag{8}$$

and may be thought of as a statement that force balances the product of mass and acceleration in the rotating frame if 'force' is interpreted to include the apparent forces which arise from the rotation. Of the terms on the left-hand side of (8), $-2m\omega \times (\delta r/\delta t)$ is called the *Coriolis* force and $-m\omega \times (\omega \times r)$ the *centrifugal* force. Since the Coriolis force is the vector product of $-2m\omega$ and $\delta r/\delta t$, it must appear, to an observer moving with the rotating frame, to act in a direction perpendicular to the particle's path at any instant. The centrifugal force is a triple vector product and by the expansion given at the end of Section 1.6:

$$-m\omega \times (\omega \times r) = -m(\omega \cdot r)\omega + m\omega^2 r = m\omega^2 r$$

remembering that $\omega = \omega k$ and therefore that $\omega \cdot r = 0$ in the special case we are considering. The direction of this force coincides with that of the position vector and acts outwards.

If the moving frame rotates with constant angular velocity then $\dot{\omega} = 0$, and

$$F - 2m\omega \times \frac{\delta r}{\delta t} - m\omega \times (\omega \times r) = m \frac{\delta^2 r}{\delta t^2}. \tag{9}$$

Example 2   A straight wire of length a rotates with constant angular velocity ω about a fixed perpendicular axis through one end of the wire. A bead of mass m is placed on the wire at its mid-point and released. If friction is negligible and the wire rotates in a horizontal plane, describe the motion of the bead.

Take rotating axes Ox along and Oy perpendicular to the wire, and let Ox coincide with the fixed axis OX when the bead is in its initial position. The scheme of axes is shown in Figure 10.4. The only horizontal force acting on the bead is the reaction due to the wire and this acts in a direction perpendicular to the wire since no friction is present. Thus the reaction is given by Rj, say. The position vector of the bead in the *rotating frame* is given by $r = xi$ where x is the distance of the bead from O at time t.

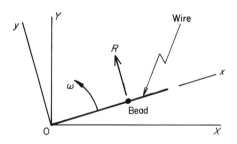

Figure 10.4

Since the angular velocity is a constant $\omega\mathbf{k}$, we apply Equation (9):

$$\mathbf{Rj} - 2m\omega\mathbf{k} \times \dot{x}\mathbf{i} - m\omega\mathbf{k} \times (\omega\mathbf{k} \times x\mathbf{i}) = m\ddot{x}\mathbf{i}$$

which simplifies to

$$\mathbf{Rj} - 2m\omega\dot{x}\mathbf{j} + m\omega^2 x\mathbf{i} = m\ddot{x}\mathbf{i}.$$

This vector equation is equivalent to the two scalar equations

$$R = 2m\omega\dot{x}, \quad \omega^2 x = \ddot{x}.$$

The second of these gives the displacement of the bead and is a standard second-order equation with solution

$$x = A\,e^{\omega t} + B\,e^{-\omega t}.$$

Initially, we are given that $x = \tfrac{1}{2}a$ and $\dot{x} = 0$. Hence the constants A and B are given by

$$\tfrac{1}{2}a = A + B, \quad 0 = A - B$$

from which we find that $A = B = \tfrac{1}{4}a$. The displacement of the bead is given by

$$x = \tfrac{1}{4}a(e^{\omega t} + e^{-\omega t}) = \tfrac{1}{2}a\cosh\omega t.$$

The bead leaves the wire at time $t = T$ where $2 = \cosh\omega T$ with velocity

$$\frac{\delta x}{\delta t} = \tfrac{1}{2}a\omega\sinh\omega T = \tfrac{1}{2}a\omega(\cosh^2\omega T - 1)^{1/2} = \frac{\sqrt{3}}{2}\,a\omega$$

relative to the wire.

10.3 ROTATION OF A RIGID BODY ABOUT A FIXED AXIS

Consider now a rigid body constrained to rotate about a fixed axis in the body. Suppose that the origin O is a point on this axis and that OZ is the axis of rotation. The axes Ox and Oy are taken to be axes fixed in the rigid body, which means that there is no relative rotation between these axes and the body. As before OX and OY specify the inertial frame (see Figure 10.5). We showed

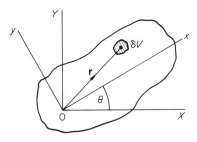

Figure 10.5

in Section 3.9 that the moment **M** of the external forces about a fixed point balances the rate of change of angular momentum $\dot{\mathbf{h}}$:

$$\mathbf{M} = \dot{\mathbf{h}}$$

where the angular momentum

$$\mathbf{h} = \int_V \mathbf{r} \times \dot{\mathbf{r}} \rho \; \mathrm{d}V$$

r being the position vector of a typical volume element δV, and ρ being the density of the rigid body. The moment **M** may also include *couples* which, although they add nothing to the resultant force on the system, do affect the moment. For example, the frictional resistance between an axle and wheel will be a couple. A couple may be thought of as a pair of equal and opposite forces which do not act in the same straight line.

We shall now examine the angular momentum in relation to a body rotating about a fixed axis. Choose the element of volume to be a cylinder parallel to the axis OZ. In Figure 10.5, δV is really a cylinder erected on the shaded element. Obviously we take moments about the axis of rotation. Consider now a point in the plane OXY with the *plane* position vector **r**. By Equation (6) the velocity of this point

$$\dot{\mathbf{r}} = \frac{\delta \mathbf{r}}{\delta t} + \boldsymbol{\omega} \times \mathbf{r}$$

where $\boldsymbol{\omega}$ is the angular velocity of the rotating frame which, since this frame is embedded in the rigid body, we describe as the angular velocity of the body. But **r** is the position vector of a *fixed* point in the rotating frame so that $\delta \mathbf{r}/\delta t = 0$, and

$$\dot{\mathbf{r}} = \boldsymbol{\omega} \times \mathbf{r}.$$

Consequently the angular momentum

$$\mathbf{h} = \int_V \mathbf{r} \times (\boldsymbol{\omega} \times \mathbf{r}) \rho \; \mathrm{d}V$$

$$= \int_V [r^2 \boldsymbol{\omega} - (\boldsymbol{\omega} \cdot \mathbf{r})\mathbf{r}] \rho \; \mathrm{d}V \quad \text{(expanding the triple vector product)}$$

$$= \boldsymbol{\omega} \int_V r^2 \rho \; \mathrm{d}V$$

since $\boldsymbol{\omega} \cdot \mathbf{r} = 0$ and $\boldsymbol{\omega}$ does not depend on the positions of individual parts of the body. We write

$$\mathbf{h} = I\boldsymbol{\omega}$$

where I, called the *moment of inertia*, is the integral $\int_V r^2\rho \, dV$, a fixed number for any selected rigid body and axis of rotation. It cannot depend on time or position since r, the distance of the element δV from the origin is a constant. The moment of inertia distinguishes the rotational behaviour of two bodies in much the same way as differences in mass distinguish the behaviour of the translational motion of particles and bodies.

The equation of motion for the body can be written

$$\mathbf{M} = I\dot{\boldsymbol{\omega}}.$$

Since $\mathbf{h}$, $\boldsymbol{\omega}$ and consequently $\mathbf{M}$ can have only one component—in the direction of $\mathbf{k}$—we can replace this equation by the single scalar equation

$$M = I\dot{\omega} = I\ddot{\theta}$$

where $\boldsymbol{\omega} = \omega\mathbf{k} = \dot{\theta}\mathbf{k}$. Having obtained the equation of motion, we can forget the rotating frame since θ is essentially the angle between a line fixed in the body and one fixed in space.

Consider a rigid body of mass m rotating freely about a fixed horizontal axis which passes through the point O of the body, the only external force being a uniform gravitational field. The total effect of this force is equivalent to a single force mg acting at the mass-centre G of the body (Figure 10.6). If $OG = b$, the moment of the weight about O is $mgb \sin \theta$ (there will be reactions on the body at the axis but these forces must have zero moment about the axis) and the equation of motion is therefore

$$-mgb \sin \theta = I\ddot{\theta}. \tag{10}$$

The angle θ is the angle between OG and the downward vertical. This equation should be compared with the equation of motion for a simple pendulum of length a:

$$-g \sin \theta = a\ddot{\theta}.$$

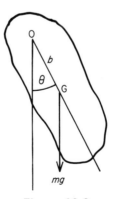

Figure 10.6

The equations of motion are similar and the rigid body behaves as a simple pendulum of length I/mb.

10.4   MOMENT OF INERTIA

The moment of inertia I of a body about a fixed axis of itself is defined by

$$I = \int_V r^2 \rho \, \mathrm{d}V$$

which is a volume integral. However, it may reduce to a surface integral for a lamina or a line integral for a wire, the density being then interpreted as mass per unit area or mass per unit length as the case may be. We shall derive some moments of inertia for simple bodies.

(i) *Rod about an axis through its mid-point.* Suppose that the axis is perpendicular to the rod and that the density per unit length ρ is constant. An element of length δx at a distance x from the mid-point of the rod will have mass $\rho \delta x$ and its moment of inertia is $x^2 \rho \delta x$ about O. The total moment of inertia is the sum of such quantities throughout the length of the rod. Thus, for a rod of length $2a$,

$$I = \int_{-a}^{a} x^2 \rho \, \mathrm{d}x = \rho(\tfrac{1}{3}x^3)^a_{-a} = \tfrac{2}{3}\rho a^3 = \tfrac{1}{3}ma^2$$

where m, the mass of the rod, is $2\rho a$.

(ii) *Circular disc about its axis.* The axis is through the centre of the disc perpendicular to its plane. We use the obvious symmetry of the disc and take a circular element of width δr and radius r as shown in Figure 10.7. Every part of the element is, to the first order, at the same distance r from the axis and

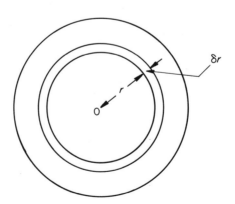

Figure 10.7

consequently $2\rho\pi r^3 \delta r$ is, to the same order, the moment of inertia of this element about the axis. If a is the radius of the disc, the total moment of inertia is given by

$$I = 2\pi\rho \int_0^a r^3 \, \mathrm{d}r = 2\pi\rho[\tfrac{1}{4}r^4]_0^a = \tfrac{1}{2}\pi\rho a^4 = \tfrac{1}{2}ma^2$$

where $m = \pi\rho a^2$ is the mass of the disc. The circular cylinder has the same moment of inertia about its axis.

(iii) *Uniform block about a central axis parallel to two faces.* Let the edges of the block have lengths $2a$, $2b$, $2c$ parallel to the axes Ox, Oy, Oz where O is the centre of the block (Figure 10.8). Take a small rectangular box with edges δx, δy, δz as the element of volume. The square of the distance of this element from the axis (say the x-axis) is $y^2 + z^2$, and the moment of inertia of the element is $\rho(y^2 + z^2)\delta x\delta y\delta z$. The total moment is therefore given by

$$I = \int_{-c}^c \int_{-b}^b \int_{-a}^a \rho(y^2 + z^2) \, \mathrm{d}x \, \mathrm{d}y \, \mathrm{d}z$$

$$= \rho \int_{-c}^c \int_{-b}^b [(y^2 + z^2)x]_{x=-a}^{x=a} \, \mathrm{d}y \, \mathrm{d}z$$

$$= 2\rho a \int_{-c}^c [\tfrac{1}{3}y^3 + z^2 y]_{y=-b}^{y=b} \, \mathrm{d}z$$

$$= 4\rho a b \int_{-c}^c (\tfrac{1}{3}b^2 + z^2) \, \mathrm{d}z$$

$$= 4\rho a b[\tfrac{1}{3}b^2 z + \tfrac{1}{3}z^3]_{z=-c}^{z=c}$$

$$= \tfrac{8}{3}\rho abc(b^2 + c^2) = \tfrac{1}{3}m(b^2 + c^2)$$

where $m = 8abc\rho$ is the mass of the block.

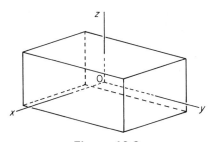

z

O

x

y

Figure 10.8

(iv) *Uniform sphere about a diameter.* Let a be the radius of the sphere. Slice a disc of thickness δx and distance x from the centre of the sphere perpendicular to the chosen axis. By (ii) the moment of inertia of this disc is, to first order, $\frac{1}{2}\pi\rho(a^2 - x^2)^2\delta x$. Summing this expression throughout the sphere, we find that the moment of inertia is

$$I = \frac{1}{2}\pi\rho \int_{-a}^{a} (a^2 - x^2)^2 \, dx$$

$$= \frac{1}{2}\pi\rho(a^4x - \frac{2}{3}a^2x^3 + \frac{1}{5}x^5)^a_{-a}$$

$$= \pi\rho a^5(1 - \frac{2}{3} + \frac{1}{5}) = \frac{8}{15}\pi\rho a^5$$

$$= \frac{2}{5}ma^2$$

where $m = \frac{4}{3}\pi a^3\rho$ is the mass of the sphere.

The moment of inertia is always the product of mass and (length)^2 and if it is written as mk^2, k is called the *radius of gyration*. Thus the radius of gyration of a disc about its axis is $a/\sqrt{2}$. A particle of mass m situated at a distance k from the axis will have the same moment of inertia as the body.

Parallel axis theorem. Moments of inertia about other axes can often be found quickly by the following useful result. If I_G is the moment of inertia of a body of mass m about an axis through its mass-centre and I is the moment of inertia about a parallel axis which is at a distance d from the first, then

$$I = I_G + md^2.$$

In Figure 10.9, GA is the axis through the mass-centre G and BC is the parallel axis. Consider an element δV of volume which is at a distance r from GA and r' from BC. By definition

$$I_G = \int_V \rho r^2 \, dV.$$

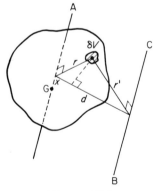

Figure 10.9

In the triangle shown we observe that $r^2 = r'^2 - d^2 + 2xd$ so that

$$I_G = \int_V \rho(r'^2 - d^2 + 2xd)\, dV$$

$$= I - d^2 \int_V \rho\, dV + 2d \int_V x\rho\, dV$$

$$= I - md^2$$

the last integral vanishing since x is the distance of δV from a fixed plane passing through the mass-centre.

Example 3   Investigate the difference between the periods of a simple pendulum and a similar pendulum in which the bob is replaced by a small sphere of the same mass for small oscillations.

If the simple pendulum has length a and a bob (idealised as a particle) of mass m, its equation for small displacements θ from the downward vertical is

$$a\ddot\theta = -g\theta.$$

The period of the oscillations is $2\pi(a/g)^{1/2}$.

Let the sphere have radius ε. By the parallel axis theorem the moment of inertia of the sphere about the point of suspension will be $\frac{2}{5}m\varepsilon^2 + ma^2$. The equation of motion is

$$(\tfrac{2}{5}\varepsilon^2 + a^2)\ddot\theta = -ga\theta$$

from Equation (10) for small θ. The period of the oscillations is therefore

$$2\pi[(\tfrac{2}{5}\varepsilon^2 + a^2)/ga]^{1/2} = 2\pi\left[\frac{a}{g}\left(1 + \frac{2\varepsilon^2}{5a^2}\right)\right]^{1/2}$$

$$\approx 2\pi\left(\frac{a}{g}\right)\cdot\left(1 + \frac{\varepsilon^2}{5a^2}\right)$$

using the binomial expansion for $\varepsilon \ll a$. The period is *increased* by a factor $\varepsilon^2/5a^2$ approximately.

The kinetic energy of a body rotating about a fixed axis is given by

$$\mathscr{T} = \tfrac{1}{2}\int_V \rho\dot{\mathbf r}\cdot\dot{\mathbf r}\, dV$$

(Section 5.4), where the velocity $\dot{\mathbf r} = \boldsymbol\omega \times \mathbf r = -\omega y\mathbf i + \omega x\mathbf j$ and $\boldsymbol\omega = \omega\mathbf k$ is the angular velocity of the body. Thus

$$\mathscr{T} = \tfrac{1}{2}\int_V \rho\omega^2(x^2 + y^2)\, dV = \tfrac{1}{2}\omega^2\int_V \rho r^2\, dV$$

$$= \tfrac{1}{2}I\omega^2$$

and, as before, I is the moment of inertia about the axis.

For a rigid body rotating freely about a fixed axis, the energy of the body will be conserved (Section 5.4). Referring back to Figure 10.6, we see that the potential energy of the body is given by

$$\mathcal{V} = -mga \cos \theta$$

which is negative because $a \cos \theta$ is the *depth* of the mass-centre below O. Energy conservation implies that

$$\tfrac{1}{2}I\dot{\theta}^2 - mga \cos \theta = \text{constant},$$

an equation which can also be obtained by integrating the equation of motion.

Example 4   One end of a uniform heavy chain is attached to a drum of radius a, and the chain is wrapped round the drum. It makes n complete turns with a small piece of chain hanging free from the horizontal diameter. The drum is mounted on a smooth horizontal axis, and the chain is allowed to unwrap. If the moment of inertia of the drum about the axis is I, find the angular speed of the drum when the chain has just unwrapped itself.

Figure 10.10 shows the drum and chain at the instant when the chain has unwrapped itself. Let the drum then have an angular speed ω. Since there is no frictional resistance in the system, energy will be conserved. Initially the mass-centre of the chain lies on the axis and finally G will be at a depth πna below the axis since the chain has length $2\pi na$. This loss of potential energy of $Mg\pi na$ is compensated by a gain in kinetic energy of $\tfrac{1}{2}I\omega^2$ for the drum and $\tfrac{1}{2}M\omega^2a^2$ for the chain (the chain will at this instant be falling vertically with speed ωa).

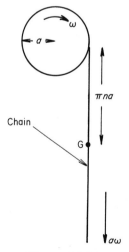

Figure 10.10

Hence

$$\tfrac{1}{2}I\omega^2 + \tfrac{1}{2}M\omega^2 a^2 = Mg\pi na$$

whence

$$\omega = [2Mg\pi na/(I + Ma^2)]^{1/2}.$$

10.5 GENERAL PLANE MOTION OF A RIGID BODY

By the term plane motion of a system, we mean that every point of the system moves parallel to a fixed plane. Suppose that OXY represents, in Figure 10.11, this fixed plane. Let G be the mass-centre of a rigid body and let GX' and GY' be axes which remain parallel to OX and OY respectively. Without loss of generality we can let the planes OXY and $GX'Y'$ coincide, since the velocity of every point of the body on a line parallel to the Z-axis (drawn out of the paper) will be the same by the definition of plane motion. Suppose that P is a point of the body with (plane) position vectors $\mathbf{r}$ relative to O and $\mathbf{r}'$ relative to G.

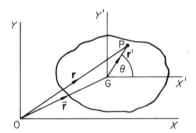

Figure 10.11   Coordinate scheme for the polar motion of a rigid body.

The motion may be considered as the combination of a translation of the mass-centre and a rotation about the mass-centre. The angular velocity of the body is $\boldsymbol{\omega}$, where $\boldsymbol{\omega} = \dot{\theta}\mathbf{k}$ and θ is an angle between a direction fixed in the body and one fixed in space, in this case the angle PGX'. By analogy with the motion of a body about a fixed axis the velocity of P *relative to* G is

$$\dot{\mathbf{r}}' = \boldsymbol{\omega} \times \mathbf{r}'.$$

The actual velocity of P will therefore be

$$\dot{\mathbf{r}} = \dot{\bar{\mathbf{r}}} + \boldsymbol{\omega} \times \mathbf{r}'. \tag{11}$$

The angular momentum of the body about O is defined by

$$\mathbf{h} = \int_V \rho\mathbf{r} \times \dot{\mathbf{r}} \, dV. \tag{12}$$

By the triangle law $\mathbf{r} = \bar{\mathbf{r}} + \mathbf{r}'$, and substituting for the velocity from (11) into (12), we find that

$$\mathbf{h} = \int_V \rho(\bar{\mathbf{r}} + \mathbf{r}') \times (\dot{\bar{\mathbf{r}}} + \boldsymbol{\omega} \times \mathbf{r}')\, dV$$

$$= \int_V \rho[\bar{\mathbf{r}} \times \dot{\bar{\mathbf{r}}} + \mathbf{r}' \times \dot{\bar{\mathbf{r}}} + \bar{\mathbf{r}} \times (\boldsymbol{\omega} \times \mathbf{r}') + \mathbf{r}' \times (\boldsymbol{\omega} \times \mathbf{r}')]\, dV$$

$$= \bar{\mathbf{r}} \times \dot{\bar{\mathbf{r}}} \int_V \rho\, dV + \int_V \rho \mathbf{r}'\, dV \times \dot{\bar{\mathbf{r}}} + \bar{\mathbf{r}} \times \left(\boldsymbol{\omega} \times \int_V \rho \mathbf{r}'\, dV\right)$$

$$+ \int_V \rho \mathbf{r}' \times (\boldsymbol{\omega} \times \mathbf{r}')\, dV.$$

Since $\mathbf{r}'$ is the position vector relative to the mass-centre, $\int_V \rho \mathbf{r}'\, dV = \mathbf{0}$ by definition. Therefore

$$\mathbf{h} = m\bar{\mathbf{r}} \times \dot{\bar{\mathbf{r}}} + \int_V \rho \boldsymbol{\omega} r'^2\, dV$$

where $m = \int_V \rho\, dV$ is the mass of the body. The expansion for the triple vector product has been used in the integrand of the last integral. Finally, since $\boldsymbol{\omega}$ depends on time only,

$$\mathbf{h} = m\bar{\mathbf{r}} \times \dot{\bar{\mathbf{r}}} + I_G \boldsymbol{\omega} \tag{13}$$

where $I_G = \int_V \rho r'^2\, dV$ is the moment of inertia of the body about an axis through its mass-centre.

The laws of motion for a rigid body are described in Section 3.8 and are equivalent to

  (i) the total force equals the product of the mass and the acceleration of the mass-centre,

  (ii) The moment of the forces about a *fixed* point equals the rate of change of angular momentum about that point.

For the plane motion considered above,

$$\mathbf{F} = m\ddot{\mathbf{r}}, \qquad \mathbf{M} = \dot{\mathbf{h}}$$

where $\mathbf{F}$ is the total force and $\mathbf{M}$ is the moment of the forces about O. The second of these equations can be expressed in a more convenient form if we substitute for $\mathbf{h}$ from (13):

$$\mathbf{M} = m\frac{d}{dt}(\bar{\mathbf{r}} \times \dot{\bar{\mathbf{r}}}) + I_G \dot{\boldsymbol{\omega}}$$

$$= m\bar{\mathbf{r}} \times \ddot{\bar{\mathbf{r}}} + I_G \dot{\boldsymbol{\omega}}$$

$$= \bar{\mathbf{r}} \times \mathbf{F} + I_G \dot{\boldsymbol{\omega}}.$$

However $\mathbf{M} - \bar{\mathbf{r}} \times \mathbf{F}$ is $\mathbf{M}_G$, the moment of the forces about G, the mass-centre. The equations of motion can therefore be represented concisely by

$$\mathbf{F} = m\ddot{\bar{\mathbf{r}}}, \qquad \mathbf{M}_G = I_G \dot{\boldsymbol{\omega}}.$$

These vector equations contain the three scalar equations

$$F_x = m\ddot{\bar{x}}, \qquad F_y = m\ddot{\bar{y}}, \qquad M_G = I_G \ddot{\theta}$$

(where we have let $\mathbf{F} = F_x \mathbf{i} + F_y \mathbf{j}$ and $\mathbf{M}_G = M_G \mathbf{k}$) to determine $\bar{x}$, $\bar{y}$ and θ; that is the position of the mass-centre and the orientation of the body about the mass-centre at any time.

As before, the kinetic energy is defined by

$$\mathscr{T} = \tfrac{1}{2} \int_V \rho \dot{\mathbf{r}} \cdot \dot{\mathbf{r}} \, dV.$$

The velocity of any point of the body can be expressed as

$$\dot{\mathbf{r}} = \dot{\bar{\mathbf{r}}} + \boldsymbol{\omega} \times \mathbf{r}' = (\dot{\bar{x}} - \omega y')\mathbf{i} + (\dot{\bar{y}} + \omega x')\mathbf{j}$$

with the result that

$$\mathscr{T} = \tfrac{1}{2} \int_V [(\dot{\bar{x}} - \omega y')^2 + (\dot{\bar{y}} + \omega x')^2]\rho \, dV$$

$$= \tfrac{1}{2} \int_V (\dot{\bar{x}}^2 + \dot{\bar{y}}^2)\rho \, dV + \tfrac{1}{2} \int_V \omega^2 (x'^2 + y'^2)\rho \, dV$$

$$- \omega \int \rho y' \, dV + \omega \int \rho x' \, dV$$

$$= \tfrac{1}{2} m\bar{v}^2 + \tfrac{1}{2} I_G \omega^2 \tag{14}$$

the last two integrals on the right-hand side vanishing by the definition of the mass-centre. The kinetic energy may be considered as the sum of two terms, one giving the translational energy and the other the rotational energy of the body. If non-conservative forces acting on the body do no work, then the energy principle will hold.

Example 5   A uniform sphere of mass m is released from rest on a rough plane inclined at an angle α to the horizontal. If no slipping occurs between the sphere and the plane, discuss the motion of the sphere.

By the phrase 'no slipping occurs' we mean that the sphere rolls on the plane, the point of contact P on the sphere in Figure 10.12 being instantaneously at rest. Suppose that O is the initial point of contact with O′ the corresponding contact point on the sphere. Let $\widehat{PGO} = \theta$ and $OP = x$. If a is the radius of the sphere, $a\theta = x$ for

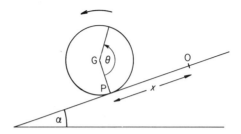

Figure 10.12 Sphere rolling down an inclined plane.

rolling contact. Since the reaction and frictional force between the sphere and the plane both act through P, which is instantaneously at rest, they do no work throughout the motion and the energy equation holds. By Equation (14)

$$\mathcal{T} = \tfrac{1}{2}m\dot{x}^2 + \tfrac{1}{2}\cdot\tfrac{2}{5}ma^2\cdot\dot{\theta}^2$$

since $\dot{\theta}$ is the angular velocity of the sphere. Note also that $\tfrac{2}{5}ma^2$ is the moment of inertia of a sphere about its diameter. Since $a\theta = x$, we can write

$$\mathcal{T} = \tfrac{1}{2}m\dot{x}^2 + \tfrac{1}{5}m\dot{x}^2 = \tfrac{7}{10}m\dot{x}^2.$$

The kinetic energy must balance the potential energy lost. In this case the centre of the sphere falls a distance $x \sin\alpha$, with a corresponding potential energy loss of $mgx \sin\alpha$. Hence

$$\tfrac{7}{10}m\dot{x}^2 = mgx \sin\alpha.$$

Taking the derivative of this equation with respect to the time, we find that

$$\ddot{x} = (5g \sin\alpha)/7$$

from which we infer that the centre of the sphere moves with constant acceleration. Since $\dot{x} = x = 0$ at $t = 0$ (say) the position of the sphere at time t is given by $x = 5gt^2 \sin\alpha/14$.

Example 6   In Figure 10.13 the planetary gear B is enmeshed between a fixed outer gear C and a free inner gear A. There is no resistance to motion. Initially the system is at rest and a constant couple G is applied to A for a time t_0 and then removed. Find the final angular speeds of the gears. (Assume that the motion takes place in a horizontal plane.)

Let the radii of A and B be a and b and let their moments of inertia be I_A and I_B respectively. In Figure 10.13, OPQ is the initial position of the gears when P_A, P_B and P, and Q_B and Q, coincide. Let the angles θ, ϕ and ψ be as shown in the figure. It is clear from the geometry of the figure that the angular velocity of A is $\dot{\theta}$ and that of B is $\dot{\phi} - \dot{\theta} + \dot{\psi}$ (remember that in two dimensions angular velocity is the rate of change of an angle between a direction in the body and one fixed in space). However, there exist relations between the angles θ, ϕ and ψ. For rolling contact, the arc lengths $P'P_A$ and $P'P_B$ must be equal and Q_BQ' must equal QQ'. Thus

$$a\psi = b\phi = (a + 2b)(\theta - \psi)$$

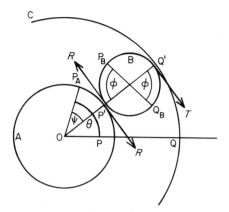

Figure 10.13

so that

$$\psi = (a + 2b)\theta/2(a + b), \qquad \phi = a(a + 2b)\theta/2b(a + b).$$

After some simplification it follows that the angular velocity of B

$$\dot\phi - \dot\theta + \dot\psi = a\dot\theta/2b.$$

For A we take moments about O, and for B we take moments about the centre of that gear and resolve transversely. With R and T the tangential reactions between the gear teeth (normal reactions are not required), the following equations result when the couple G is acting:

$$G - Ra = I_A\ddot\theta \tag{15}$$

$$-Rb - Tb = -I_B(\ddot\phi - \ddot\theta + \ddot\psi) = -I_B a\ddot\theta/2b \tag{16}$$

$$R - T = m_B(a + b)(\ddot\theta - \ddot\psi) = \tfrac{1}{2}m_B a\ddot\theta \tag{17}$$

where in the last equation m_B is the mass of gear B and $\tfrac{1}{2}a\ddot\theta$ is the transverse acceleration of its centre.

The behaviour of θ supplies the information we require. Eliminate R and T between Equations (15), (16) and (17) to give

$$G = \frac{\ddot\theta}{4b^2}(4b^2 I_A + m_B a^2 b^2 + a^2 I_B).$$

Integration of this equation with respect to t yields

$$Gt = \frac{\dot\theta}{4b^2}(4b^2 I_A + m_B a^2 b^2 + a^2 I_B).$$

Putting $\dot\theta = \omega$ when $t = t_0$, we find that the inner gear A ultimately rotates with angular speed

$$\omega = 4b^2 Gt_0/(4b^2 I_A + m_B a^2 b^2 + a^2 I_B).$$

The reader should calculate the corresponding angular speed of B.

Example 7   A uniform circular cylinder of radius a and mass m rolls without slipping on the inner surface of a fixed circular cylinder of radius 4a. Find the period of small oscillations of the rolling cylinder.

Let O′ coincide with O when the rolling cylinder is in equilibrium in Figure 10.14. Let θ and ϕ be the angles shown in Figure 10.14. The conditions for rolling contact implies that $a\phi = 4a\theta$. We shall apply the energy principle to the rolling cylinder; this is justifiable since both the normal reaction and the frictional force acting at the point of contact P between the cylinders do no work.

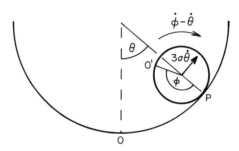

Figure 10.14

The centre of the rolling cylinder must move in a circle with speed $3a\dot\theta$ and its angular speed is given by $a(\dot\phi - \dot\theta)$, the sense of both being shown in the figure. By Equation (14),

$$\mathcal{T} = \tfrac{1}{2}m(9a^2\dot\theta^2) + \tfrac{1}{2}\cdot\tfrac{1}{2}ma^2(\dot\phi - \dot\theta)^2.$$

Relative to the centre of the fixed cylinder,

$$\mathcal{V} = -3mga \cos\theta.$$

The energy principle asserts that

$$\tfrac{9}{2}ma^2\dot\theta^2 + \tfrac{1}{4}ma^2(\dot\phi - \dot\theta)^2 - 3mga \cos\theta = \text{constant}$$

or, by writing ϕ in terms of θ, that

$$\tfrac{9}{2}ma^2\dot\theta^2 + \tfrac{9}{4}ma^2\dot\theta^2 - 3mga \cos\theta = \text{constant}$$

that is

$$\tfrac{27}{4}ma^2\dot\theta^2 - 3mga \cos\theta = \text{constant}.$$

Differentiating this equation with respect to t, we deduce that

$$\tfrac{27}{2}ma^2\ddot\theta + 3mga \sin\theta = 0$$

which for small values of θ can be approximated by

$$\ddot\theta + \frac{2g}{9a}\theta = 0.$$

Thus to the first order the cylinder rocks with simple harmonic motion of period $3\pi(2a/g)^{1/2}$. Note that the methods of Section 7.2 could have been applied to this problem.

10.6 GENERAL ROTATING FRAME OF REFERENCE

Figure 10.15 shows a fixed frame of reference specified by the unit vectors **I, J, K** and a rotating frame whose axes are in the directions of the unit vectors **i, j, k**, the frames having a coincident origin O. The rotating frame may be imagined fixed in a rigid body which is pivoted at the point O. The unit vectors **i, j, k**, which will be time dependent, must satisfy the relations

$$\mathbf{i}\cdot\mathbf{i} = \mathbf{j}\cdot\mathbf{j} = \mathbf{k}\cdot\mathbf{k} = 1, \qquad \mathbf{i}\cdot\mathbf{j} = \mathbf{j}\cdot\mathbf{k} = \mathbf{k}\cdot\mathbf{i} = 0.$$

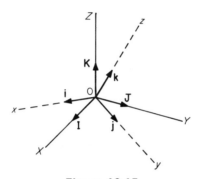

Figure 10.15

Taking derivatives of these relations with respect to the time, we must have

$$\mathbf{i}\cdot\frac{d\mathbf{i}}{dt} = 0, \qquad \mathbf{j}\cdot\frac{d\mathbf{j}}{dt} = 0, \qquad \mathbf{k}\cdot\frac{d\mathbf{k}}{dt} = 0 \tag{18}$$

$$\frac{d\mathbf{i}}{dt}\cdot\mathbf{j} + \mathbf{i}\cdot\frac{d\mathbf{j}}{dt} = 0, \qquad \frac{d\mathbf{j}}{dt}\cdot\mathbf{k} + \mathbf{j}\cdot\frac{d\mathbf{k}}{dt} = 0, \qquad \frac{d\mathbf{k}}{dt}\cdot\mathbf{i} + \mathbf{k}\cdot\frac{d\mathbf{i}}{dt} = 0. \tag{19}$$

The three relations (18) imply that the derivative of the unit vector is perpendicular to that unit vector in each case. Therefore there exist functions $\alpha_2(t)$, $\alpha_3(t)$, $\beta_3(t)$, $\beta_1(t)$, $\gamma_1(t)$, $\gamma_2(t)$ such that

$$\frac{d\mathbf{i}}{dt} = \beta_1\mathbf{j} + \gamma_1\mathbf{k}, \qquad \frac{d\mathbf{j}}{dt} = \gamma_2\mathbf{k} + \alpha_2\mathbf{i}, \qquad \frac{d\mathbf{k}}{dt} = \alpha_3\mathbf{i} + \beta_3\mathbf{j}$$

but conditions (19) must also be satisfied, and they imply that

$$\beta_1 = -\alpha_2 = \omega_3 \text{ (say)}, \quad \gamma_2 = -\beta_3 = \omega_1 \text{ (say)}, \quad \alpha_3 = -\gamma_1 = \omega_2 \text{ (say)}.$$

Thus

$$\frac{d\mathbf{i}}{dt} = \omega_3\mathbf{j} - \omega_2\mathbf{k}, \qquad \frac{d\mathbf{j}}{dt} = \omega_1\mathbf{k} - \omega_3\mathbf{i}, \qquad \frac{d\mathbf{k}}{dt} = \omega_2\mathbf{i} - \omega_1\mathbf{j}.$$

We now introduce the vector $\boldsymbol{\omega} = \omega_1\mathbf{i} + \omega_2\mathbf{j} + \omega_3\mathbf{k}$ and observe by the definition of a vector product that

$$\frac{d\mathbf{i}}{dt} = \boldsymbol{\omega} \times \mathbf{i}, \qquad \frac{d\mathbf{j}}{dt} = \boldsymbol{\omega} \times \mathbf{j}, \qquad \frac{d\mathbf{k}}{dt} = \boldsymbol{\omega} \times \mathbf{k}. \qquad (20)$$

One immediate conclusion from this equation is that each of the points with position vectors $\mathbf{i}$, $\mathbf{j}$ and $\mathbf{k}$ is moving in a direction perpendicular to a certain direction indicated by the vector $\boldsymbol{\omega}$.

Consider now a point which is rigidly embedded in the rotating frame with position vector $\mathbf{r} = a\mathbf{i} + b\mathbf{j} + c\mathbf{k}$, where, of course, a, b and c must be constants. The velocity of this point is

$$\frac{d\mathbf{r}}{dt} = a\frac{d\mathbf{i}}{dt} + b\frac{d\mathbf{j}}{dt} + c\frac{d\mathbf{k}}{dt}$$

$$= a\boldsymbol{\omega} \times \mathbf{i} + b\boldsymbol{\omega} \times \mathbf{j} + c\boldsymbol{\omega} \times \mathbf{k}$$

$$= \boldsymbol{\omega} \times \mathbf{r}$$

using the relations above. With the frame considered attached to a rigid body we deduce that

$$\boldsymbol{\omega} \cdot \frac{d\mathbf{r}}{dt} = \boldsymbol{\omega} \cdot (\boldsymbol{\omega} \times \mathbf{r}) = 0$$

which means that at any instant every point of the body is moving in a direction perpendicular to $\boldsymbol{\omega}$. Furthermore if the point $\mathbf{r}$ lies on the line through O in the direction of $\boldsymbol{\omega}$, so that $\mathbf{r} = \boldsymbol{\omega}k/|\boldsymbol{\omega}|$, where k is a constant, then

$$\frac{d\mathbf{r}}{dt} = \boldsymbol{\omega} \times \frac{\boldsymbol{\omega}k}{|\boldsymbol{\omega}|} = \mathbf{0}.$$

At any instant there exists a straight line which is momentarily at rest and such that every point of the body is turning about that line. In other words at any instant the body is spinning about an *instantaneous axis of rotation*. This axis will, in general, be fixed neither in space nor in the body.

That the magnitude of $\boldsymbol{\omega}$ represents an angular rate of rotation can be seen from Figure 10.16. If θ is the angle between $\boldsymbol{\omega}$ and $\mathbf{r}$, then the magnitude of the velocity of the body-fixed point P is given by

$$|\dot{\mathbf{r}}| = |\boldsymbol{\omega} \times \mathbf{r}| = \omega r \sin \theta$$

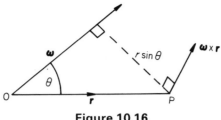

Figure 10.16

and clearly $r \sin \theta$ is the perpendicular distance from P on to ω. Thus P is moving instantaneously in a circle of radius $r \sin \theta$ at an angular rate ω. Note that the direction of $\dot{\mathbf{r}}$ is such that ω, $\mathbf{r}$ and $\dot{\mathbf{r}}$ form a right-handed set of vectors.

Let us consider now a point with position vector

$$\mathbf{r} = X\mathbf{I} + Y\mathbf{J} + Z\mathbf{K} = x\mathbf{i} + y\mathbf{j} + z\mathbf{k}$$

being not necessarily stationary in either frame. The velocity of the point is

$$\dot{\mathbf{r}} = \dot{X}\mathbf{I} + \dot{Y}\mathbf{J} + \dot{Z}\mathbf{K}$$

$$= \dot{x}\mathbf{i} + \dot{y}\mathbf{j} + \dot{z}\mathbf{k} + x\frac{d\mathbf{i}}{dt} + y\frac{d\mathbf{j}}{dt} + z\frac{d\mathbf{k}}{dt}.$$

If the frame $Oxyz$ is spinning with angular velocity ω, Equations (20) hold and the velocity becomes

$$\dot{\mathbf{r}} = \frac{\delta \mathbf{r}}{\delta t} + \omega \times x\mathbf{i} + \omega \times y\mathbf{j} + \omega \times z\mathbf{k}$$

$$= \frac{\delta \mathbf{r}}{\delta t} + \omega \times \mathbf{r} \tag{21}$$

the generalisation of the previous two-dimensional formula in Section 10.2.

10.7   CORIOLIS FORCE IN THREE DIMENSIONS

Consider now a particle of mass m with position vector $\mathbf{r}$ subject to a force $\mathbf{F}$. Suppose we wish to find the equation of motion relative to a frame rotating with angular velocity ω. The equation of motion of the particle will be $\mathbf{F} = m\ddot{\mathbf{r}}$

and we can express $\ddot{\mathbf{r}}$ in terms of the relative acceleration by using the result (21) twice. Thus

$$\mathbf{F} = m\ddot{\mathbf{r}} = m\left(\frac{\delta}{\delta t} + \boldsymbol{\omega} \times \right)\left(\frac{\delta \mathbf{r}}{\delta t} + \boldsymbol{\omega} \times \mathbf{r}\right)$$

$$= m\left(\frac{\delta^2 \mathbf{r}}{\delta t^2} + 2\boldsymbol{\omega} \times \frac{\delta \mathbf{r}}{\delta t} + \frac{\delta \boldsymbol{\omega}}{\delta t} \times \mathbf{r} + \boldsymbol{\omega} \times (\boldsymbol{\omega} \times \mathbf{r})\right)$$

$$= m\left(\frac{\delta^2 \mathbf{r}}{\delta t^2} + 2\boldsymbol{\omega} \times \frac{\delta \mathbf{r}}{\delta t} + \dot{\boldsymbol{\omega}} \times \mathbf{r} + \boldsymbol{\omega} \times (\boldsymbol{\omega} \times \mathbf{r})\right) \qquad (22)$$

since by replacing $\mathbf{r}$ by $\boldsymbol{\omega}$ in (21), $\dot{\boldsymbol{\omega}} = \delta\boldsymbol{\omega}/\delta t$. Equation (22) should be compared with Equation (7). As before, Equation (22) can be written as a balance between actual and apparent forces and the product of mass and relative acceleration:

$$\mathbf{F} - 2m\boldsymbol{\omega} \times \frac{\delta \mathbf{r}}{\delta t} - m\dot{\boldsymbol{\omega}} \times \mathbf{r} - m\boldsymbol{\omega} \times (\boldsymbol{\omega} \times \mathbf{r}) = m\frac{\delta^2 \mathbf{r}}{\delta t^2}.$$

The *centrifugal* force, $-m\boldsymbol{\omega} \times (\boldsymbol{\omega} \times \mathbf{r})$, acts towards the instantaneous axis of rotation and the *Coriolis* force, $-2m\boldsymbol{\omega} \times (\delta\mathbf{r}/\delta t)$, acts in a direction perpendicular to the particle's path as seen by an observer moving with the rotating frame.

We shall illustrate the Coriolis effect by investigating the motion of a particle on the rotating earth. Let $OXYZ$ in Figure 10.17 be the inertial frame with its Z-axis along the polar axis of the Earth. The Earth is assumed to be a sphere. The rotating frame $Oxyz$ is embedded in the Earth with the z-axis inclined at

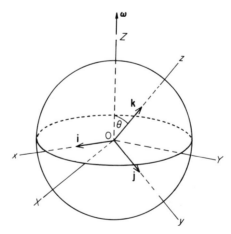

Figure 10.17

an angle θ to the axis of rotation and with the x-axis in the Earth's equatorial plane. The Earth's spin can be resolved into components in the rotating frame:

$$\boldsymbol{\omega} = -\omega \sin \theta \mathbf{j} + \omega \cos \theta \mathbf{k}$$

where ω is the angular speed in a sidereal frame. Let us restrict our attention to motion in the neighbourhood of the Earth's surface so that the gravitational force experienced by a particle of mass m is mg in magnitude and directed towards the centre of the Earth. In the vicinity of the point P, we can approximate this gravitational force by $-m g \mathbf{k}$. The equation of motion (22) will become

$$-mg\mathbf{k} = m\left(\frac{\delta^2 \mathbf{r}}{\delta t^2} + 2\boldsymbol{\omega} \times \frac{\delta \mathbf{r}}{\delta t} + \boldsymbol{\omega} \times (\boldsymbol{\omega} \times \mathbf{r})\right)$$

noting that $\dot{\boldsymbol{\omega}} = \mathbf{0}$.

We could now write the vector equation of motion in its three components, but before we do this we will consider some approximations which can be made.

(i) We have

$$\mathbf{r} = x\mathbf{i} + y\mathbf{j} + (a + z)\mathbf{k},$$

where a is the mean radius of the Earth. For z of the order of a few hundred metres, $a + z \approx a$, since $a = 6370$ km.

(ii) In the term $\boldsymbol{\omega} \times (\boldsymbol{\omega} \times \mathbf{r})$ all the terms apart from $-a\omega^2 \sin \theta \cos \theta \mathbf{j}$ are of magnitude $x\omega^2$ or $y\omega^2$. From the second term on the right-hand side of the equation of motion, we obtain terms of magnitude $\dot{x}\omega$ and $\dot{y}\omega$. Now to fall a distance of a few hundred metres a particle will take a few seconds. Therefore $\dot{x} \approx x/(\text{few seconds})$ whereas $x\omega = 2\pi x/(1\text{ day})$ since ω is the angular velocity of the Earth. Thus $\dot{x}\omega \gg x\omega^2$ and $\dot{y}\omega \gg y\omega^2$.

(iii) For the Earth $g \gg a\omega^2 \sin^2 \theta$.

(iv) Provided θ is not small, $2\dot{x}\omega \cos \theta \ll a\omega^2 \sin \theta \cos \theta$.

With these approximations the equation of motion resolved into its components becomes

$$\ddot{x} - 2\omega(\dot{z} \sin \theta + \dot{y} \cos \theta) = 0 \qquad (23)$$

$$\ddot{y} - a\omega^2 \sin \theta \cos \theta = 0 \qquad (24)$$

$$\ddot{z} + 2\dot{x}\omega \sin \theta = -g. \qquad (25)$$

Suppose a body at rest relative to the Earth is dropped from a height h above the Earth's surface. The initial conditions become $\dot{x} = \dot{y} = \dot{z} = 0$, $x = y = 0$, $z = h$ at $t = 0$. Equations (24) and (25) can be integrated immediately to give

$$\dot{y} = a\omega^2 t \sin \theta \cos \theta, \qquad \dot{z} + 2x\omega \sin \theta = -gt.$$

Substituting for $\dot{y}$ and $\dot{z}$ in (23) and using the approximations again we find that x satisfies

$$\ddot{x} + 2\omega g t \sin \theta = 0$$

which, subject to the initial conditions, has the solution

$$x = -\tfrac{1}{3}\omega g t^3 \sin \theta.$$

Similarly

$$y = \tfrac{1}{2} a \omega^2 t^2 \sin \theta \cos \theta, \qquad z = h - \tfrac{1}{2} g t^2.$$

The body hits the Earth when $z = 0$ which must occur in a time $(2h/g)^{1/2}$. The body therefore reaches the Earth at a point *east* of the vertical at an approximate distance

$$\tfrac{1}{3}\omega g \left(\frac{2h}{g}\right)^{3/2} \sin \theta,$$

and *south* of the true vertical at an approximate distance

$$a h \omega^2 \sin \theta \cos \theta / g$$

from the point where it would have fallen in the absence of rotation. Note that the curvature of the Earth has been ignored. Since

$$\omega = 7.29 \times 10^{-5} \text{ rad/s},$$

the deflection of a particle dropped at $\theta = 45°$ from a height of 100 m is about 1.6 cm to the east and 16 cm to the south.

The Coriolis force is of importance in atmospheric motions. It is found that for a particle of air the term $2\boldsymbol{\omega} \times \delta \mathbf{r}/\delta t$, or $2\boldsymbol{\omega} \times \mathbf{v}_w$ where $\mathbf{v}_w$ is the horizontal velocity of the wind, is the most dominant term on the right-hand side of Equation (22). The horizontal component of the force $\mathbf{F}$ is the pressure force $\mathbf{F}_p$ directed from high to low pressure. We get therefore

$$\mathbf{F}_P \approx 2\boldsymbol{\omega} \times \mathbf{v}_w.$$

We see from the definition of the vector product that $\mathbf{v}_w$ will be perpendicular to the pressure force. This implies that the wind will blow *between* high and low pressure centres to the first approximation rather than *from* high to low pressure; that is along the lines of constant pressure, or *isobars*, as is indicated in Figure 10.18. By further considering the direction of $\boldsymbol{\omega}$ the reader will be able to deduce that in the northern hemisphere the winds blow anti-clockwise around centres of low pressure, whilst in the southern hemisphere they blow in the opposite sense.

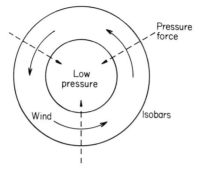

Figure 10.18 Anticlockwise wind around low pressure region in the northern hemisphere.

Exercises

1. A pendulum consists of a bob suspended by a wire of length 30 m. The pendulum is set up at a point on the Earth's surface with geographic latitude of 30°. Assuming the Earth to be a sphere, estimate the displacement of the bob from the true vertical. (Use the following data: radius of the Earth = 6370 km; $g = 9.81$ m/s²; $\omega = 7.29 \times 10^{-5}$ rad/s.)

2. A point has a position vector

$$\mathbf{r} = (a + b \cos \Omega t)\mathbf{I} + b \sin \Omega t \mathbf{J}$$

in an inertial frame (a, b and Ω are constants). Verify that the point describes a circle with constant speed. Show that the coordinates (x, y) of the point referred to a frame rotating with constant angular velocity $\omega \mathbf{k}$ satisfy

$$\dot{x} - \omega y = -b\Omega \sin \Omega t, \qquad \dot{y} + \omega x = b\Omega \cos \Omega t.$$

Show further that these equations can be combined into the single equation

$$\dot{z} + i\omega z = ib\Omega \, e^{i\Omega t}.$$

where $z = x + iy$. Assuming that the axes of the two frames coincide at time $t = 0$, obtain the solution of this equation in the case $\Omega + \omega \neq 0$:

$$z = \left(a + \frac{b\omega}{\Omega + \omega}\right) e^{-i\omega t} + \frac{b\Omega}{\Omega + \omega} e^{i\Omega t}.$$

3. A circular wire of radius a rotates with angular speed ω in its own plane about an axis which passes through a point of the wire. A bead slides on the wire with constant speed V relative to the wire. Find the actual velocity and acceleration of the bead in suitable rotating and fixed axes.

4. A satellite moves round the Earth in a circular orbit of radius a. The angular speed ω of the radius to the satellite is given by $a^3\omega^2 = \gamma m$ where γ is the gravitational constant and m is the mass of the Earth. An origin is taken in the satellite with the axis

Ox along the radius from the Earth and the axis Oy perpendicular to it in the direction in which the satellite is moving. Show that the equations of motion of an object in the vicinity of the satellite are given approximately by

$$\ddot{x} - 2\omega\dot{y} - 3\omega^2 x = 0$$
$$\ddot{y} + 2\omega\dot{x} = 0$$

with respect to the rotating frame. (Ignore the gravitational effect of the satellite.)

5. A smooth elliptic tube of eccentricity e rotates with constant angular velocity ω about a vertical axis through its centre and perpendicular to its plane. Show that a particle can remain at rest at an end of the major axis. If slightly disturbed show that it oscillates with period $2\pi(1 - e^2)^{1/2}/e\omega$.

6. A horizontal turntable rotates with constant angular velocity ω about a vertical axis which intersects a smooth circular groove of radius a in the turntable. A particle of mass m in the groove moving in the same sense as that of the rotation just makes complete revolutions. Show that the force exerted by the groove on the particle at its greatest distance from the axis of rotation is of magnitude $10ma\omega^2$.

7. A rocking cylinder is pivoted at O and drives, through a linkage, a crankshaft B. The appropriate lengths are given in Figure 10.19. If the rod BA rotates about B with a constant angular velocity Ω show that the magnitude of the reaction of the piston on the wall of the cylinder when BA is perpendicular to OA, i.e. when angle $A\hat{O}B$ has a maximum value, is $m a \Omega^2 (2 - \sqrt{2})/6$ where m is the mass of the piston.

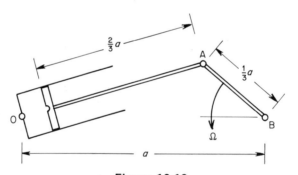

Figure 10.19

8. A turntable is rotating with angular velocity ω. A bullet of mass m is fired from a rifle barrel of length ℓ, fixed relative to the turntable, with a constant acceleration a. The bullet starts from the centre of the turntable. Show that, if $\omega^2 \ll a/\ell$, the reaction of the bullet on the barrel as it leaves the end is approximately

$$2m\omega(2a\ell)^{1/2}.$$

9. A smooth straight tube rotates in a plane with constant angular velocity ω about a perpendicular axis through a point O of it. Inside the tube is a particle of mass m joined to O by spring of stiffness $5m\omega^2$ and natural length a. The particle is released

from its position of relative rest with the spring at its natural length. Show that the particle describes a simple harmonic motion of period π/ω and amplitude $\frac{1}{4}a$ relative to the tube. What is the largest reaction on the tube?

10. OA, OB are two perpendicular lines fixed in a smooth horizontal plane which rotates with constant angular velocity ω about a vertical axis OC (OA, OB, OC form a right-handed system). There is a smooth groove in the plane parallel to OB and meeting OA at L, where $OL = a$. A particle of mass m is projected from L along the groove at time $t = 0$ with speed $a\omega$ relative to the plane. Find the force, due to the constraint of the groove, on the particle at time t.

11. Find the moments of inertia of the following bodies:

   (i) a spherical shell of radius a about a diameter,
   (ii) a uniform disc of radius a about a diameter,
   (iii) a composite body consisting of a circular cylinder of radius a and length h with two hemispheres of the same radius a attached to the plane faces of the cylinder about the axis of symmetry, the density of the body being uniform,
   (iv) a circular cylinder of radius a in which the density varies as distance from the axis about its axis,
   (v) a uniform circular cylinder of radius a and length h about an axis through its centre perpendicular to its axis of symmetry,
   (vi) a uniform square plate of side a about a diagonal.

12. Show that the moment of inertia of a plane lamina about a perpendicular axis is equal to the sum of its moments of inertia about two perpendicular axes in the plane which intersect the perpendicular axis.

13. Find the kinetic energy of a uniform circular cylinder of mass m and radius a which is rolling without slipping on a horizontal plane with speed V.
   The cylinder is released from rest on a plane inclined at an angle α to the horizontal with its axis horizontal. Obtain the angular speed of the cylinder when it has rolled a distance x down the plane.

14. A flywheel has a mass of 900 kg and a radius of gyration of 85 cm and is rotating at 150 r.p.m. Calculate its kinetic energy. What is the maximum weight which the flywheel could lift through 1 m if 14% of the transmitted energy is lost through friction?

15. Find the period of small oscillations of a uniform sphere about a horizontal axis which is a tangent to the sphere.

16. A rigid body is moving in general plane motion with angular velocity $\omega\mathbf{k}$ and such that its mass-centre has velocity $\bar{\mathbf{v}} = \bar{u}\mathbf{i} + \bar{v}\mathbf{j}$. Show that the point with position vector (see Figure 10.11)

$$\mathbf{r} = \bar{\mathbf{r}} - (\bar{v}\mathbf{i} - \bar{u}\mathbf{j})/\omega$$

is instantaneously at rest (this point is known as the *instantaneous centre of rotation* and at any instant the body is rotating about this point; note that, in general, it is fixed neither in the body nor in space). Show that the kinetic energy of the body can be written as $\frac{1}{2}I'\omega^2$ where I' is the moment of inertia about the instantaneous centre of rotation.

Find the instantaneous centres in the following cases:

(i) a circular cylinder rolling on a plane;
(ii) a rod moving with its ends sliding on two straight intersecting perpendicular wires.

17. A uniform rod is smoothly pivoted at one end to a fixed point. The rod is released from rest in a horizontal position. Find the reaction at the pivot when the rod is inclined at an angle θ to the vertical.

18. A uniform rod of length $2a$ stands on a smooth horizontal table and is slightly displaced. Verify that the rod hits the plane with angular speed $(3g/2a)^{1/2}$.

19. A uniform solid circular cylinder C of mass m and diameter a is placed between, and in contact with, two coaxial circular cylindrical surfaces of radii a and $2a$. The inner and outer surfaces rotate with angular velocities ω_1 and ω_2, respectively, about their common axis. If C rolls in contact with the two surfaces and no slipping occurs, show that the centre C describes a circle at an angular rate $\frac{1}{3}(2\omega_2 + \omega_1)$, and find the kinetic energy of C.

20. A particle of mass m is fixed to a point P of the rim of a uniform circular disc, centre O, mass m and radius a. The vertical disc is released from rest in contact with a rough horizontal table with OP inclined at $60°$ to the upward vertical. If no slipping occurs, show that the angle θ between OP and the upward vertical satisfies

$$a(7 + 4 \cos \theta)\left(\frac{d\theta}{dt}\right)^2 = 2g(1 - 2 \cos \theta).$$

21. Kinetic energy can be represented by

$$\mathcal{T} = \frac{1}{2} \int_V \rho \dot{\mathbf{r}} \cdot \dot{\mathbf{r}} \, dV$$

in the usual notation. A rigid body rotates about a fixed point with angular velocity $\boldsymbol{\omega}$. Show that

$$\mathcal{T} = \frac{1}{2}\omega_x^2 I_{xx} + \frac{1}{2}\omega_y^2 I_{yy} + \frac{1}{2}\omega_z^2 I_{zz} + \omega_y \omega_z I_{yz} + \omega_z \omega_x I_{zx} + \omega_x \omega_y I_{xy}$$

where

$$I_{xx} = \int_V \rho(y^2 + z^2) \, dV \quad \text{etc.}$$

$$I_{yz} = -\int_V \rho yz \, dV \quad \text{etc.}$$

The quantities I_{yz}, I_{zx}, I_{xy} are called the products of inertia of the body about the chosen set of rotating axes.

22. Show that the angular momentum **h** of a rigid body turning about a fixed point is given by

$$\mathbf{h} = \int_V \rho[r^2\boldsymbol{\omega} - (\mathbf{r}\cdot\boldsymbol{\omega})\mathbf{r}] \, dV$$

in the usual notation. Deduce that

$$\mathscr{T} = \tfrac{1}{2}\mathbf{h}\cdot\boldsymbol{\omega}.$$

23. Calculate the position of the point of impact of a body dropped from a height of 200 m above the Earth's surface at the equator.

24. A particle is projected with a small speed V at an angle α to the horizontal from a point on the Earth's surface whose angular displacement from the north pole is θ. With origin at the centre of the Earth, take the z-axis through the point of projection and the x-axis in the equatorial plane of the Earth as shown in Figure 10.17. Using Equations (23), (24) and (25), show that the projectile's path is given by

$$x = -\tfrac{1}{3}\omega g t^3 \sin\theta + V\omega t^2(\sin\alpha\sin\theta + \cos\alpha\sin\beta\cos\theta) + Vt\cos\alpha\cos\beta$$
$$y = Vt\cos\alpha(\sin\beta - \omega t\cos\beta\cos\theta)$$
$$z = a - \tfrac{1}{2}g t^2 + Vt\sin\alpha - V\omega t^2\cos\alpha\cos\beta\sin\theta$$

where β is the angle between the vertical plane through the initial velocity vector and the x-axis (ignore terms in ω^2, where ω is the angular speed of the Earth, and variations in g).

25. A spacecraft, considered to be a solid cylinder of uniform density with mass 1000 kg and radius 1 m, is rotating about its axis of symmetry once every 12 s. A compressed air jet, with muzzle velocity 150 m/s and cross-sectional area $\tfrac{1}{3}$ cm², is used tangentially at the extremity of a radius to arrest the rotation. For how many seconds should such a jet be activated if the density of the air in the jet is taken as 1 kg/m³?

26. A smooth circular wire of mass M and radius a can turn freely about a horizontal axis which passes through a point P of its circumference. A bead B also of mass m slides on the wire. If Q is the centre of the wire find the equations of motion which determine θ and ϕ, the angles between PQ and QB and the downward vertical. Linearise the equations and use the methods of Section 7.7 to find the periods of the normal modes of oscillation of the system about its stable equilibrium.

Appendix
Differential Equations and Numerical Methods

This appendix contains some brief notes on analytic and numerical methods of solving ordinary differential equations. The notes are fairly self-contained but are directed towards the dynamical applications encountered in this text. It is not our intention to include a comprehensive treatment of the subject.

A.1 ORDINARY DIFFERENTIAL EQUATIONS

Theoretical problems in particle and rigid body mechanics generally reduce to the solution of one or more differential equations. For example, Newton's laws of motion for a particle can be written, in vector form, as

$$\mathbf{F} = m\ddot{\mathbf{r}}$$

(Section 3.2). This vector equation is equivalent to the three scalar equations

$$F_x = m \frac{d^2 x}{dt^2}, \qquad F_y = m \frac{d^2 y}{dt^2}, \qquad F_z = m \frac{d^2 z}{dt^2} \tag{1}$$

where $\mathbf{F} = F_x \mathbf{i} + F_y \mathbf{j} + F_z \mathbf{k}$. Whenever possible our aim is to *solve* these (and similar) equations for x, y and z in terms of t, given the behaviour of the force components in these variables.

Usually the equations of motion are *ordinary differential equations* as in (1) above, that is, x, y and z depend only on the single variable t. For this reason we shall concentrate only on equations of this type. The equations in (1) have the added complication that they are *simultaneous* differential equations in that the dependent variables x, y and z could appear in all three equations. We shall look first at equations with just one dependent variable.

The *order* of a differential equation is the order of the derivative of highest order in the equation. Thus

$$\frac{dx}{dt} = t \quad \text{and} \quad \frac{d^3 x}{dt^3} = \left(\frac{dx}{dt}\right)^4 + x \sin t$$

are respectively of first and third orders. Any relation between the variables which does not contain derivatives and satisfies a differential equation is called a *solution* of that equation. For example, it is easy to verify that

$$x = e^{5t}$$

is a solution of

$$\frac{dx}{dt} = 5x. \tag{2}$$

However,

$$x = C\,e^{5t} \tag{3}$$

is a solution of (1) for *all* values of the real constant C. We say that (3) is the *general solution* of the equation whilst (1) is a *particular* solution. Usually the number of unknown constants in the general solution equals the order of the differential equation.

The differential equation governing vertical motion under gravity is (Section 2.3)

$$\ddot{z} = -g \tag{4}$$

where z is measured upwards and g is the acceleration due to gravity. It can be verified by direct substitution that the general solution is

$$z = A + Bt - \tfrac{1}{2}gt^2 \tag{5}$$

where A and B are constants. The constants are usually determined either by *initial* conditions on z and $\dot{z}$ at the same time or by *boundary* conditions on z at two different times. For example, for the particular solution of (4) which satisfies $z = 0$ and $\dot{z} = U$ at time $t = 0$, we must let $A = 0$ and $B = U$ in (5) resulting in

$$z = Ut - \tfrac{1}{2}gt^2.$$

A.2 FIRST-ORDER EQUATIONS

There are a number of methods of solving certain first-order equations. The choice of method depends on the classification of the particular first-order equation into one of a number of categories. We shall briefly list the types and their solutions. The general first-order equation is

$$\frac{dx}{dt} = f(x, t) \tag{6}$$

(i) *Variables separable.* If

$$f(x, t) = g(x)h(t)$$

that is, the derivative is the product of functions of x and t, then the variables in

$$\frac{dx}{dt} = g(x)h(t)$$

can be separated formally as

$$\frac{dx}{g(x)} = h(t) \, dt.$$

We can now integrate both sides to give the general solution

$$\int \frac{dx}{g(x)} = \int h(t) \, dt + C \tag{7}$$

where C is a constant. By

$$P(t) = \int h(t) \, dt$$

we mean an *indefinite* integral of $h(t)$, that is, any function $P(t)$ such that $dP(t)/dt = h(t)$. All indefinite integrals differ by a constant. We assume that in any specific problem a particular indefinite integral is chosen and not changed in any subsequent manipulations.

Example 1   Solve

$$\frac{dx}{dt} = x \, e^t. \tag{8}$$

The right-hand side of (8) is separable with $g(x) = x$ and $h(t) = e^t$ in the previous notation. Hence the solution is

$$\int \frac{dx}{x} = \int e^t \, dt + C.$$

In the answer

$$\int \frac{dx}{x} = \ln |x| \quad \text{and} \quad \int e^t \, dt = e^t.$$

Suppose that $x > 0$. Then the solution is

$$\ln x = e^t + C \tag{9}$$

or

$$x = e^{C + e^t} = D\,e^{e^t}$$

where $D = e^C$. What is the solution for $x < 0$? Observe that not all solutions are really given by (9): $x = 0$ clearly satisfies (8) but not (9) due to the singularity in ln x at $x = 0$. Particular solutions of this type are known as *singular solutions*.

(ii) *Homogeneous equations.* The first-order equation

$$\frac{dx}{dt} = f(x,\, t)$$

can be reduced to an equation of the variables separable type if $f(x, t)$ is dependent only on the ratio x/t. In the equation

$$\frac{dx}{dt} = g\!\left(\frac{x}{t}\right) \tag{10}$$

we make the substitution $x = vt$. Now

$$\frac{dx}{dt} = t\,\frac{dv}{dt} + v$$

so that Equation (4) becomes the separable equation

$$t\,\frac{dv}{dt} = g(v) - v$$

for the new variable v. When v has been found in terms of t the substitution can be reversed to find x.

Example 2   Solve

$$2t^2\,\frac{dx}{dt} = x^2 + t^2$$

for $t \geq 1$ given that $x = 1$ at $t = 1$.

This is a homogeneous first-order equation. Put $x = vt$. We have

$$2t^2\!\left(t\,\frac{dv}{dt} + v\right) = v^2 t^2 + t^2$$

or

$$2\!\left(t\,\frac{dv}{dt} + v\right) = v^2 + 1.$$

By separating the variables, the solution is given by

$$2 \int \frac{dv}{(v-1)^2} = \int \frac{dt}{t} + C, \qquad (11)$$

so that

$$-\frac{2}{v-1} = \ln |t| + C.$$

Putting $v = x/t$, the general solution becomes

$$-2t = (x - t)(\ln |t| + C).$$

Putting $x = 1$ when $t = 1$ we see that, according to this equation, no solution is possible for these initial conditions. The reason for this apparent breakdown in the method can be seen if we retrace our steps to Equation (11). The integral on the left-hand side of (11) is meaningless (we say that it does not converge) in any interval containing $v = 1$ because of the singularity there in the integrand $1/(v - 1)^2$. We note also that $v = 1$ initially. However there is, in fact, a solution satisfying the initial conditions since if $v = 1$, it follows that $x = t$. It is a simple matter to verify that $x = t$ satisfies the original equation and is such that $x = 1$ when $t = 1$. Thus $x = t$ is the required solution.

This example is significant since it again reveals that differential equations may have isolated particular solutions which are not displayed by the formal evaluation of the indefinite integrals.

(iii) *Integrating factor type.* The integrating factor method is applicable to first-order equations of the type

$$\frac{dx}{dt} + p(t)x = q(t) \qquad (12)$$

where $p(t)$ and $q(t)$ are sufficiently well-behaved functions of t. We attempt to write the left-hand side of (12) as the derivative of a product:

$$\frac{d}{dt}[r(t)x] = r(t)\frac{dx}{dt} + xr'(t). \qquad (13)$$

Multiply both sides of (6) by $r(t)$:

$$r(t)\frac{dx}{dt} + r(t)p(t)x = r(t)q(t). \qquad (14)$$

If we put

$$r(t)p(t) = r'(t) \qquad (15)$$

and use the identity (13), Equation (14) may be expressed as

$$\frac{d}{dt}[r(t)x] = r(t)q(t)$$

an equation essentially of separable type. On integration

$$r(t)x = \int r(t)q(t)\, dt + C$$

which will be the solution. The unknown $r(t)$ is given by (15):

$$\frac{r'(t)}{r(t)} = p(t)$$

which can be integrated to give

$$\ln |r(t)| = \int p(t)\, dt + D$$

where D is a constant. If we take the exponential of both sides

$$r(t) = e^D \exp\left(\int p(t)\, dt\right).$$

We can drop the modulus sign since the exponential on the right-hand side must be positive. The constant e^D can also be ignored since each term in the differential equation is multiplied by $r(t)$.

Example 3   Solve

$$\frac{dx}{dt} + 2x = t.$$

In this case $p(t) = 2$ and $q(t) = t$, and the integrating factor

$$r(t) = \exp\left(\int 2\, dt\right) = e^{2t}.$$

The differential equation can be rewritten in the form

$$\frac{d}{dt}(e^{2t}\, x) = t\, e^{2t}.$$

Integrating,

$$x\, e^{2t} = \int t\, e^{2t}\, dt + C$$

$$= \tfrac{1}{2} t\, e^{2t} - \tfrac{1}{4} e^{2t} + C$$

or

$$x = \tfrac{1}{2} t - \tfrac{1}{4} + C\, e^{-2t}.$$

A.3   HIGHER-ORDER EQUATIONS

A general second-order equation is

$$\frac{d^2x}{dt^2} = f\left(x, \frac{dx}{dt}, t\right).$$
(16)

One approach is to express this equation as a first-order system. We let

$$\frac{dx}{dt} = y$$
(17)

so that

$$\frac{dy}{dt} = f(x, y, t).$$
(18)

The first-order system (17) and (18) is equivalent to (16).

If x is absent from f then we might be able to solve (18) for y in terms of t, in which case x can be obtained by integrating (17).

Example 4   Solve

$$\frac{d^2x}{dt^2} = \left(\frac{dx}{dt}\right)^2 e^t.$$

Let $y = dx/dt$. Then

$$\frac{dy}{dt} = y^2 e^t$$

which is first-order separable with general solution

$$\int \frac{dy}{y^2} = \int e^t \, dt + C.$$

Hence

$$-\frac{1}{y} = e^t + C.$$

Further

$$\frac{dx}{dt} = y = -\frac{1}{e^t + C}.$$

Thus

$$x = -\int \frac{dt}{e^t + C} = \frac{1}{C} \ln |1 + C e^{-t}| + D$$

which is the general solution. Observe that the differential equation also has the singular solution $x = 0$.

If t is absent from f then Equations (17) and (18) define an *autonomous system* (see Section 9.3). In this case we can eliminate t by forming the quotient of the derivatives. Hence

$$\frac{dy}{dx} = \frac{dy/dt}{dx/dt} = \frac{f(x, y)}{y} \tag{19}$$

which is a first-order equation relating x and y. Once this relation is known then x can be obtained in terms of t by integrating the separable equation $dx/dt = y$.

Example 5   Solve

$$\frac{d^2x}{dt^2} = \frac{dx}{dt}\, x. \tag{20}$$

Let $y = dx/dt$. Then

$$\frac{dy}{dt} = yx.$$

Hence

$$\frac{dy}{dx} = \frac{yx}{y} = x \quad y \neq 0.$$

(Note that $y = 0$ satisfies (20), which implies that $x = $ constant is a solution.) This first-order separable equation has the general solution

$$y = \tfrac{1}{2}x^2 + C.$$

Hence

$$\frac{dx}{dt} = y = \tfrac{1}{2}x^2 + C$$

so that

$$x = (2C)^{1/2}\, \tan\,[(\tfrac{1}{2})(2C)^{1/2}t + D].$$

A.4   SECOND-ORDER LINEAR DIFFERENTIAL EQUATIONS

The general second-order linear differential equation with constant coefficients can be written as

$$\frac{d^2x}{dt^2} + a_1 \frac{dx}{dt} + a_0 x = f(t) \tag{21}$$

where a_1 and a_0 are constants, and $f(t)$ is a given function. The mechanical interpretation of the various terms in (21) can be found in Chapter 7. Equation (21) is said to be *homogeneous* if $f(t) \equiv 0$ and *inhomogeneous* otherwise. The general solutions of the homogeneous and inhomogeneous equations have a common part as we shall see.

Let us consider the homogeneous equation first. In the equation

$$\frac{d^2x}{dt^2} + a_1 \frac{dx}{dt} + a_0 x = 0 \tag{22}$$

the second derivative is a linear combination of x and its derivative, and the only function which has the property that all its derivatives are proportional to itself is the exponential function. For this reason we attempt a solution $x = e^{mt}$ where m is a constant to be determined. Thus the left-hand side of (22) becomes

$$\frac{d^2x}{dt^2} + a_1 \frac{dx}{dt} + a_0 x = (m^2 + a_1 m + a_0) \exp(mt)$$

and this vanishes if and only if m satisfies the quadratic equation

$$m^2 + a_1 m + a_0 = 0. \tag{23}$$

Suppose that its roots are $m = m_1$ and $m = m_2$ with $m_1 \neq m_2$ for the moment. We have found two solutions $\exp(m_1 t)$ and $\exp(m_2 t)$. Since the equation is *linear* any linear combination of these solutions is also a solution. We conclude that the general solution is

$$x = A \exp(m_1 t) + B \exp(m_2 t)$$

where A and B are constants. Equation (23) is known as the *characteristic equation* of (22).

Example 6   Obtain the general solution of

$$\frac{d^2x}{dt^2} + \frac{dx}{dt} - 2x = 0.$$

The characteristic equation is

$$m^2 + m - 2 = 0 \quad \text{or} \quad (m + 2)(m - 1) = 0.$$

Hence we can select $m_1 = 1$ and $m_2 = -2$. The general solution is therefore

$$x = Ae^t + B e^{-2t}.$$

Obviously a repeated root could occur as in the following example.

Example 7   Solve

$$\frac{d^2x}{dt^2} + 2\frac{dx}{dt} + x = 0.$$

The characteristic equation is

$$m^2 + 2m + 1 = 0 \quad \text{or} \quad (m + 1)^2 = 0.$$

At this point we have the solution $A\,e^{-t}$ only, with just one arbitrary constant. We expect a further solution to exist. Try $x = f(t)\,e^{-t}$. Then

$$\frac{d^2x}{dt^2} + 2\frac{dx}{dt} + x = [(f'' - 2f' + f)\,e^{-t} + 2(f' - f)\,e^{-t} + f\,e^{-t}]$$

$$= f''\,e^{-t}$$

which vanishes if $f'' = 0$. Hence $f = A + Bt$, and the general solution is

$$x = (A + Bt)\,e^{-t}.$$

This has been worked through for a specific example but the factor $A + Bt$ appears for a repeated root as a general rule.

The roots of the characteristic equation can also be *complex numbers* as in the following example for simple harmonic motion.

Example 8   Solve

$$\frac{d^2x}{dt^2} + \omega^2 x = 0 \qquad \omega > 0.$$

The characteristic equation is

$$m^2 + \omega^2 = 0$$

with roots

$$m = \pm\omega\sqrt{-1} = \pm\omega i.$$

Hence the general solution is, in complex form,

$$x = A\exp(i\omega t) + B\exp(-i\omega t).$$

We now use the identities

$$\exp(i\omega t) = \cos\omega t + i\sin\omega t \quad \exp(-i\omega t) = \cos\omega t - i\sin\omega t$$

so that, in real form, the solution can be written

$$x = A(\cos\omega t + i\sin\omega t) + B(\cos\omega t - i\sin\omega t)$$
$$= (A + B)\cos\omega t + (Ai - Bi)\sin\omega t$$
$$= C\cos\omega t + D\sin\omega t.$$

Example 9   Obtain the general solution of

$$\frac{d^2x}{dt^2} - 2\frac{dx}{dt} + 5x = 0$$

in both complex and real forms.

The characteristic equation is

$$m^2 - 2m + 5 = 0$$

with roots

$$m_1, m_2 = 1 \pm 2i.$$

Hence the general solution can be expressed as

$$x = A \exp\left[(1 + 2i)t\right] + B \exp\left[(1 - 2i)t\right] = e^2 \left[A \exp(2it) + B \exp(-2it)\right].$$

We now use the identities

$$\exp(\pm 2it) = \cos 2t \pm i \sin 2t$$

so that, in real form, the solution can be written *a*

$$x = e^t (C \cos 2t + D \sin 2t).$$

Consider now the inhomogeneous Equation (21). Let $x = p(t)$ be a *particular* solution of the equation. Consider the function $x = u(t) + p(t)$. Substitution into (21) implies that $u(t)$ must satisfy the homogeneous equation

$$\frac{d^2u}{dt^2} + a_1 \frac{du}{dt} + a_0 u = 0 \tag{24}$$

and, as we have shown, it will contain two arbitrary constants. In this context $u(t)$ is known as the *complementary function* and $p(t)$ the *particular integral*. We know how to construct the complementary function but the particular integral will be dependent on $f(t)$. It can be verified by direct differentiation that, if $m_1 \neq m_2$, then (21) has the particular solution

$$p(t) = \frac{1}{m_1 - m_2} \left(\int_{t_0}^t \{\exp\left[m_1(t - s)\right] - \exp\left[m_2(t - s)\right]\} f(s) \, ds \right) \tag{25}$$

where t_0 is a constant.

Example 10   Find a particular solution of (21) if $m_1 \neq m_2$ and $f(t) = 1$.

From (25)

$$p(t) = \frac{1}{m_1 - m_2} \left(\int_{t_0}^t \{\exp[m_1(t-s)] - \exp[m_2(t-s)]\} \, ds \right)$$

$$= \frac{1}{m_1 - m_2} \left(-\frac{\exp(m_1 t)}{m_1} [\exp(-m_1 t) - \exp(-m_1 t_0)] \right.$$

$$\left. + \frac{\exp(m_2 t)}{m_2} [\exp(-m_2 t) - \exp(-m_2 t_0)] \right)$$

$$= \frac{1}{m_1 m_2} + \frac{1}{m_1 m_2 (m_1 - m_2)} \{-m_2 \exp[m_1(t-t_0)] + m_1 \exp[m_2(t-t_0)]\}.$$

In fact since the second term is made up of multiples of the homogeneous solutions, then the particular integral can be taken to be

$$p(t) = \frac{1}{m_1 m_2} = \frac{1}{a_0}.$$

The constant t_0 plays no part in the particular integral. The general solution becomes

$$x = A \exp(m_1 t) + B \exp(m_2 t) + \frac{1}{a_0}.$$

Whilst the integrals in (25) define a general formula for the particular integrals, it is often easier in practice to fit given functions as solutions. For example, it can be seen from (25) that, if $f(t)$ is a polynomial in t of degree n, then there exists a particular solution which is also a polynomial of degree n (assuming $m_1 \neq m_2$). We insert a polynomial with arbitrary coefficients into the equation, and find these coefficients by matching like powers of t on both sides of the equation. This approach is known as the *method of undetermined coefficients*. The following example illustrates the method.

Example 11   Find a particular integral of

$$\frac{d^2 x}{dt^2} - 2 \frac{dx}{dt} + 5x = t^2 + 1.$$

The right-hand side is a polynomial of degree 2. Hence we try a solution $p(t) = Mt^2 + Nt + P$ where M, N and P are coefficients to be determined. Thus

$$\frac{d^2 p}{dt^2} - 2 \frac{dp}{dt} + 5p - t^2 - 1 = 2M - 2(2Mt + N) + 5(Mt^2 + Nt + P) - t^2 - 1$$

$$= (5M - 1)t^2 + (-4M + 5N)t + (2M - 2N + 5P - 1)$$

$$= 0$$

for all t if

$$5M - 1 = 0, \qquad -4M + 5N = 0, \qquad 2M - 2N + 5P - 1 = 0.$$

This set of linear equations has the solution

$$M = \tfrac{1}{5}, \qquad N = \tfrac{4}{25}, \qquad P = \tfrac{23}{125}.$$

From Example 9, it follows that the general solution is

$$x = e^t(C \cos 2t + D \sin 2t) + (25t^2 + 20t + 23)/125.$$

There are exceptional cases if any component of the trial solution is part of the complementary function. For example, in the equation

$$\frac{d^2x}{dt^2} + 3\frac{dx}{dt} = 4$$

$p(t) = M$ will not be a solution for any M since the complementary function is $A + B \, e^{-3t}$. If this trial function is used then an inconsistency will appear. We try $p(t) = Mt$ instead. It then follows that $M = 4/3$.

The following table gives some trial functions with unknown coefficients (capital letters) to be determined in the third column, for given forcing functions $f(t)$ in the first column. Restrictions on the characteristic roots are indicated in column 2. The table is not intended to be exhaustive but to suggest possible trial solutions. Some compound functions such as $f(t) = e^{at} \cos bt$ can also be handled in this way. Since the equations are linear, the particular integral of the sum of two functions from column 1 will be the sum of the particular integrals in column 3.

The methods described for second-order constant coefficient equations can be readily generalised to higher order equations. For example, every third-order equation will have a cubic characteristic equation. Similarly, the table of trial

Table A.1

$f(t)$	Restrictions on m_1, m_2	Trial function, $p(t)$
$b_0 + b_1 t + \cdots + b_n t^n$	$m_1 \neq 0$ and $m_2 \neq 0$	$B_0 + B_1 t + \cdots + B_n t^n$
$b_0 + b_1 t + \cdots + b_n t^n$	$m_1 = 0, \, m_2 \neq 0$ or $m_1 \neq 0, \, m_2 = 0$	$B_1 t + B_2 t^2 + \cdots + B_{n+1} t^{n+1}$
e^{bt}	$m_1 \neq b$ and $m_2 \neq b$	$B \, e^{bt}$
e^{bt}	$m_1 = b, \, m_2 \neq b$	$Bt \, e^{bt}$
e^{bt}	$m_1 = m_2 = b$	$Bt^2 \, e^{bt}$
$\begin{cases} \cos \omega t \\ \sin \omega t \end{cases}$	$m_1 \neq \pm i\omega$	$A \cos \omega t + B \sin \omega t$

functions will have the same first and third columns but there will be changes to the restrictions in column 2 to accommodate the three characteristic roots.

A.5 NUMERICAL METHODS OF SOLUTION

Many simple ordinary differential equations can now be readily solved on personal computers and computer terminals, and solutions either delivered as data, or represented on screen graphically. The necessary equipment is now so widespread that numerical or graphical output is an important component in our understanding of the behaviour of mechanical systems. As is explained in Chapter 9, systems which seem to be governed by quite simple differential equations can exhibit complex outputs.

We shall describe some methods for solving initial value problems for ordinary differential equations. Consider the first-order equation $dx/dt = f(x, t)$ with $x(0) = x_0$ given. We require the solution for $t > 0$. Let time be divided into equal time intervals of duration h, and let x_1 be an approximation to the true value $x(h)$ of the solution. Approximately, the derivative at $t = 0$ can be represented by the quotient

$$\frac{x_1 - x_0}{h}.$$

From the differential equation the slope of the solution at $t = 0$ must be $f(x_0, 0)$. Hence the basis of our approximation is

$$\frac{x_1 - x_0}{h} = f(x_0, 0)$$

so that

$$x_1 = x_0 + hf(x_0, 0). \tag{26}$$

The difference between the approximation and the true value at $t = h$ will be $x_1 - x(h)$ (Figure A.1).

In a similar manner we can calculate x_2 at $t = 2h$ by repeating the procedure with the new initial data $x = x_1$ at $t = h$. Hence

$$\frac{x_2 - x_1}{h} = f(x_1, h) \quad \text{or} \quad x_2 = x_1 + hf(x_1, h).$$

After n steps

$$x_n = x_{n-1} + hf(x_{n-1}, (n-1)h).$$

This is known as *Euler's method* of solution and is probably the simplest method of solving initial value problems. A simple loop at equal time steps will generate the sequence $\{x_n\}$.

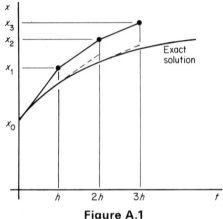

Figure A.1

We can now estimate the *local error* $|E(h)|$ in using this approximation by expanding $x(t + h)$ as a Taylor series about $t = 0$. Thus

$$x(h) = x(0) + hx'(0) + E(h)$$

where the remainder $E(h) = \frac{1}{2}h^2 x''(\xi)$ for some ξ such that $0 \le \xi \le h$. Hence the local error at each step is $O(h^2)$ (we say of order h^2).

Example 12   Solve the equation

$$\frac{dx}{dt} + 2x = \sin t$$

using Euler's method over $(0, 1)$ with step length $h = 0.1$. Compare the answers with the exact solution.

Table A.2, column 2, lists the values for x obtained by the Euler iteration at $t = 0, 0.1, \ldots, 1.0$. The exact solution is

$$x = 1.2 \, e^{-2t} - 0.2 \cos t + 0.4 \sin t$$

and the values of x given by this solution at the same steps are shown in column 3. The *global errors*, that is the absolute values of the differences between columns 2 and 3 are shown in column 4. The same information is presented graphically in Figure A.2.

Higher order equations can be reduced to first-order systems. For example, the equation

$$\frac{d^2x}{dt^2} + \left(\frac{dx}{dt}\right)^2 + x \, e^{-t} = 0, \qquad x(0) = x_0, \qquad dx(0)/dt = y_0$$

Table A.2

t	Computed value, x_n	Exact value, $x(t)$	Global error, $|x_n - x(t)|$
0	1.000 00	1.000 00	
0.1	0.800 00	0.823 41	0.023 41
0.2	0.649 98	0.687 84	0.037 86
0.3	0.539 85	0.585 71	0.045 86
0.4	0.461 43	0.510 75	0.049 32
0.5	0.408 09	0.457 71	0.049 62
0.6	0.374 41	0.422 22	0.047 81
0.7	0.356 00	0.400 63	0.044 63
0.8	0.349 22	0.389 88	0.040 66
0.9	0.351 11	0.387 37	0.036 26
1.0	0.359 22	0.390 93	0.031 71

is equivalent to the first-order system

$$\frac{dx}{dt} = y, \qquad \frac{dy}{dt} = -y^2 - x\,e^{-t}, \qquad x(0) = x_0, \qquad y(0) = y_0$$

or, in vector notation,

$$\frac{d\mathbf{x}}{dt} = \mathbf{f}(\mathbf{x}, t), \qquad \mathbf{x} = \begin{bmatrix} x \\ y \end{bmatrix}, \qquad \mathbf{f} = \begin{bmatrix} y \\ -y^2 - x\,e^{-t} \end{bmatrix}.$$

We can still apply Euler's method to each component, which results in the iteration

$$\mathbf{x}_n = \mathbf{x}_{n-1} + h\mathbf{f}(\mathbf{x}_{n-1}, (n-1)h).$$

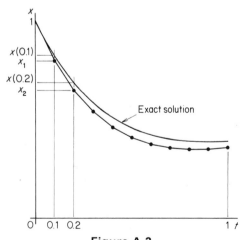

Figure A.2

302   *Appendix*

As can be judged from Table A.2 or from Figure A.2, the accuracy of the Euler method is not great with this step length for this particular equation. We can reduce the step length or develop more accurate iterations. We can reduce the local error at each step by using the formula

$$x_1 = x_0 + \tfrac{1}{2}h\{f(x_0, 0) + f(x_1, h)\} \tag{27}$$

in which the derivative at $t = 0$ is replaced by the mean of the derivatives at $t = 0$ and $t = h$. However Equation (27) still has to be solved for x_1, and it is given implicitly by this equation. However, we can construct an explicit approximation formula if we replace x_1 on the right-hand side of (27) by the Euler formula (26). Thus

$$x_1 = x_0 + \tfrac{1}{2}h\{f(x_0, 0) + f(x_0 + hf(x_0, 0), h)\}.$$

The general term is

$$x_n = x_{n-1} + \tfrac{1}{2}h\{f(x_{n-1}, (n-1)h) + f(x_{n-1} + hf(x_{n-1}, (n-1)h), nh\}.$$

This is an example of what is known as a *Runge–Kutta* formula for initial value problems. The local error at each step is now $O(h^3)$. This procedure has been applied to Example 12. The computations are shown in Table A.3: column 4 indicates a considerable reduction in the global errors.

Table A.3

t	Computed value, x_n	Exact value, $x(t)$	Global error, $\lvert x_n - x(t)\rvert$
0	1.000 00	1.000 00	
0.1	0.824 99	0.823 41	0.001 58
0.2	0.690 42	0.687 84	0.002 58
0.3	0.588 87	0.585 71	0.003 16
0.4	0.514 16	0.510 75	0.003 41
0.5	0.461 16	0.457 71	0.003 45
0.6	0.425 56	0.422 22	0.003 34
0.7	0.403 76	0.400 63	0.003 13
0.8	0.392 72	0.389 88	0.002 84
0.9	0.389 89	0.387 37	0.002 52
1.0	0.393 12	0.390 93	0.002 19

We can write the simple Runge–Kutta scheme in the form

$$x_1 = x_0 + \tfrac{1}{2}(k_1 + k_2) \tag{28}$$

where $k_1 = hf(x_0, h)$, $k_2 = hf(x_0 + k_1, h)$. This is known as a second-order formula, and it can be generalised. Consider the general formula

$$x_1 = x_0 + \alpha_1 k_1 + \alpha_2 k_2 \tag{29}$$

where

$$k_1 = hf(x_0, 0), \qquad k_2 = hf(x_0 + \beta k_1, \gamma h) \tag{30}$$

and α_1, α_2, β and γ are constants. We now compare the expansion of (29) and (30) in powers of h with the general expansion of $x(h)$ in powers of h by its Taylor series. Thus, from (30),

$$h_2 = hf(x_0, 0) + \beta h k_1 f_x(x_0, 0) + \gamma h^2 f_t(x_0, 0) + O(h^3)$$
$$= hf(x_0, 0) + \beta h^2 f(x_0, 0) f_x(x_0, 0) + \gamma h^2 f_t(x_0, 0) + O(h^3).$$

Hence (29) becomes

$$x_1 = x_0 + (\alpha_1 + \alpha_2)hf(x_0, 0)$$
$$+ \alpha_2 h^2 \{\beta f(x_0, 0)f_x(x_0, 0) + \gamma f_t(x_0, 0)\} + O(h^3). \tag{31}$$

The Taylor expansion of $x(h)$ is also given by

$$x(h) = x(0) + hx'(0) + \tfrac{1}{2}h^2 x''(0) + O(h^3)$$
$$= x_0 + hf(x_0, 0) + \tfrac{1}{2}h^2\{f(x_0, 0)f_x(x_0, 0) + f_t(x_0, 0)\} + O(h^3). \tag{32}$$

Comparison of (31) and (32) implies

$$\alpha_1 + \alpha_2 = 1, \qquad \alpha_2 \beta = \tfrac{1}{2}, \qquad \alpha_2 \gamma = \tfrac{1}{2}$$

which is a set of three equations in four unknowns. Hence, there is some flexibility built into the Runge–Kutta scheme. The simple one given by (28) and used in Table A.3 uses $\beta = 1$, $\alpha_1 = \alpha_2 = \tfrac{1}{2}$, $\gamma = 1$. All these schemes have local error.

It is possible to design higher order Runge–Kutta iterations. The one most commonly used is the fourth-order scheme which we shall state but not derive. In this case

$$x_1 = x_0 + \tfrac{1}{6}(k_1 + 2k_2 + 2k_3 + k_4) \tag{33}$$

where

$$k_1 = hf(x_0, 0), \quad k_2 = hf(x_0 + \tfrac{1}{2}k_1, \tfrac{1}{2}h) \tag{34}$$

$$k_3 = hf(x_0 + \tfrac{1}{2}k_2, \tfrac{1}{2}h), \quad k_4 = hf(x_0 + k_3, h). \tag{35}$$

The local error is $O(h^5)$. This Runge–Kutta scheme has been used in the computation of most of the initial-value problems in this book. Table A.4 contains a re-calculation of Example 12 using this method. The global errors now only just affect the fifth decimal place over the interval.

The iterative scheme defined by (33), (34) and (35) can be easily adapted to nth-order systems. All that happens in practice is that $x_1, x_0, f, k_1, k_2, k_3$ and k_4 are replaced by vectors with n components.

Table A.4

t	Computed value, x_n	Exact value, $x(t)$	Global error, $|x_n - x(t)|$
0	1.000 00	1.000 00	
0.1	0.823 41	0.823 41	0.000 00
0.2	0.687 84	0.687 84	0.000 00
0.3	0.585 72	0.585 71	0.000 01
0.4	0.510 76	0.510 75	0.000 01
0.5	0.457 72	0.457 71	0.000 01
0.6	0.422 23	0.422 22	0.000 01
0.7	0.400 64	0.400 63	0.000 01
0.8	0.389 88	0.389 88	0.000 00
0.9	0.387 37	0.387 37	0.000 00
1.0	0.390 93	0.390 93	0.000 00

Phase diagrams in the plane can be computed by solving the first-order system

$$\frac{dx}{dt} = P(x, y, t), \qquad \frac{dy}{dt} = Q(x, y, t)$$

by the Runge–Kutta method and then either outputting the coordinates (x, y) or by displaying them directly on screen or by plotting for hard copy diagrams. Poincaré maps can be obtained in the same way except that the programs are adapted to plot a single point in the (x, y) plane at a fixed period in t.

Exercises

1. Obtain general solutions of the following first-order differential equations:

(i) $\dfrac{dx}{dt} = t^3 x^2$

(ii) $\dfrac{dx}{dt} = e^{-x} \sin t$

(iii) $(2t - x) \dfrac{dx}{dt} = 2x - t$

(iv) $t \dfrac{dx}{dt} = (x^2 + t^2)^{1/2} + x$

(v) $t \dfrac{dx}{dt} - 3x = t^5$

(vi) $\dfrac{dx}{dt} - tx = t.$

2. Obtain general solutions of

   (i) $\dfrac{d^2x}{dt^2} + 4x = 0$

   (ii) $\dfrac{d^2x}{dt^2} + 7\dfrac{dx}{dt} + 12x = 0$

   (iii) $\dfrac{d^2x}{dt^2} + 6\dfrac{dx}{dt} + 9x = 0.$

3. Obtain general solutions of

   (i) $\dfrac{dx}{dt} + x = t^3$

   (ii) $\dfrac{d^2x}{dt^2} - \dfrac{dx}{dt} - 2x = 44 - 76t - 48t^2$

   (iii) $\dfrac{d^2x}{dt^2} + 6\dfrac{dx}{dt} + 25x = 104\ e^{3t}$

   (iv) $\dfrac{d^2x}{dt^2} - k^2x = \sinh kt$

   (v) $\dfrac{d^2x}{dt^2} + 8\dfrac{dx}{dt} + 25x = 48 \cos t - 16 \sin t.$

4. Find the solution of

$$5\frac{dx}{dt} + 4x - 2\frac{dy}{dt} - y = e^{-t}$$

$$\frac{dx}{dt} + 8x - 3y = 5\ e^{-t}$$

such that $x = 2$, $y = 4$ when $t = 0$.

5. Obtain complementary functions and particular integrals for each of the following differential equations:

   (i) $\dfrac{d^2x}{dt^2} + 4x = 3 + \sin t$

   (ii) $\dfrac{d^2x}{dt} + x = e^{2t} \sin 2t$

   (iii) $\dfrac{d^3x}{dt^3} - x = 1$

   (iv) $\dfrac{d^2x}{dt^2} + \dfrac{dx}{dt} + x = \sin^2 t$

   (v) $\dfrac{d^3x}{dt^3} - 3\dfrac{d^2x}{dt^2} + 3\dfrac{dx}{dt} - x = 1.$

6. Obtain general solutions of

(i) $x \dfrac{d^2x}{dt^2} = \left(\dfrac{dx}{dt}\right)^2$

(ii) $x \dfrac{d^2x}{dt^2} + \left(\dfrac{dx}{dt}\right)^2 = \dfrac{dx}{dt}$

(iii) $1 + \left(\dfrac{dx}{dt}\right)^2 = x \dfrac{d^2x}{dt^2}.$

7. Solve the first-order system

$$\frac{dx}{dt} + x = e^{-t}$$

both analytically and numerically using the Euler method over $0 \le t \le 1$ with step-length $h = 0.1$. Compare graphs of the two answers. Repeat the procedure using the Runge–Kutta fourth-order method.

8. Find the forced part of the solution of the equation

$$\frac{d^2x}{dt^2} + x = K \cos \Omega t, \qquad \Omega > 0$$

for (i) $\Omega \ne 1$, (ii) $\Omega = 1$. Plot the solutions in both cases.

Answers and Comments on the Exercises

Chapter 1

2. (i) $10.46\mathbf{i} + 12\mathbf{j}$.
    (ii) 15.9 km.
3. (i) 0; (ii) $2\mathbf{i} - 2\mathbf{j} + 2\mathbf{k}$; (iii) 2; (iv) $-4\mathbf{i} + 4\mathbf{k}$.
4. $(-2\mathbf{i} + \mathbf{j} + \mathbf{k})/\sqrt{6}$; an alternative unit vector could be formed by changing the sign of all the components.
5. A is the position vector of any point on the plane. **B** is any vector perpendicular to the plane.
8. $\mathbf{v} = -a\omega \sin \omega t\mathbf{i} + a\omega \cos \omega t\mathbf{j} + b\mathbf{k}$,
    $\mathbf{f} = -a\omega^2 \cos \omega t\mathbf{i} - a\omega^2 \sin \omega t\mathbf{j}$.
10. (iii) represents twice the area swept out by the radius vector in time t_1.
12. (i) $(2t + 1)\mathbf{i} + \mathbf{j} + \mathbf{k}$; (ii) $3t^2 - 6t$; (iii) $2(1 + t)\mathbf{i} - \mathbf{j} + (1 - 4t - 3t^2)\mathbf{k}$.
13. The radial and transverse components of velocity and acceleration are respectively (e^t, e^t) and $(0, 2e^t)$.
14. (i) $\frac{3}{2}$; (ii) $-\frac{1}{2}\mathbf{i} - \frac{1}{5}\mathbf{j} + \frac{1}{3}\mathbf{k}$; (iii) $\frac{62}{15}$; (iv) 0.
15. (i) $\frac{11}{10}$; (ii) $-\frac{1}{12}\mathbf{i} - \frac{3}{10}\mathbf{j} + \frac{5}{12}\mathbf{k}$.
16. $(yz\mathbf{i} + zx\mathbf{j} + xy\mathbf{k}) e^{xyz}$.
17. (i) and (iii) conservative, (ii) and (iv) not conservative.

Chapter 2

1. 12.32 hours; 8 km.
2. $16°$ west of north; 47 km out from port.
3. $8° 8'$ west of north; 2 hr 21 min.
5. 117 km/h; $59°$ south of east.
7. $R(v^2 - n^2)^{1/2}/v$; east or west.
8. 54.
9. 3.3 s.
10. $\mathbf{v} = v\mathbf{e}_r + r\omega\mathbf{e}_\theta$; $r = v\theta/\omega$.
11. 8 170 km/h; 35 220 km.
12. (a) $50/27$ m/s^2; (b) 208.3 m; (c) 66.7 km/h.
13. 45 m.
15. 41 cm/s^2.
17. First-second gear, 30 km/h; second-third gear, 80 km/h; 9.95 s.

20. 166 s—the curved sections are taken at the maximum speed of 180 km/h, acceleration and braking take place on the straight.

21. $V\left\{\mathbf{i}\left[1 - \cos\left(\dfrac{Vt}{a} + \alpha\right)\right] + \mathbf{j}\sin\left(\dfrac{Vt}{a} + \alpha\right)\right\}$, where α is a constant denoting an initial position on the circumference. To prove perpendicularity show that the scalar product of the appropriate vectors is zero.

22. (i) 11.3 km/h; (ii) 17.0 km/h, (iii) 22.6 km/h; (iv) 45.2 km/h.

24. $a + b \cos \omega t$.

27. Displacements are: $x = 0$, $(t < t_0)$; $x = \frac{1}{2}(t - t_0)^2$, $(t_0 \le t \le t_1)$, $x = (t_1 - t_0)t + \frac{1}{2}(t_1 - t_0)^2$, $(t > t_1)$.

28. $\alpha > 1$: $v_0^{1-\alpha}/(\alpha - 1)k$.

30. Path is $y = x \sinh[k \ln(d/x)]$.

31. $4aV_0/3V$ downstream from the starting point.

Chapter 3

1. 2,000 N.
2. $\mathbf{r} = (t - a)\mathbf{i} - b\mathbf{j} - t^2\mathbf{k}$.
3. 1.23 m/s².
4. 11° 18′.
5. 23.9 m/s.
6. $M(f_1 + f_2)/(g + f_2)$.
7. Distance $3a/20$ along the axis of the cylinder from the circular face of the hemisphere.
8. 30 m.
9. 233.9 m.
10. 2.00×10^{30} kg.
11. (i) 127.5 s; (ii) 14.7 N.
12. 18.2 N.
13. 10^3 km/h.
14. Upper hinge: bMg/l; lower hinge: $Mg(l^2 + b^2)^{1/2}/l$.
15. Rear axle: 4,989 N; front axle: 4,821 N.
16. 19.2 km/h.
17. $2m(2gh)^{1/2}/(M - m)g$.
18. 1.8 s.
19. $s(r + 1)[2^{n-1} - (1 + e)^{n-1}]/rV(1 - e)(1 + e)^{n-1}$.
21. Velocity of D has magnitude $U(1 + \sin^2 2\alpha)^{1/2}/(1 + 2\sin^2 \alpha)$ in a direction making an angle $\tan^{-1}(\sin 2\alpha)$ with CA. Velocity of B may be found from the velocity of D by symmetry. Velocity of C has magnitude $U \cos 2\alpha/(1 + 2\sin^2 \alpha)$ in the direction CA.
22. (i) $U(1 - e^n)/2$, $U(1 + e^n)/2$; (ii) $2\pi a(1 - e^n)/Ue^{n-1}(1 - e)$.
23. (i) 1.1 m kg/s; (ii) 219.5 N.
24. (i) $\frac{2}{3}$ m; (ii) No. The jump would be equivalent to a standing jump of 6 m; (iii) 145 m kg/s.

Chapter 4

1. $y^3 + 15x = 0$.
2. $(1 - e^{-t})\mathbf{i} + (1 - \cos t)\mathbf{j} + \sin t\mathbf{k}$.
4. 35.3 m; 5.1 m.
6. 22.1 m/s.

10. 5.05 s.
11. 3.57 m.
14. $(0, 2/\pi)$.
16. 140 m.
17. 111.5 m.
19. (i) 58.6 m. The transcendental equation can be solved with sufficient accuracy by expanding the exponential function to its fourth term. (ii) 104 m/s.
21. $\theta = \dfrac{1}{\beta a} \ln [(\beta a V + \alpha)/\alpha]$.
23. (i) mV/c; (ii) it never comes to rest: it may seem paradoxical but the resistance at low speeds is insufficient to dissipate energy quickly enough.
24. 37.2°.
25. $2\pi[(a \cos \alpha)/g]^{1/2}$.
26. $\tan \alpha = v^2/(\alpha g)$; 11.12°.

Chapter 5

1. 8.48×10^6 J.
2. 2070 J.
3. Approximately 2.5×10^8 W.
4. 87%.
5. (i) 0.022 m/s²; (ii) 61 km/h.
6. 716 kW.
7. $m(v_1 - v_2)^2(1 - e^2)/4$.
8. 9.6 s.
11. 22.5 cm above the point from which it was released. Note that it traverses the final part of its path as a projectile.
12. Stiffness $k = 4.9 \times 10^4$ N/m.
13. $(w + W)/5$.
14. Equilibrium position given by $\cos \theta = (b/a)^{1/3}$; stable.
15. Equilibrium position given by $\tan \theta = m \sin \phi/(m \cos \phi + M)$; unstable.
17. Other position of equilibrium given by $\cos \theta = a/c$ where θ is the angle between the rod and the vertical.
18. Radius to R makes with the vertical (i) 15°, stable, (ii) 75°, unstable.
19. (i) $\lambda = 12mg$; $\theta = \frac{1}{2}\pi$, unstable, $\theta = \sin^{-1}(6/7)$, stable (spring subtends an angle of 2θ at the centre).
    (ii) $\lambda = mg$; $\theta = \frac{1}{2}\pi$, stable; no other position of equilibrium.
20. $2kd^2 > mgl$.

Chapter 6

1. The variables v and m separate giving a final velocity $-c \ln (1 - \varepsilon)$.
4. Height attained whilst the fuel is burning is about 56.8 km and in free flight about 75.8 km giving a total distance of 132.3 km.
5. (i) 12.3 m/s; (ii) 10.5 m/s. The rate at which momentum is removed by the rope falling on the ground exactly balances the force due to air resistance. Initially, therefore, since the weight exceeds the buoyancy force, the balloon accelerates until the buouancy force is the larger. The balloon is then retarded.

7. The parametric equations of the path are:

$$x = t(V + c) + \frac{c}{k}(M - kt)\ln\left[(M - kt)/M\right], \qquad z = -\tfrac{1}{2}gt^2.$$

9. The equilibrium point is 3.5×10^5 km from the Earth. This large distance has two implications:
   (i) the velocity required is virtually the escape velocity,
   (ii) the height attained during burning is a negligible proportion of the total distance.
   By using the formula for speed at burn-out we find that ε would be 0.997, a figure far in excess of anything presently attainable.

10. (i) The second stage would have mass 900 kg; (ii) The final speed of the satellite would be 5480 m/s (compared with 3500 m/s for a single rocket).

11. $\dfrac{1}{\alpha}(ck - mg)\left[(1 - \left(1 - \dfrac{kt}{M}\right)^{a/k}\right]$.

12. $Mg\{-2\ln\left[\tfrac{1}{2}(1 + \varepsilon)\right] - 1 + \varepsilon\}/k$.

Chapter 7

1. (i) 1.4 s; (ii) 0.71 oscillations/s. (iii) 4.46 rad/s.
2. (i) Distance a vertically below the beam; (ii) $4\pi(a/5g)^{1/2}$.
3. Arrangement in series gives the longer period.
4. 0.0493 N/m.
5. $4\pi(m/k)^{1/2}$.
7. Note that, if $x_0 \leq 3F/k$, the mass comes to rest after $t = \pi(m/k)^{1/2}$.
8. $\left[\dfrac{1}{m}\left(\dfrac{\lambda_1}{a_1} + \dfrac{\lambda_2}{a_2}\right)\right]^{1/2}$.
10. $F_0 t \sin \Omega t/2\Omega$.
13. Stiffness should be less than 4.57×10^7 N/m.
14. The length should be decreased by 0.2 mm.
15. $2(\pi + 4)(am/\lambda)^{1/2}$.
16. <7.95 rad/s.
17. If x_1 is the distance of m_1 from its original position and x_2 is the distance of m_2 from the original position of m_1, then

$$x_1 = -m_2 a(1 - \cos \omega t)/4(m_1 + m_2),$$
$$x_2 = a(4m_1 + 3m_2 - m_1 \cos \omega t)/4(m_1 + m_2),$$

where $\omega^2 = \lambda(m_2 + m_2)/m_1 m_2 a$. Note that the centre of mass of m_1 and m_2 remains fixed.

18. $I/(mk)^{1/2}$.
19. $2\pi(h\rho_1/g\rho_2)^{1/2}$.
20. (i) Strongly damped; (ii) weakly damped; (iii) weakly damped; (iv) strongly damped.
21. If x is measured from the equilibrium position,
    (i) $m\ddot{x} + c\dot{x} + 2kx = 0$
    (ii) $m\ddot{x} + (c_1 + c_2)\dot{x} + kx = 0$
    (iii) $m\ddot{x} + c\dot{x} + (k_1 + k_2)x = 0$.
22. 0.867 N/m.
24. 12.
25. (i) $2 \times$ (velocity) N; (ii) 308.7 N/m.

27. (i) 4.1 rad/s, (ii) 73°.
28. Note that the mass of water hitting the ball per second when it is moving with speed $\dot{x}$ is $kM(v - \dot{x})/v$.
30. Speed of rotation > 273 r.p.m.
31. $v \exp(-ct/2m) \sin \beta t/\beta$.
32. (i) $F(t) = F_0 + \dfrac{2F_0}{\pi}(-\cos \omega t + \tfrac{1}{3}\cos 3\omega t - \cdots)$;

    (ii) $x = \dfrac{F_0}{k} + \dfrac{2F_0}{\pi}\left(\dfrac{-\cos(\omega t - \alpha_1)}{[(k - m\omega^2)^2 + c^2\omega^2]^{1/2}} + \dfrac{\cos(3\omega t - \alpha_3)}{3[(k - 9m\omega^2)^2 + 9c^2\omega^2]^{1/2}} - \cdots\right)$

    where $\tan \alpha_n = cn\omega/(k - mn^2\omega^2)$.
33. $2\pi P^2 r\omega/\{m[(5r^2 - \omega^2)^2 + 4r^2\omega^2]\}$.
35. $[k(n + 2)/(mn)]^{1/2}$; $(k/m)^{1/2}$.
37. $\omega_r = (k/m)^{1/2}\cos \tfrac{1}{2}\theta_r$, $\theta_r = r\pi/(n + 1)$, $r = 1, 2, \ldots, n$.
38. $\begin{bmatrix} x \\ y \end{bmatrix} = -\alpha\begin{bmatrix} 1 \\ 1 \end{bmatrix}\cos t + \alpha\begin{bmatrix} 3 \\ 1 \end{bmatrix}\cos 2t$.
39. $(3g/2a)^{1/2}$; $(g/2a)^{1/2}$; $(3g/2a)^{1/2}$.
42. The eigenfrequencies are $[k(b - a)/m]^{1/2}$, $[3k(b - a)/m]^{1/2}$: in the normal modes $x = y$ and $x = -y$.

Chapter 8

2. Since the differential equation for x is oscillatory for $\alpha < 3$, the circular orbit is stable for $\alpha < 3$.
3. (i) 2030 km; (ii) 0.12; (iii) 1 h 48 min.
7. (i) $a = 2.7 \times 10^9$ km; (ii) 8×10^7 km. Note that the orbit of Halley's comet does not extend beyond the orbit of Pluto.
10. 42 100 km; 3.
12. 2.71×10^4 m/s (that is, 2700 m/s less than the orbital speed of the Earth); (ii) 145 days.
14. Inequalities for V^2 imply that if V is too small the rocket will not reach the satellite orbit, and if V is too large the rocket trajectory cannot have an apogee at such a small height.
16. 110 min.
18. 2.45×10^4 km.
19. The two orbits each have semi-latus rectum and eccentricity given respectively by $\mu(V + v)^2/V^4$, $v(2V + v)/V^2$ and $\mu(V - v)^2/V^4$, $v|2V - v|/V^2$.
20. (i) $|V^2d - 3\gamma m|/3\gamma m$; (ii) $6\pi(d^3/\gamma m|V^2d/\gamma m - 6|^3)^{1/2}$ if $V^2 < 6\gamma m/d$. Otherwise no period since orbit not closed.
22. (i) 10.3 m/s²; (ii) 6% reduction.

Chapter 9

3. Equilibrium at $\theta = \tfrac{1}{2}(2n - 1)\pi$ $(n = 0, \pm 1, \ldots)$.
4. Equilibrium at (i) $x = 0$, $x = \pm 1$; (ii) all values of x; (iii) $x = 0$; (iv) $x = 0$.
6. (i) saddle; (ii) saddle; (iii) unstable critical node; (iv) stable spiral.
7. $b > a$: $x = 0$ is a node or spiral depending on the sign of $k^2(b - a)^2 - 4cb^2$, $b < a$: $x = 0$ is a saddle; $x = (a^2 - b^2)^{1/2}$ is a spiral or node depending on the sign of $4k(a^2 - b^2)^{1/2} - c^2a^2$.

9. (i) Equilibrium at $x = 0$ (saddle), $x = 1$ (centre), $x = -1$ (centre);
   (ii) Equilibrium at $x = 0$ (centre), $x = 1$ (saddle), $x = -1$ (saddle);
   (iii) Equilibrium at $x = 0$ (saddle), $x = 1$ (stable spiral), $x = -1$ (stable spiral);
   (iv) Equilibrium at $x = n\pi$ (saddle if n odd, stable spiral if n even);
   (v) Equilibrium at $x = 0$ (stable spiral), $x = 1$ (saddle).
10. Equilibrium where $\sin \theta = aF_0\omega/mg$—only possible if $|aF_0\omega/mg| \le 1$.
13. Both equations have unstable equilibrium points at $x = 0$. The phase diagram should show a stable limit cycle in both cases. Plot graphs of the phase diagrams for $\mu = 0.1$, 1, 10 in each case.
14. Period is $2\left(\dfrac{m}{k}\right)^{1/2}\left(\pi + \dfrac{b}{A-b}\right)$.
16. $g < a\omega^2$: equilibrium where $\sin \theta = 0$, $\cos \theta = g/a\omega^2$ $g > a\omega^2$: equilibrium where $\sin \theta = 0$. Bifurcation value where $g = a\omega^2$.
18. Origin is an unstable spiral.
19. Fixed point at $\Gamma\omega^2[g - a\omega^2, c\omega]/[(g - a\omega^2)^2 + c^2\omega^2]$.
20. Computer investigations should reveal the following orbits, (i) $\Gamma = 2$, symmetric period 1, (ii) $\Gamma = 4$, two of period 1, (iii) $\Gamma = 10$, period 3, (iv) $\Gamma = 114$, chaotic solution, (iv) $\Gamma = 14$, period 1, (vi) $\Gamma = 24$, two of period 1.
23. Amplitude $Z = 2.1$.
25. The initial data $x_0 = 2.085$, $y_0 = 0$, $t = 0$ generates the synchronous orbit.
31. $\pi(m/k_1)^{1/2} + \pi(m/k_2)^{1/2}$.
32. The next term in the series for T is $11\pi a^4/1536$.

Chapter 10

1. 4.48 cm.
3. (i) In the rotating frame, $\mathbf{v} = -V \sin (Vt/a)\mathbf{i} + V \cos (Vt/a)\mathbf{j}$, and
$$\mathbf{f} = -V^2[\cos (Vt/a)\mathbf{i} + \sin (Vt/a)\mathbf{j}]/a;$$
   (ii) In the fixed frame,
   $\mathbf{v} = -(V + a\omega) \sin \{(v + a\omega)t/a\}\mathbf{I} + (V + a\omega) \cos \{(V + a\omega)t/a\}\mathbf{J}$,
   and $\mathbf{f} = -[(V + a\omega)^2]\{\cos [(V + a\omega)t/a]\mathbf{I} + \sin [(V + a\omega)t/a]\mathbf{J}\}/a$.
9. $m\omega^2 a$.
10. $m\omega^2 a(1 + e^{\omega t} - e^{-\omega t})$.
11. (i) $2Ma^2/3$; (ii) $\frac{1}{4}Ma^2$; (iii) $\pi a^4\rho(h/2 + 8a/15)$; (iv) $3Ma^2/5$; (v) $M(a^2/4 + h^2/12)$; (vi) $Ma^2/12$.
13. (i) $\frac{3}{4}mV^2$; (ii) $(4gx \sin \alpha/3a^2)^{1/2}$.
14. (i) 8×10^4 J; (ii) 7×10^3 kg.
15. $2\pi(7a/5g)^{1/2}$.
16. (i) Point of contact of cylinder with plane.
   (ii) With the perpendicular wires as axes and with $(\bar{x}, \bar{y})$ the coordinates of the centre of mass, the instantaneous centre has coordinates $(2\bar{x}, 2\bar{y})$.
17. The reaction has components $5 mg \cos \theta/2$ along the rod and $mg \sin \theta/4$ perpendicular to the rod.
19. $[Ma^2(3\omega_1^2 + 4\omega_1\omega_2 + 12\omega_2^2)/16]$.
23. 6.2 cm east
25. 126 s. Note that the variation of mass of the spacecraft due to the loss of approximately 1 kg of air may be ignored.
26. $2\pi(2a/g)^{1/2}$, $2\pi[Ma/g(M + m)]^{1/2}$.

Appendix

1. (i) $x = 4/(C - t^4)$, (ii) $x = \ln(C - \cos t)$, $C > 1$, (iii) $x - t = C(x + t)^3$; (iv) $x + (x^2 + t^2)^{1/2} = Bt^2$; (v) $x = \frac{1}{2}t^5 + Ct^3$; (vi) $x = C e^{1/2t^2} - 1$.

2. (i) $x = C_1 \cos 2t + C_2 \sin 2t$; (ii) $x = C_1 e^{-3t} + C_2 e^{-4t}$;
   (iii) $x = (C_1 + C_2 t) e^{-3t}$.

3. (i) $x = t^2 - 3t^2 + 6t - 6 + C e^{-t}$;
   (ii) $x = C_1 e^{-t} + C_2 e^{2t} + 24t^2 + 14t - 5$;
   (iii) $x = e^{-3t}(C_1 \cos 4t + C_2 \sin 4t) + 2 e^{3t}$;
   (iv) $x = C_1 \sinh kt + (C_2 + t/2k) \cosh kt$;
   (v) $x = e^{-4t}(C_1 \cos 3t + C_2 \sin 3t) + 2 \cos t$.

4. $x = e^t - e^{-2t} + 2 e^{-t}$, $y = 3 e^t - 2 e^{-2t} + 3 e^{-t}$

5. (i) $x = A \cos 2t + B \sin 2t + \frac{3}{4} + \frac{1}{3} \sin t$;
   (ii) $x = A \cos t + B \sin t + e^{2t}(-8 \cos 2t + \sin 2t)/65$;
   (iii) $x = A e^t + e^{-1/2t}[B \cos(\sqrt{3}t/2) + C \sin(\sqrt{3}t/2)]$;
   (iv) $x = e^{-1/2t}[A \cos(\sqrt{3}t/2) + B \sin(\sqrt{3}t/2)] + \frac{1}{2} + \frac{3}{26} \cos 2t - \frac{1}{13} \sin 2t$;
   (v) $x = e^t(A + Bt + Ct^2) - 1$.

6. (i) $x = A e^{Bt}$;
   (ii) $x - c + c \ln|x - c| = t + D$;
   (iii) $x = \dfrac{1}{a} \cosh(At + B)$.

8. (i) $x = \dfrac{K}{1 - \Omega^2} \cos \Omega t$; (ii) $x = \frac{1}{2}Kt \sin t$.

Further Reading

Mechanics has a long history with a very extensive literature, which we could not begin to cover in this bibliography. However, some suggestions for further reading are discussed and listed here by chapter to provide alternative viewpoints, more details where required, historical perspectives and guides to more advanced work.

INTRODUCTION

Background to the development of mechanics by Galileo, Kepler and Newton can be found in *The History of Mathematics—A Reader* edited by J. Fauvel and J. Gray (MacMillan, London, 1987). Newton's *Principia*, edited by F. Cajori (University of California Press, Berkeley, 1962) is available in translation.

CHAPTER 1: VECTORS

There are many standard introductions to vectors. *Vector Analysis* by M. R. Spiegel (Schaum/McGraw-Hill, New York, 1974) contains a large number of worked examples.

CHAPTER 2: KINEMATICS: GEOMETRY OF MOTION

Many specific engineering applications are given in *Engineering Mechanics, Volume 2: Dynamics* by J. L. Meriam and L. G. Kraige (Wiley, New York, 1987), for this and later chapters.

CHAPTER 3: PRINCIPLE OF MECHANICS

A rigorous and original introduction to dynamical principles can be found in *Principles of Dynamics* by R. Hill (Pergamon Press, Oxford, 1964). The author approaches the subject through the gravitational theory of planetary systems.

CHAPTER 4: APPLICATIONS IN PARTICLE DYNAMICS

Modelling with Projectiles by D. Hart and A. Croft (Ellis Horwood, Chichester, 1988) is devoted exclusively to the subject of projectiles. The book also contains lists of computer programs for the numerical computation of trajectories. More applications of dynamics in sport are presented in *The Physics of Ball Games* by C. B. Daish (English Universities Press, London, 1972).

CHAPTER 6: VARIABLE MASS PROBLEMS: ROCKET MOTION

More information on space vehicle motion and satellite orbits is given in *Introduction to Space Dynamics*, by W. T. Thomson (Wiley, New York, 1961).

CHAPTER 7: MECHANICAL OSCILLATIONS: LINEAR THEORY

Many applications of the linear theory of oscillations including normal modes are included in *Theory of Vibrations with Applications* by W. T. Thomson (Unwin–Hyman, London, Third edition, 1988).

CHAPTER 8: ORBITS

Extensive data concerning the planets and their satellites in the solar system (and also other physical constants) are listed in Kaye and Laby's *Tables of Physical and Chemical Constants* (Longman, London) and in *Handbook of Chemistry and Physics* (CRC Press, Florida). Details of possible Hohmann transfer orbits between planets are also given in the latter publication.

CHAPTER 9: NON-LINEAR DYNAMICS

More details and a more advanced approach to the phase plane, limit cycles, forced nonlinear systems, Poincaré maps, etc. can be found in *Nonlinear Ordinary Differential Equations* by D. W. Jordan and P. Smith (Oxford University Press, Oxford, Second edition, 1987). *Chaotic Vibrations* by F. C. Moon (Wiley-Interscience, New York, 1987) contains descriptions of many mechanical (and other) systems which lead to chaotic dynamics. With particular reference to Section 9.8, 'Steady motions exhibited by Duffing's equation: a picture book of regular and chaotic motions' by Y. Ueda in *New Approaches to Nonlinear Problems in Dynamics* edited by P. J. Holmes (SIAM, Philadelphia, 1980) contains further phase diagrams of a variety of periodic and chaotic oscillations for the forced, damped oscillator with cubic restoring force.

APPENDIX

There are many introductory textbooks on ordinary differential equations. *Modern Introductory Differential Equations* by R. Bronson (Schaum/McGraw-Hill, New York, 1973) contains many worked examples. Further explanation of numerical methods for initial-value problems in differential equations can be found in *A Simple Introduction to Numerical Analysis* by R. D. Harding and D. A. Quinney (Adam Hilger, Bristol, 1986) together with its associated software.

Index